METAL IONS IN BIOLOGICAL SYSTEMS

edited by

Helmut Sigel

Institute for Inorganic Chemistry
University of Basel
Basel, Switzerland

VOLUME 5
Reactivity of Coordination Compounds

MARCEL DEKKER, INC. New York and Basel

MARCEL DEKKER, INC.
270 Madison Avenue, New York, New York 10016

LIBRARY OF CONGRESS CATALOG CARD NUMBER: 75-18372

ISBN: 0-8247-6032-8

Current printing (last digit):
10 9 8 7 6 5 4 3 2 1

PRINTED IN THE UNITED STATES OF AMERICA

CONTENTS

Chapter 3

METAL ION-CATALYZED DECARBOXYLATIONS OF BIOLOGICAL INTEREST

R. W. Hay

Chapter 4

METAL ION-PROMOTED HYDROLYSIS OF AMINO ACID ESTERS AND PEPTIDES

R. W. Hay and P. J. Morris

Chapter 5

CATALYTIC ACTIVITY OF POLY-L,α-AMINO ACID-METAL ION COMPLEXES: NEW APPROACHES TO ENZYME MODELS

Masahiro Hatano and Tsunenori Nozawa

LIST OF CONTRIBUTORS

Numbers in parentheses indicate the pages on which the authors' contributions begin.

B. S. COOPERMAN, Chemistry Department, University of Pennsylvania, Philadelphia, Pennsylvania (79)

M. HATANO, Chemical Research Institute of Non-Aqueous Solutions, Tohoku University, 2-1-1 Katahira, Sendai, Japan (245)

R. W. HAY, Department of Chemistry, University of Stirling, Stirling, Scotland (127, 173)

D. L. LEUSSING, Department of Chemistry, Ohio State University, Columbus, Ohio (1)

P. J. MORRIS, Department of Chemistry, University of Waikato, Hamilton, New Zealand (173)

T. NOZAWA, Chemical Research Institute of Non-Aqueous Solutions, Tohoku University, 2-1-1 Katahira, Sendai, Japan (245)

J. T. SPENCE, Department of Chemistry and Biochemistry, Utah State University, Logan, Utah (279)

A. D. ZUBERBÜHLER, Institute of Inorganic Chemistry, University of Basel, Basel, Switzerland (325)

CONTENTS OF OTHER VOLUMES

Volume 3. High Molecular Complexes

Volume 4. Metal Ions as Probes

Volume 6. Biological Action of Metal Ions

Other volumes are in preparation.

Comments and suggestions with regard to contents, topics, etc., for future volumes of the series would be greatly welcome.

PREFACE

Recently, the importance of metal ions to the vital functions of living organisms, hence their health and well-being, has become increasingly apparent. As a result, the long-neglected field of "bioinorganic chemistry" is now developing at a rapid pace. The research centers on the synthesis, stability, formation, structure, and reactivity of biological metal ion-containing compounds of low and high molecular weight. The metabolism and transport of metal ions and their complexes is being studied, and new models for complicated natural structures and processes are being devised and tested. The focal point of our attention is the connection between the chemistry of metal ions and their role for life.

No doubt, we are only at the brink of this process. Thus, it is with the intention of linking coordination chemistry and biochemistry in their widest sense that the series METAL IONS IN BIOLOGICAL SYSTEMS reflects the growing field of "bioinorganic chemistry." We hope, also, that this series will help to break down the barriers between the historically separate spheres of chemistry, biochemistry, biology, medicine, and physics, with the expectation that a good deal of the future outstanding discoveries will be made in the interdisciplinary areas of science.

Should this series prove a stimulus for new activities in this fascinating "field" it would well serve its purpose and would be a satisfactory result for the efforts spent by the authors.

Helmut Sigel

Chapter 1

THE FORMATION OF SCHIFF BASES IN THE COORDINATION SPHERE OF METAL IONS

D. L. Leussing
Department of Chemistry
Ohio State University
Columbus, Ohio

1. SCHIFF BASES

Schiff bases play an important role in many biological processes involving amino acids and keto acids: transamination, decarboxylation, condensation, β-elimination, and racemization are just some of the reactions which may involve Schiff base intermediates. An essential cofactor in many of these enzymic processes is pyridoxal phosphate. It has been found that many of the pyridoxal reactions occur nonenzymically and are catalyzed by metal ions [1-11]. These metal ion model systems serve as excellent reference points for assessing the important factors in the reaction mechanisms and, as a result, have received much attention in recent years. This chapter will be concerned with the influence of metal ions on Schiff base stabilities and on their rates of formation and transamination.

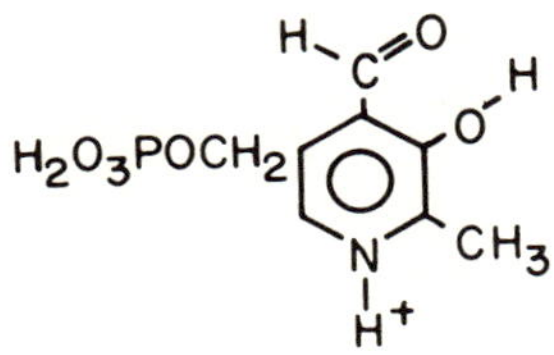

H_4PLP^+ (I)

2. SCHIFF BASE EQUILIBRIA IN THE ABSENCE OF METAL IONS

Primary amines react with carbonyl compounds to form adducts which in aqueous media may generally be comprised of hydrated and dehydrated forms:

$$\rangle C{=}O + H_2NR \rightleftarrows \rangle C\langle^{OH}_{NHR} \rightleftarrows \rangle C{=}N^{\diagup R} + H_2O \qquad (1)$$

Because it is usually not possible to determine through equilibrium measurements the number of solvent molecules eliminated in a reaction, the adduct formation constants which are determined through UV absorption spectroscopy, potentiometry, or similar methods represent the sum of the individual constants for the formation of carbinolamine and imine.

NMR methods, in those situations where chemical exchange is slow and where it is possible to see the individual proton resonances, may yield direct information pertaining to the carbinolamine-imine ratio of an adduct, and permit the evaluation of the microscopic constants. In the absence of measurements of this type estimates of the concentration of imine may be made from a comparison of the intensities of the adduct $n-\pi^*$ absorption bands to those of the parent carbonyl compound. The magnitudes of a polarographic diffusion current have also been used to estimate imine levels [12]. Pertinent to pyridoxal and its analogs, adducts of other aromatic aldehydes that have a hydroxy group ortho to the azomethine group are usually found to have a low proportion of carbinolamine to imine [13-18].

The basicity of the imine nitrogen atom in a Schiff base formed between a primary amine and an aliphatic carbonyl compound is considerably lower than that of the parent amine. The pK_a of N-pyruvideneglycine, for one example, is sufficiently lower than that of glycine to permit the evaluation of the Schiff base formation constant by potentiometric pH titration methods [19], and the pK_a of the N-isobutylidenemethylammonium ion, for another example, has been directly determined through stopped-flow spectroscopy to be 6.9 [20]. This corresponds to a reduction of almost 4 pK_a units below the value 10.7 for free $CH_3NH_3^+$. Schiff bases derived from aromatic carbonyl compounds on the other hand, can give markedly different behavior depending on the nature and position of ring substituents. The presence of an ortho oxy group, for example, tends to cause the basicity of the imine nitrogen atom to become considerably greater that that of the parent amine. This point has important implications regarding Schiff base stability and is discussed in greater detail below.

In Table 1 are presented formation and protonation constants which have been reported for some typical Schiff base systems in aqueous media.

TABLE 1

Equilibrium Constants for Schiff Base Formation in Aqueous Solution

Reaction	log K[a]	Ref.
$Pyu^- + Gly^- \rightleftarrows (Pyu\cdot Gly)^{2-}$	0.4	12,19
$Pyu^- + Ala^- \rightleftarrows (Pyu\cdot Ala)^{2-}$	0.32	21
	0.06	29
$Pyu^- + \beta\text{-}Ala^- \rightleftarrows (Pyu\cdot \beta\text{-}Ala)^{2-}$	0.91	21
$Pyu^- + i\text{-}Leu^- \rightleftarrows (Pyu\cdot i\text{-}Leu)^{2-}$	1.32	21
$Pyu^- + Sarc^- \rightleftarrows (Pyu\cdot Sarc)^{2-}$	0.7	21
$Pyu^- + L\text{-}Glu^{2-} \rightleftarrows (Pyu\cdot Gly)^{3-}$	0.03	29
$Glyox^- + Gly^- \rightleftarrows (Glyox\cdot Gly)^{2-}$	1.75	24
$Glyox^- + Ala^- \rightleftarrows (Glyox\cdot Ala)^{2-}$	1.27	24
$Glyox^- + Aib^- \rightleftarrows (Glyox\cdot Aib)^{2-}$	1.00	24
$Glyox^- + en \rightleftarrows (Glyox\cdot en)^{1-}$	5.01	26
$(Glyox\cdot en) + H^+ \rightleftarrows H(Glyox\cdot en)$	8.85	26
$H(Glyox\cdot en) + H^+ \rightleftarrows H_2(Glyox\cdot en)^+$	4.86	26
$Glyox^- + tetrameen \rightleftarrows (Glyox\cdot tetrameen)$	6.70	26
$(Glyox^-\cdot tetrameen)^{1-} + H^+ \rightleftarrows H(Glyox\cdot tetrameen)$	8.66	26
$H(Glyox\cdot tetrameen) + H^+ \rightleftarrows H_2(Glyox\cdot tetrameen)^{1+}$	2.69	26

Reaction	Value	Ref.
$\alpha\text{-Kg}^{2-} + \text{Ala}^{1-} \rightleftarrows (\alpha\text{-Kg}\cdot\text{Ala})^{3-}$	-0.11	29
i-Butal + Mea $\rightleftarrows$ (i-Butal•Mea)	0.93(Ca), 2.07(Im)(35°)	23,58
i-Butal + $PrNH_2$ $\rightleftarrows$ (i-Butal•$PrNH_2$)	0.71(Ca), 2.004(Im)(35°)	23,58
i-Butal + $CH_3O(CH_2)_3NH_2$ $\rightleftarrows$ adduct	0.48(Ca), 1.95(Im)(35°)	23,58
i-Butal + $CH_3O(CH_2)_2NH_2$ $\rightleftarrows$ adduct	0.38(Ca), 1.72(Im)(35°)	23,58
i-Butal + $(CH_3O)_2CHCH_2NH_2$ $\rightleftarrows$ adduct	0.04(Ca), 1.48(Im)(35°)	23,58
i-Butal + $F_3CCH_2NH_2$ $\rightleftarrows$ adduct	-0.62(Ca), 0.78(Im)(35°)	23
$(\text{i-Butal}\cdot\text{Mea})_{Im} + H^+ \rightleftarrows H(\text{i-Buta}\cdot\text{Mea})^+_{Im}$	6.88 (35°)	20
i-Butal + Mea $\rightleftarrows$ (i-Butal•Mea)	3.70 (35°)	22
i-Butal + Eta $\rightleftarrows$ (i-Butal•Eta)	3.56 (35°)	22
i-Butal + n-$PrNH_2$ $\rightleftarrows$ (i-Buta = NPr)	3.62 (35°)	22
i-Butal + i-$PrNH_2$ $\rightleftarrows$ (i-Buta = N-i-Pr)	3.26 (35°)	22
i-Butal + n-$BuNH_2$ $\rightleftarrows$ (i-Buta = N-n-Bu)	3.61 (35°)	22
i-Butal + t-$BuNH_2$ $\rightleftarrows$ (i-Buta = N-t-Bu)	2.30 (35°)	22
$\text{Sal}^- + \text{Mea} \rightleftarrows (\text{Sal}\cdot\text{Mea})^{1-}$	1.0	34
$(\text{Sal}\cdot\text{Mea})^{1-} + H^+ \rightleftarrows H(\text{Sal}\cdot\text{Mea})$	11.7	34
$\text{Sal}^- + \text{Eta} \rightleftarrows (\text{Sal}\cdot\text{Eta})^{1-}$	0.86	35
	1.05	31
$(\text{Sal}\cdot\text{Eta})^{1-} + H^+ \rightleftarrows H(\text{Sal}\cdot\text{Eta})$	11.8	35
	11.55	31
	11.81	38

TABLE 1 (Continued)

Reaction	log K[a]	Ref.
$H(Sal{\cdot}Eta) + H^{+} \rightleftarrows H_2(Sal{\cdot}Eta)^{+}$	4.96	38
$Sal^{-} + n\text{-}BuNH_2 \rightleftarrows (Sal = N\text{-}n\text{-}Bu)^{1-}$	0.97	72
$(Sal{=}N\text{-}n\text{-}Bu)^{-} + H^{+} \quad H(Sal{=}n\text{-}Bu)$	12.0	72
$Sal^{-} + en \rightleftarrows (Sal{\cdot}en)^{1-}$	1.34	33
$(Sal{\cdot}en)^{1-} + H^{+} \rightleftarrows H(Sal{\cdot}en)$	11.04	33
$H(Sal{\cdot}en) + H^{+} \rightleftarrows H_2(Sal{\cdot}en)^{+}$	8.85	33
$H_2(Sal{\cdot}en)^{+} + H^{+} \rightleftarrows H_3(Sal{\cdot}en)^{2+}$	4.87	33
$Sal^{-} + Gly^{-} \rightleftarrows (Sal{\cdot}Gly)^{2-}$	0.48	30
$(Sal{\cdot}Gly)^{2-} + H^{+} \rightleftarrows H(Sal{\cdot}Gly)^{1-}$	11.21	30
$Sal^{-} + \alpha\text{-}Ala^{-} \rightleftarrows (Sal{\cdot}\alpha\text{-}Ala)^{2-}$	0.78	31
$(Sal{\cdot}\alpha\text{-}Ala)^{2-} + H^{+} \rightleftarrows H(Sal{\cdot}\alpha\text{-}Ala)^{1-}$	9.95	31
$Sal^{-} + \beta\text{-}Ala^{-} \rightleftarrows (Sal{\cdot}\beta\text{-}Ala)^{2-}$	0.82	31
$(Sal{\cdot}\beta\text{-}Ala)^{2-} + H^{+} \rightleftarrows H(Sal{\cdot}\beta\text{-}Ala)^{1-}$	11.41	31
$SSal^{2-} + Eta \rightleftarrows (SSal{\cdot}Eta)^{2-}$	1.3	38
$(SSal{\cdot}Eta)^{2-} + H^{+} \rightleftarrows H(SSal{\cdot}Eta)^{1-}$	11.1	38
$H(SSal{\cdot}Eta)^{1-} + H^{+} \rightleftarrows H_2(SSal{\cdot}Eta)$	3.79	38
$m\text{-}Oxybenzal^{1-} + Mea \rightleftarrows (m\text{-}Oxybenzal{\cdot}Mea)^{1-}$	2.16	35
$(m\text{-}Oxybenzal{\cdot}Mea)^{1-} + H^{+} \rightleftarrows H(m\text{-}Oxybenzal{\cdot}Mea)$	9.6 (I = 0.2)	35
$p\text{-}Oxybenzal^{1-} + Mea \rightleftarrows (p\text{-}Oxybenzal{\cdot}Mea)^{1-}$	0.74	35

$(\text{p-Oxybenzal}\cdot\text{Mea})^{1-} + H^+ \rightleftarrows H(\text{p-Oxybenzal}\cdot\text{Mea})$	10.76(I = 0.29)	35
$\text{3-Oxypy-4-al}^{1-} + \text{Gly}^- \rightleftarrows (\text{3-Oxypy-4-al}\cdot\text{Gly})^{2-}$	1.69(30°)	13
$(\text{3-Oxypy-4-al}\cdot\text{Gly})^{2-} + H^+ \rightleftarrows H(\text{3-Oxypy-4-al}\cdot\text{Gly})^{1-}$	9.12(30°)	13
$H(\text{3-Oxypy-4-al}\cdot\text{Gly})^{1-} + H^+ \rightleftarrows H_2(\text{3-Oxypy-4-al}\cdot\text{Gly})$	5.50(30°)	13
$\text{3-Oxypy-4-al}^{1-} + \text{Val}^{1-} \rightleftarrows (\text{3-Oxypy-4-al}\cdot\text{Val})^{2-}$	1.75(30°)	13
$(\text{3-Oxypy-4-al}\cdot\text{Val})^{2-} + H^+ \rightleftarrows H(\text{3-Oxypy-4-al}\cdot\text{Val})^{1-}$	9.48(30°)	13
$H(\text{3-Oxypy-4-al}\cdot\text{Val})^{1-} + H^+ \rightleftarrows H_2(\text{3-Oxypy-4-al}\cdot\text{Val})$	5.16	13
$\text{3-Oxypy-4-al}^{1-} + \text{Glu}^{2-} \rightleftarrows (\text{3-Oxypy-4-al}\cdot\text{Glu})^{3-}$	1.39(30°)	13
$(\text{3-Oxypy-4-al}\cdot\text{Glu})^{3-} + H^+ \rightleftarrows H(\text{3-Oxypy-4-al}\cdot\text{Glu})^{2-}$	9.28(30°)	13
$H(\text{3-Oxypy-4-al}\cdot\text{Glu})^{2-} + H^+ \rightleftarrows H_2(\text{3-Oxypy-4-al}\cdot\text{Glu})^-$	5.15(30°)	13
$\text{PL}^- + \text{Gly}^- \rightleftarrows (\text{PL}\cdot\text{Gly})^{2-}$	1.28	39
$(\text{PL}\cdot\text{Gly})^{2-} + H^+ \rightleftarrows H(\text{PL}\cdot\text{Gly})^{1-}$	10.86	39
$H(\text{PL}\cdot\text{Gly})^{1-} + H^+ \rightleftarrows H_2(\text{PL}\cdot\text{Gly})$	5.88	48
$H_2(\text{PL}\cdot\text{Gly}) + H^+ \rightleftarrows H_3(\text{PL}\cdot\text{Gly})^+$	2.3	48
$\text{PL}^- + \text{Val}^- \rightleftarrows (\text{PL}\cdot\text{Val})^{2-}$	1.64	6
$(\text{PL}\cdot\text{Val})^{2-} + H^+ \rightleftarrows H(\text{PL}\cdot\text{Val})^{1-}$	10.49	6
$H(\text{PL}\cdot\text{Val})^{1-} + H^+ \rightleftarrows H_2(\text{PL}\cdot\text{Val})$	5.9	6
$\text{PL} + \gamma\text{-ABA} \rightleftarrows H_i(\text{PL}\cdot\gamma\text{-ABA})$	0.53(pH 7.4, 30°)	73
$\text{PL}^- + \alpha\text{-Ala}^- \rightleftarrows (\text{PL}\cdot\alpha\text{-Ala})^{2-}$	1.16(50°)	74
$(\text{PL}\cdot\alpha\text{-Ala})^{2-} + H^+ \rightleftarrows H(\text{PL}\cdot\alpha\text{-Ala})^{1-}$	9.91(50°)	74
$H(\text{PL}\cdot\alpha\text{-Ala})^{1-} + H^+ \rightleftarrows H_2(\text{PL}\cdot\alpha\text{-Ala})$	6.26(50°)	74
$\text{deox PL}^- + \text{L-Leu}^- \rightleftarrows (\text{deox PL}\cdot\text{L-Leu})^{2-}$	1.0	74

TABLE 1 (Continued)

Reaction	log K^a	Ref.
$(\text{deox PL·L-Leu})^{2-} + H^+ \rightleftarrows H(\text{deox PL·L-Leu})^{1-}$	11.73	74
$H(\text{deox PL·L-Leu})^{1-} + H^+ \rightleftarrows H_2(\text{deox PL·L-Leu})$	6.51	74
$PLP^{3-} + Gly^{1-} \rightleftarrows (PLP·Gly)^{4-}$	0.67	41
$(PLP·Gly)^{4-} + H^+ \rightleftarrows H(PLP·Gly)^{3-}$	11.88	41
$H(PLP·Gly)^{3-} + H^+ \rightleftarrows H_2(PLP·Gly)^{2-}$	6.65	41
$PLP^{3-} + \alpha\text{-Ala}^{1-} \rightleftarrows (PLP·\alpha\text{-Ala})^{4-}$	0.53	41
$(PLP·\alpha\text{-Ala})^{4-} + H^+ \rightleftarrows H(PLP·\alpha\text{-Ala})^{3-}$	11.77	41
$H(PLP·\alpha\text{-Ala})^{3-} + H^+ \rightleftarrows H_2(PLP·\alpha\text{-Ala})^{2-}$	6.33	41
$PLP + Gly \rightleftarrows H_i[PLP·Gly]$	1.99 (pH 7.5)	75
$PLP + \alpha\text{-Ala} \rightleftarrows H_i[PLP·Gly]$	1.76 (pH 7.5)	75
PLP + many other amines	(pH 7.5)	75
PLP + leu•gly•gly	3.41 (pH 7.5)	32
PLP + gly•gly•gly•gly	2.73 (pH 7.5)	32

PLP + leu•leu	2.63(pH 7.5)	32
PLP + ala•gly	2.18(pH 7.5)	32
PLP + γ-ABA	2.06(pH 7.4, 30°)	73
$PM^{1-} + Pyu^{1-} \rightleftarrows (PM{\cdot}Pyu)^{2-}$	0.95	45
$(PM\ Pyu)^{2-} + H^{+} \rightleftarrows H(PM{\cdot}Pyu)^{-}$	10.3	45
$H(PM\ Pyu)^{1-} + H^{+} \rightleftarrows H_2(PM{\cdot}Pyu)$	6.9	45
$PMP^{3-} + Pyu^{1-} \rightleftarrows (PMP{\cdot}Pyu)^{4-}$	1.1	44
$(PMP\ Pyu)^{4-} + H^{+} \rightleftarrows H(PMP{\cdot}Pyu)^{3-}$	10.6	44

[a] 25° unless otherwise specified.

[b] This value corrects an error in Reference 26.

[c] Ca, carbinolamine; Im, imine.

2.1. Schiff Bases Derived from Aliphatic Carbonyl Compounds

Pyruvate reacts with glycinate to form a somewhat weak Schiff base, N-pyruvideneglycinate [eq. (2)] [12,19]. An increase in the

$$CH_3\overset{O}{\overset{\|}{C}}CO_2^- + H_2NCH_2CO_2^- \underset{}{\overset{\beta_{SBO} = 2.5}{\rightleftharpoons}} (H_3C)(^-O_2C)C{=}NH_2CO_2^- + H_2O \qquad (2)$$

basicity of the amine β-alanine) brings about an increase in the Schiff base stability [21], showing that an increase in the electron-donating ability of the amine group strengthens the carbon-nitrogen double bond. A bulky side chain on the amino acid (i-leucine) has the same effect. Aliphatic aldehydes are typically more reactive than aliphatic ketones and, consistent with this, isobutyraldehyde forms Schiff bases with primary amines that are appreciably more stable than those formed by pyruvate [22,23]. Studies of the influence of polar substituents on the stabilities of the isobutyraldimines further demonstrate the correlation between amine basicity and Schiff base stability. A plot of the logarithm of the isobutyraldimine formation constants vs the pK_a of the corresponding ammonium ion shows a linear relation between these two quantities: the weaker the amine basicity, the weaker the Schiff base [22,23].

Glyoxylate forms Schiff bases, the formation constants of which are more typical of those of isobutyraldehyde rather than those of pyruvate [24]. Glyoxylate is extensively hydrated in aqueous solutions as $HC(OH)_2CO_2^-$ [25] and there is a corresponding tendency for its amino acid adducts also to be hydrated as the carbinolamines. Adjacent bulky substituents interfere with the hydration of the C=N bond, however, and the fraction of carbinolamine in a glyoxylate-amino acid adduct appears to decrease with increasing alkyl substitution on the amino acid α-carbon atom. [See equation (3).] Evidence for this is obtained from the observations that the UV absorption bands of the zinc(II) complexes of the glyoxylate-α-alaninate adduct are about twice as intense as those shown for the glycinate adduct, and α-aminoisobutyrate reacts with glyoxylate to

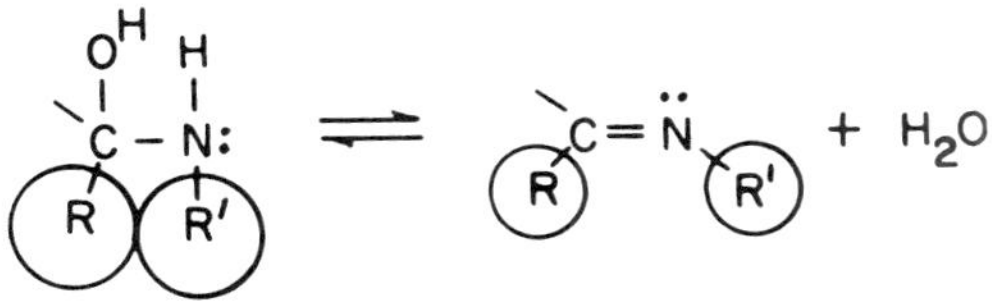 (3)

form a comparatively strong adduct having intense absorption in the UV region [24]. With this last amino acid steric repulsion between the methyl group of pyruvate and the two α-methyl groups of the amino acid is sufficiently large to prevent adduct formation, but the glyoxylate Schiff base has a normal stability.

Exceptionally strong interactions have been found between glyoxylate and the 1,2-diamines, ethylenediamine (en), and C,C,C',C'-tetramethylethylenediamine (tetrameen) [26]. The reaction products are cyclic imidazolidines of the type shown in equation (4).

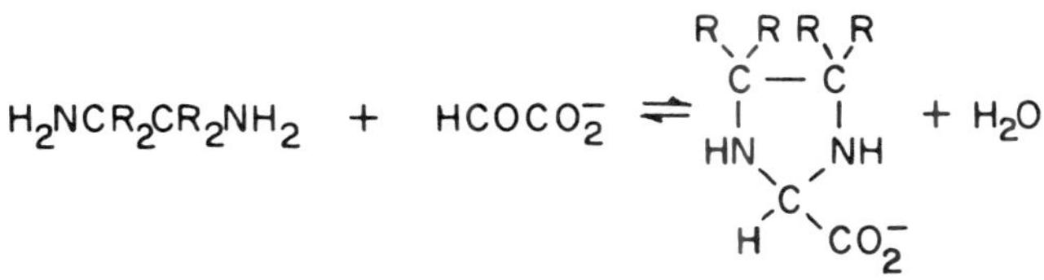 (4)

II

Interestingly, these ring structures remain in a fairly rapid dynamic equilibrium with the unassociated reactants so that the pH titration technique is well suited for the determination of their formation constants [26]. This is also feasible because ring formation produces a large decrease in the basicities of the diamine nitrogen atoms.

The imidazolidine formation constants are 4 to 5 orders of magnitude greater than those found for imine formation:

$$\text{en} + \text{glyox}^- \underset{}{\overset{\beta_{SBO} = 10^{5.04}}{\rightleftharpoons}} (\text{en}\cdot\text{glyox}^-) + H_2O \qquad (5)$$

$$\text{tetrameen} + \text{glyox}^- \underset{}{\overset{\beta_{SBO} = 10^{6.70}}{\rightleftharpoons}} (\text{tetrameen}\cdot\text{glyox})^{1-} + H_2O \qquad (6)$$

The first protonation constant [eq. (7)] is $10^{8.7}$ for R = CH_3 and $10^{8.85}$ for R = H. These values are appreciably lower than the amine protonation constant, $10^{10.52}$, reported [27] for the analogous cyclic amino acid proline, owing to the greater electronegativity of the ring N-H group of the imidazolidine than the corresponding CH_2 group of proline.

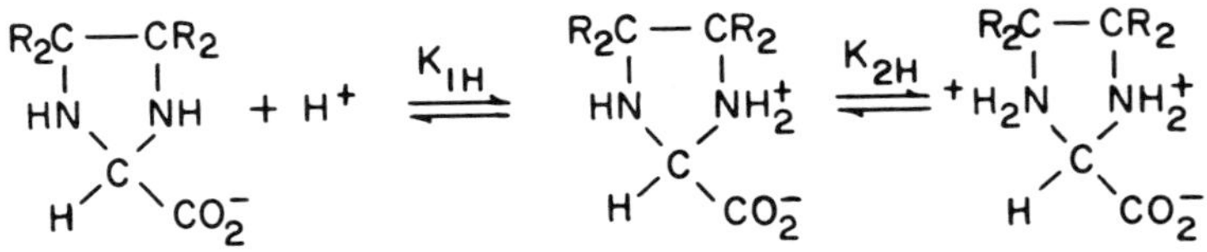

II III IV (7)

The second amino protonation constants are very low, being $10^{4.7}$ and $10^{2.69}$ for the en and tetrameen derivatives, respectively. The carboxylate group has very weakly basic properties owing to the proximity of the two positive imminium groups, and acquires a proton at even lower pH.

Transamination of α-ketoglutarate and α-alanine to give glutamic acid and pyruvate is an important biological reaction. Krebs [28] has determined the equilibrium constant of transamination to be 0.67 at pH 7.4 for the overall reaction:

$$\alpha\text{-Kg}^{2-} + \text{HAla}^{\pm} \rightleftharpoons {}^{\pm}\text{HGlu}^{-} + \text{Pyu}^{1-} \qquad (8)$$

Alaninate and pyruvate react to form a weak Schiff base, $(\text{Pyu}\cdot\text{Ala})^{2-}$ $\beta_{SBO} = 10^{0.06}$, but those formed between pyruvate and glutamate, $(\text{Pyu}\cdot\text{Glu})^{3-}$ $\beta_{SBO} = 10^{0.03}$, and alaninate and α-ketoglutarate, $(\alpha\text{-Kg}\cdot\text{Ala})^{3-}$ $\beta_{SBO} = 10^{-0.11}$, are slightly weaker owing to greater steric repulsions in the adducts [29]. These data and the pK_a values of the amino acids permit the evaluation of the Schiff base tautomerization constant for reaction (9).

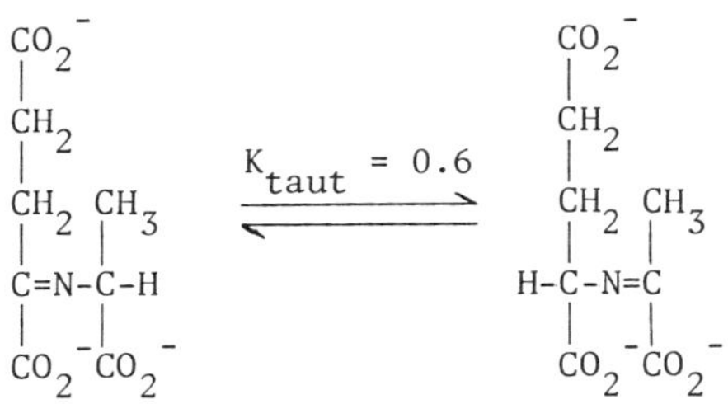

(9)

K_{taut} does not differ significantly from Krebs' conditional transamination constant at pH 7.4 because differences in the amino acid basicities are also reflected by the differences in the Schiff base formation constants. These differences tend to cancel out in K_{taut}. This transamination reaction proceeds very slowly in the absence of the enzyme and, even in the presence of metal ions, is accompanied by extensive decarboxylation.

2.2. Schiff Bases Derived from Aromatic Carbonyl Compounds

Salicylaldehyde is a simple aromatic analog of pyridoxal 5'-phosphate and serves as a good starting point from which to build up an understanding of vitamin B_6 model reactions. Salicylaldehyde anion reacts with primary amines to form 1:1 adducts, which, from the intensities of their UV absorption bands, appear to exist predominantly in the imine form (V) [14,30].

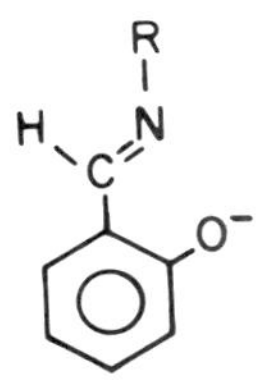

V

The stability of the glycinate Schiff base $(Sal \cdot Gly)^{2-}$ ($\beta_{SBO} = 10^{0.48}$) [30] is practically the same as that of N-pyruvideneglycinate ($\beta_{SBO} = 10^{0.42}$) and, judging from this, salicylaldehyde Schiff base stabilities are expected to run slightly lower than those derived from aliphatic aldehydes.

In contrast to pyruvate, Sal^- reacts with α-Ala^- to give a Schiff base that has a greater stability than that of Gly^-, $(Sal\cdot Ala)^{2-}$, $\beta_{SBO} = 10^{0.78}$ [31]. This is a consequence of both slightly greater α-Ala^{1-} basicity compared to Gly^- and lower steric interference between the α-Ala^- methyl side chain and the planar Sal^- molecule than exists in the pyruvate adduct. The influence of amine basicity on adduct stabilities is also evident with other amines. Sal^- Schiff base stabilities were found to follow the order Eta > En > β-Ala^- > α-Ala^- > Gly^-, which is in the same order as decreasing pK_a of the conjugate acids, with the exception of an inversion between En and β-Ala^-. The inversion of these two amines may be explained on the basis of electrostatic repulsion between the negatively charged Sal^- and β-Ala^- moieties of the Schiff base, which does not occur with en.

Unlike glyoxylate, salicylaldehyde does not form a cyclic adduct with ethylenediamine, but a normal linear imine [32] having a typical stability [33].

The position of the oxy group on the aromatic ring with respect to the azomethine group has a large influence on the Schiff base stability. Green and coworkers have determined the formation constants for Schiff base formation between methylamine (CH_3NH_2) and salicylaldehyde [34], m-hydroxybenzaldehyde [35], or p-hydroxybenzaldehyde [35]. From their results, the formation constants for reactions (10) are calculated to be $10^{1.0}$, $10^{2.16}$, and $10^{0.74}$ for ortho, meta, and para oxy substitution, respectively. The ortho

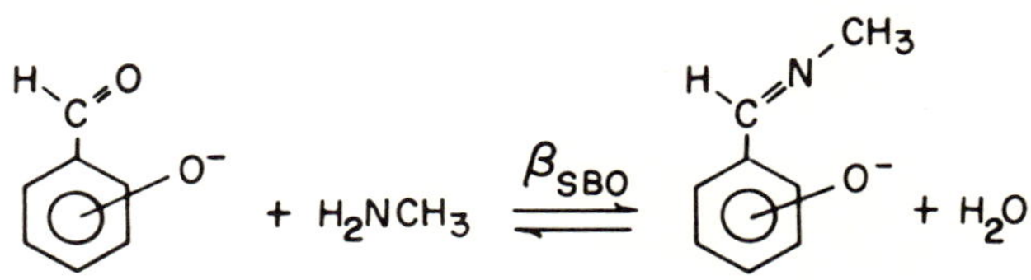

(10)

VI VII

and para oxy-substituted benzaldehydes are seen to have Schiff base formation constants that are similar in magnitude to each other, with the para derivative being slightly smaller. The formation constant for the meta derivative, however, is over an order of magnitude greater than that of the ortho derivative and lies in the range of values typical of those observed for aliphatic aldehydes. In either the ortho or para positions, the oxy group, through conjugation, increases the π-electron density at the carbonyl group. This reduces the driving force by which the carbonyl oxygen atom is replaced by a less electronegative nitrogen atom. With meta substitution, conjugation of these functional groups is absent and a normal aldehyde formation constant is obtained. The somewhat higher Schiff base stability observed for salicylaldehyde anion compared to p-oxybenzaldehyde may possibly arise from stabilization through a hydrogen-bonded, bridging H_2O molecule (VIII).

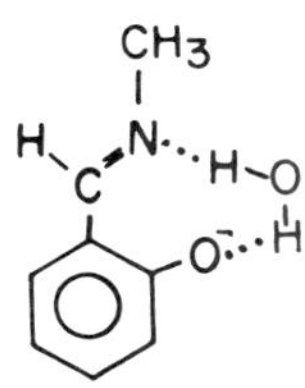

VIII

In aqueous media, the protonation site of Schiff bases derived from ortho-oxy aromatic aldehydes is the imine group and not the phenolate group (IX). Evidence for this has been obtained from UV [14,30,36] and NMR [17] spectroscopic data and molecular orbital calculations [30,36].

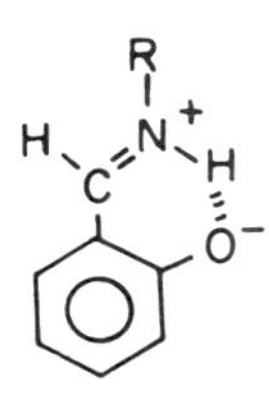

IX

Imine protonation in these aromatic Schiff bases gives rise to a π-π^* absorption band lying in the vicinity of 400 nm. As the dielectric constant of the medium is decreased this band diminishes in intensity and one at shorter wavelengths grows as the proton is transferred from the imine nitrogen to the phenolate oxygen atom [14].

It is a characteristic of this class of Schiff bases that the imine nitrogen atom has a higher basicity than that of the primary amine from which it is derived, for example,

$$(\mathrm{Sal\cdot Gly})^{2-} + \mathrm{H}^+ \rightarrow \mathrm{H(Sal\cdot Gly)}^{1-} \quad K_H = 10^{11.21} \tag{11}$$

compared to [30]

$$\mathrm{Gly}^{1-} + \mathrm{H}^+ \rightleftarrows \mathrm{H\ Gly}^{\pm} \quad K_H = 10^{9.70} \tag{12}$$

This change in basicity on Schiff base formation is in the opposite direction to that observed for aliphatic carbonyl compounds. Furthermore, the position of the oxy group relative to the imine group has a large influence on the nitrogen atom basicity. Protonation constants for the o-, m-, and p-oxybenzylidene-methyl amines are $10^{11.7}$, $10^{9.6}$, and $10^{10.8}$ compared to $10^{10.51}$ for free methyl amine [34,35]. Thus, the meta-substituted Schiff base shows the decrease in basicity relative to the free amine that is expected from the behavior of the aliphatic Schiff bases, but in this case the magnitude of the change is smaller. It is very likely that of the two basic groups available in the molecule it is the meta phenolate oxygen which is actually protonated rather than the imine nitrogen, since the observed pK_a of the Schiff base is close to that of m-hydroxybenzaldehyde itself (pK_a = 9.0). The para-substituted Schiff base, on the other band, exhibits a slightly augmented basicity over the free amine owing to conjugation. The still greater increase observed with the salicylaldehyde Schiff base (IX) arises from the influence of a strong iminium-phenolate hydrogen bond.

The constant for phenoxide protonation of the imminium

salicylidene Schiff bases appears to lie in the vicinity of $10^{5.0}$ [37,38], and represents a decrease of over 3 pK_a units from the value observed for phenolate protonation in salicylaldehyde. This low affinity of the phenolate group for a proton arises from the repulsive influence of the nearby imminium ion coupled with the energy required to break the imminium-phenolate hydrogen bond. Protonation of the phenolate group of IX causes the absorption band centered around 395 nm to disappear, while simultaneously a new band due to X appears at about 330 nm [33]. This new band lies close to the one for H(Sal) itself at 323 nm.

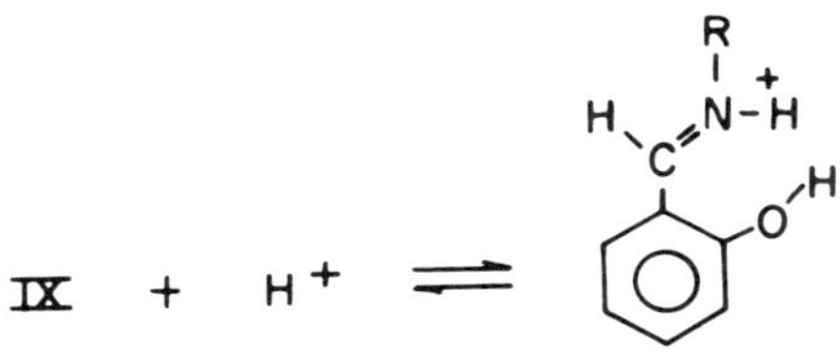

X

(13)

The Schiff bases derived from aliphatic carbonyl compounds are extensively dissociated in those pH regions where the amino groups of the parent amines are protonated. Owing to the stabilization brought about by imine protonation, association of o-oxyaromatic Schiff bases, such as N-salicylideneglycinate, is actually more pronounced in intermediate pH ranges than in strongly alkaline or more acidic solutions. Extensive dissociation usually occurs in more acidic regions where both the phenolate and amine groups are protonated, however. This behavior is quantitatively expressed in the values of the equilibrium constants for the principal reactions in the various pH regions:

$$\mathrm{Sal^- + Gly^{1-} \rightleftarrows Sal{\cdot}Gly^{2-}} \qquad \beta_{SBO} = 3\ M^{-1} \tag{14}$$

$$\mathrm{Sal^- + HGly^{\pm} \rightleftarrows H(Sal{\cdot}Gly)^{1-}} \qquad \beta'_{SB1} = 1 \times 10^{2}\ M^{-1} \tag{15}$$

$$\mathrm{HSal + HGly^{\pm} \rightleftarrows H_2(Sal{\cdot}Gly)} \qquad \beta''_{SB2} = 6 \times 10^{-2}\ M^{-1} \tag{16}$$

Thus, for a solution which contains a total of 0.01 M Gly and Sal in equilibrium, 5% of the Sal is converted to Schiff base at pH > 12, about 40% at pH ∿ 9 and 6 x 10^{-4}% at pH < 5. Distribution curves which illustrate the influence of pH on the extent of Schiff base formation are presented in Figure 1 for some systems of present interest.

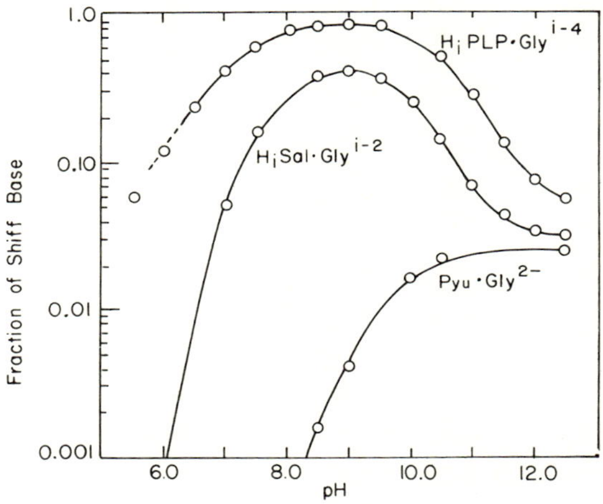

FIG. 1. Fraction of carbonyl compound converted to Schiff base as a function of pH by an excess of 0.010 M glycine.

3-Oxypyridine-4-carboxaldehyde anion (XI), a close analog of pyridoxal anion, forms a Schiff base with glycinate which has a formation constant over 10 times as large as that for salicylaldehyde anion [13]. Thus, this pyridine derivative forms Schiff bases that are as stable as those of aliphatic aldehydes. Earlier it was pointed out that an increase in the electron density of the carboxyaldehyde group by an ortho or para oxy substituent reduces the stability of the Schiff base. Here, it is seen that replacing the para ⟩CH group of the aromatic ring by a more electronegative pyridine nitrogen atom increases the Schiff base stability, since a more electronegative aromatic ring favors replacement of the carboxaldehyde oxygen atom by a less electronegative nitrogen atom.

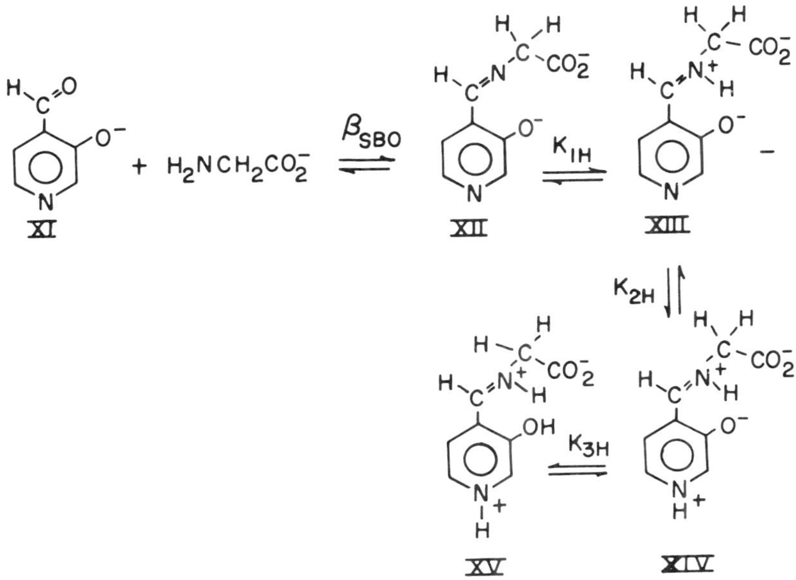

(17)

Thus, ring nitrogen substitution compensates to some extent for the weakening influence of the ortho oxy atom on the Schiff base stability with this aromatic aldehyde.

The electron-withdrawing effect of the pyridine nitrogen is also manifested in the basicity of the imine group: the protonation constant of Schiff base XII lies close to that of glycinate, unlike the situation with Sal^- where a higher imine protonation constant is observed. Protonation of XIII occurs in the region of pH 5.5, and is probably at the site of the pyridine nitrogen to give XIV.

Pyridoxal anion (XVI) reacts with glycinate to form a Schiff base which may be somewhat less stable than that formed by 3-oxy-pyridine-4-carboxaldehyde [39]. This decrease may arise from the inductive effects of the pyridoxal ring side chains; however, it is important to keep in mind that the experimental conditions differed somewhat in the two studies being compared, so firm conclusions cannot be drawn from the data.

There is evidence of another sort that indicates that the ring methyl group does indeed exert an appreciable influence on the Schiff base. The protonation constant of the imine nitrogen atom

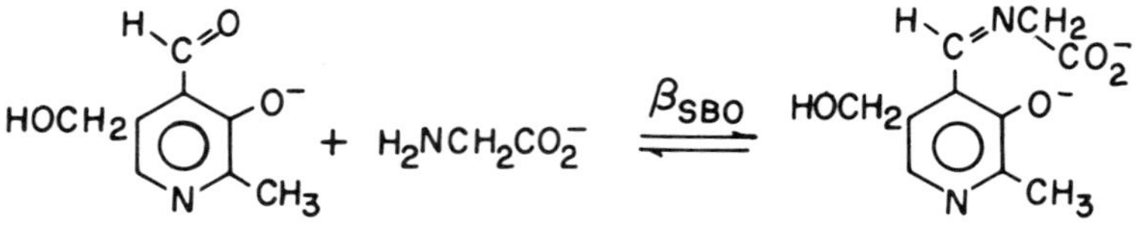

XVI XVII (18)

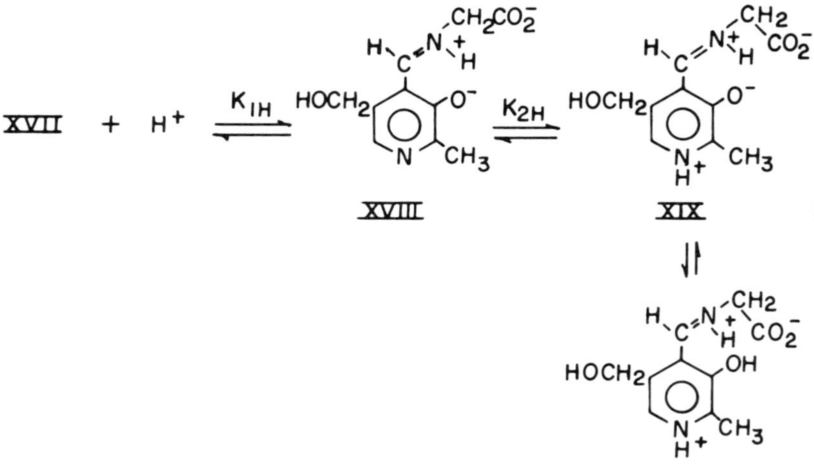

XVII XVIII XIX XX (19)

of XVII is greater than that of the amine group of the parent glycinate ion, contrary to the situation that exists with XII, and the basicity of the heterocyclic nitrogen atom is greater with pyridoxal and its derivatives than with the corresponding 3-hydroxypyridine-4-carboxaldehyde compounds. Because transamination activity is closely associated with the presence of the pyridine protonated Schiff base [8,40] these differences would tend to promote pyridoxal transamination activity at neutral pH over that exhibited by the less substituted analogs through (i) increasing the amount of Schiff base present, and (ii) increasing the proportion of pyridinium to pyridine form.

Pyridoxal 5'-phosphate is a tetraprotic acid [41-43] and this ionization behavior is carried over into its Schiff base [eqs. (20) and (21)].

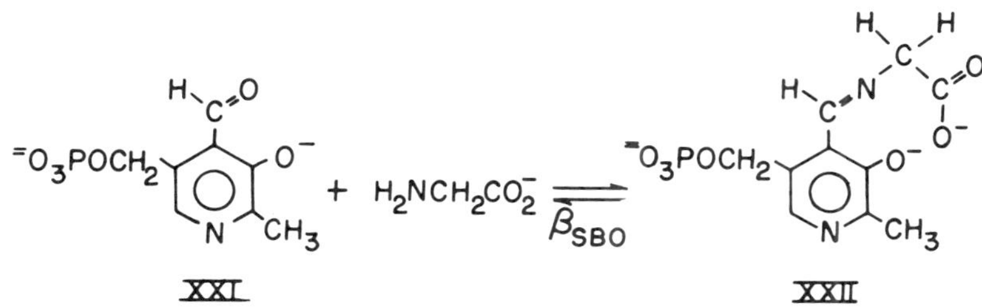

(20)

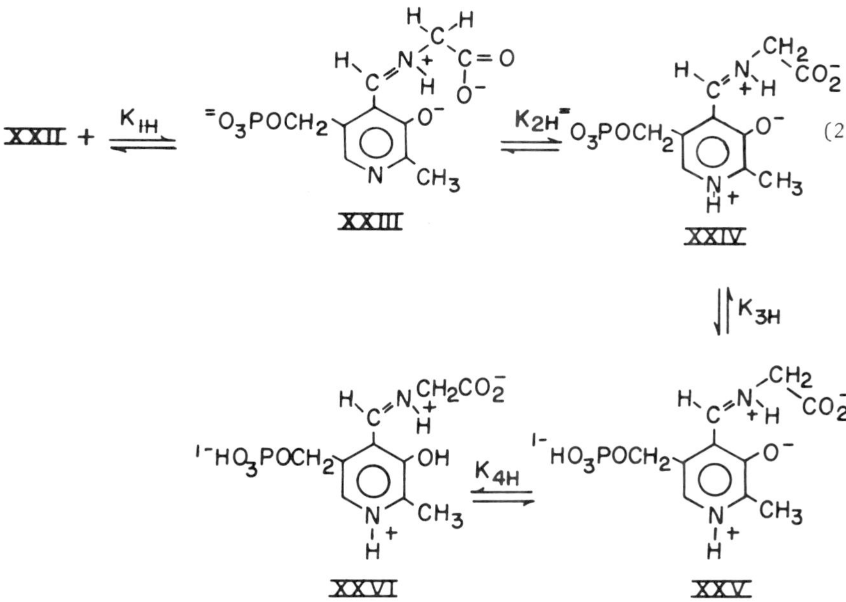

(21)

Values of β_{SBO} for Schiff base formation with glycinate or alaninate are smaller with PLP^{3-} than with PL^{1-} [41]. Interaction between the pyridoxal $-CH_2OH$ side chain and the azomethine group is not a factor in enhancing the stability of the PL Schiff base [17] so a true decrease in the PLP system seems to result from the presence of the phosphate group. A plausible explanation for this effect is electrostatic repulsion between PLP^{3-} and the negatively charged aminoacidate ions. This electrostatic phosphate effect also serves to increase the affinity of the PLP-aminoacidate Schiff bases for protons. The latter influence is greater than the former, so that the protonated PLP Schiff base, XXIII, is actually more stable than the corresponding PL Schiff base, XVIII. Thus, the phosphate group of PLP, although remote and not involved in direct bonding with those groups which are associated with the transamination process, exerts a definite and favorable influence on this reaction.

The pyridoxamine 5'-phosphate-pyruvate system forms the tautomeric ketimine isomer of the pyridoxal 5'-phosphate-alanine aldimine [eqs. (22) and (23)].

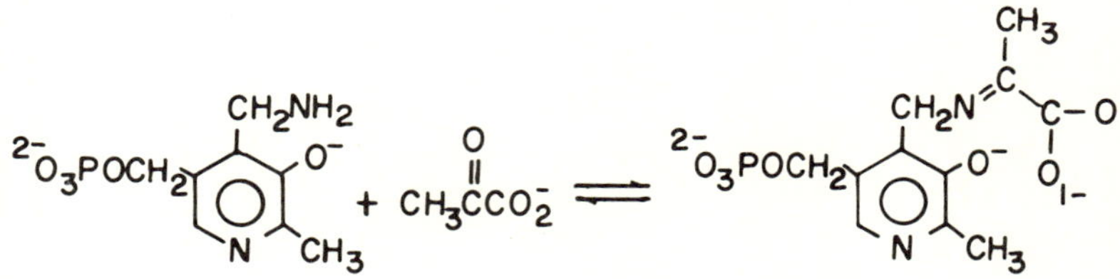

XXVII XXVIII (22)

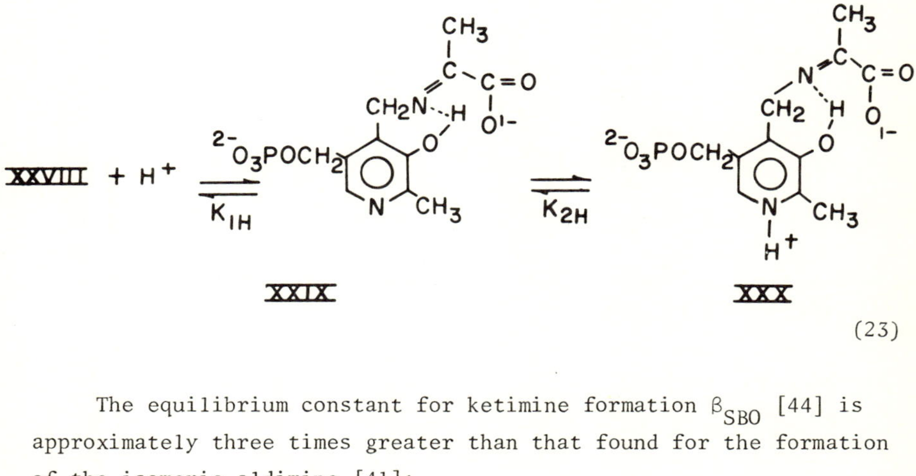

XXIX XXX

(23)

The equilibrium constant for ketimine formation β_{SBO} [44] is approximately three times greater than that found for the formation of the isomeric aldimine [41]:

$$PLP^{3-} + Ala^{1-} \underset{}{\overset{\beta_{SBO} = 3.4}{\rightleftharpoons}} (PLP \cdot Ala)^{4-} + H_2O \qquad (24)$$

As far as stability constants range this is not a great difference and seems to reflect more the greater reactivities of PMP^{3-} and Pyu^{1-} compared to PLP^{3-} and Ala^{1-} than an intrinsically greater stability of the ketimine over the aldimine. The transamination equilibrium constant for reaction (25) very slightly favors the ketimine.

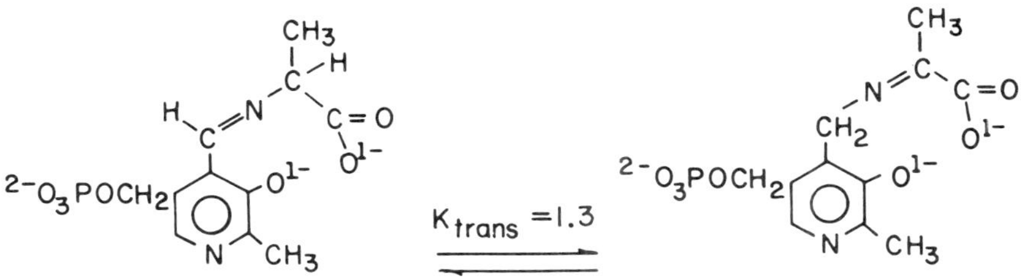

(25)

Banks et al. [45] have determined the value for the formation of the pyridoxamine ketimine:

$$PM^{1-} + Pyu^{1-} \rightleftarrows PM{\cdot}Pyu^{2-} \tag{26}$$

to be 9.0 M^{-1}. Thus, within the experimental error of these two determinations, the phosphate group appears to have little influence on the ketimine stabilities.

The basicity of the ketimine is appreciably lower than that of the aldimine

$$PMP{\cdot}Pyu^{4-} + H^{+} \xrightleftharpoons{K_H = 10^{10.6}} H(PMP{\cdot}P)^{3-} \quad [44] \tag{27}$$

$$PLP{\cdot}Ala^{4-} + H^{+} \xrightleftharpoons{K_H = 10^{11.8}} H(PLP{\cdot}Ala)^{3-} \quad [41] \tag{28}$$

One reason for this is that conjugation of the imine group and the phenolate group is absent in the ketimine, but another is that the site of protonation in the two isomers is different. While it is the azomethine nitrogen that is the protonation site with the aldimine, $H(PLP{\cdot}Ala)^{3-}$ (XXIII), it is the phenolate group that is protonated with the ketimine, $H(PMP{\cdot}Pyu)^{3-}$ (XXIX). As was discussed above, Schiff base formation between primary amines and aliphate carbonyl compounds characteristically causes a large decrease in the nitrogen basicity: with pyrividenepyridoxamine the imine basicity is depressed below that of the phenoxide group, which then acquires the proton.

Protonation stabilizes the ketimine through a phenol-imine hydrogen bond, but the effect is not as great as it is in the aldimine as evidenced from a lower ketimine basicity. The difference in proton stabilization has a large effect on tautomerization, which now favors the aldimine configuration of the monoprotonated isomers [eq. (29)].

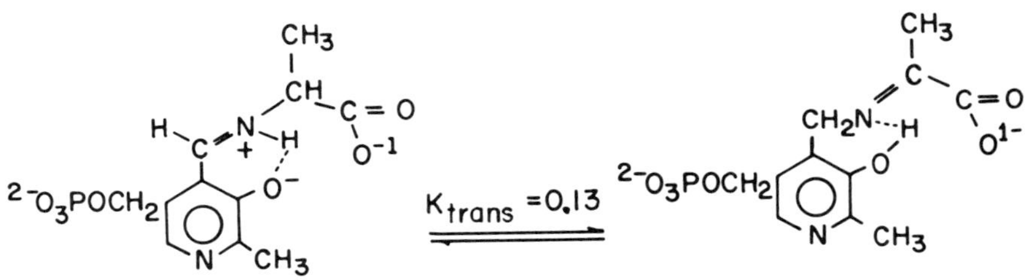

(29)

The aldimine has been found to accept a second proton at the site of the pyridine nitrogen to give XXIV, but direct evidence for pyridine protonation of XXIX to give XXX was not obtained owing to extensive dissociation of the Schiff base in the pH region where the protonation of the heterocyclic nitrogen occurs. Studies on PM-Pyu systems, however, indicate that the pyridine protonation constant of this ketimine lies in the pK_a range 6.3 to 6.9 [17,45].

3. METAL ION SCHIFF BASE COMPLEXES

Schiff bases may often function as ligands of a higher denticity than the components from which they are constituted, e.g., salicylaldehyde anion and glycinate are both bidentate ligands, but N-salicylideneglycinate is a tridentate ligand. Owing to this property, complexing metal ions are often able to form highly stable complexes with Schiff bases in aqueous media and stabilize them toward hydrolysis even under acidic conditions where the uncomplexed Schiff base is highly dissociated. Formation constants for some metal ion-Schiff base complexes are given in Table 2.

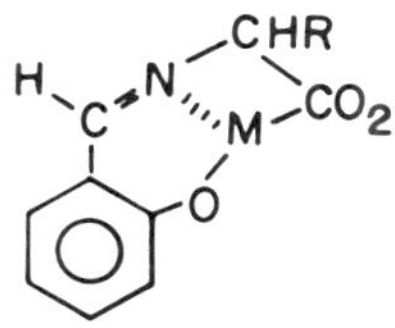

XXXI

In general, the equilibrium expressions for metal ion-Schiff base systems are quite complicated because hydrolysis and protonation reactions may be present. A generalized scheme for a typical system may be represented as shown in reaction (30) where O and A represent carbonyl and amine donors and the Σ represent summations over all the forms that are present. This scheme does not include

$$
\begin{array}{ccccccc}
\Sigma H_pO & & \Sigma H_qA & & & & \Sigma H_n(O\cdot A) \\
\upharpoonleft\downharpoonright & & \upharpoonleft\downharpoonright & & & & \upharpoonleft\downharpoonright \quad + iH_2O \\
iO & + & jA & + & M^m & \rightleftharpoons & M(O\cdot A)_i(A)_{j-i} \\
\downharpoonleft\upharpoonright & & \downharpoonleft\upharpoonright & & & & \\
\Sigma MO_r & & \Sigma MA_s & & & &
\end{array}
\qquad (30)
$$

the formation of hydroxy complexes which may be important with some metal ions even at low pH, e.g., Al(III) or Fe(III). Numerical solution of the species distributions that arise in such complicated systems has been successfully performed for a number of years in the author's laboratory using an algolrithm based on the Newton-Raphson iteration.

Early investigations of the aqueous solution reactions of Schiff base complexes were performed spectrophotometrically. While much useful information was obtained, the number of species having similar spectra proved to be too large to permit a detailed and

TABLE 2

Metal Ion-Schiff Base Complex Formation Constants in Aqueous Media

$$M + iL' + jL'' \overset{\beta_{ij}}{\rightleftharpoons} M(L'\cdot L'')_i(L'')_{j-i}$$

		log equilibrium constant	Ref.
L' = pyruvate			
L" = glycinate	Ca^{2+}	β_{11} 2.9(25°)	19
	Mn^{2+}	β_{11} 5.51(10°) 5.36(25°), β_{12} 7.7(10°) 6.9(25°), β_{22} 10.25(10°) 9.79(25°)	19
	Ni^{2+}	β_{11} 8.19(10°) 8.09(25°), β_{12} 13.57(10°) 13.00(25°), β_{22} 15.76(10°) 15.29(25°)	19
		β_{11} 8.068, β_{12} 12.95, β_{22} 15.28(25°)	21
	Zn^{2+}	β_{11} 7.61(10°) 7.53(25°), 11.54(10°) 12.0(25°) β_{22}14.54(10°) 14.25(25°)	19
		β_{11} 7.601, β_{12} 11.87, β_{22} 14.35(25°)	21
L" = α-alaninate	Ni^{2+}	β_{11} 7.14, β_{12} 11.63, β_{22} 14.02(25°)	21
	Zn^{2+}	β_{11} 6.78, β_{12} 9.36, β_{22} 13.01(25°)	21
		β_{11} 7.09, β_{12} 11.18, β_{22} 13.28(25°)	29
L" = i-leucinate	Ni^{2+}	β_{11} 6.21, β_{12} 11.08, β_{22} 13.05(25°)	21
	Zn^{2+}	β_{11} 6.76, β_{12} 9.58, β_{22} 12.34(25°)	21

Ligand	Metal ion / equilibrium	Constants	Ref.
L'' = β-alaninate	Ni^{2+}	β_{11} 8.34, β_{12} 11.95, β_{22} 15.17(25°)	21
	Zn^{2+}	β_{11} 7.08, β_{12} pptn., β_{22} 12.1(25°)	21
L'' = sarcosinate	Ni^{2+}	β_{11} 5.97, β_{12} 7.79, β_{22} 8.74(25°)	21
	Zn^{2+}	β_{11} 4.60, β_{12} 9.85, β_{22} 10.1ppt.(25°)	21
L'' = glutamate	Zn^{2+}	β_{11} 6.97, β_{12} 10.64, β_{22} 12.38(25°)	29
L'' = pyridoxamine 5'-phosphate, PMP^{3-}	Zn^{2+}	β_{11} 8.5, β_{22} 15.49(25°)	56
	$Zn(PMP\cdot Pyu)^{2-} + H^+ \rightleftarrows ZnH(PMP\cdot Pyu)^{1-}$	8.4(25°)	56
	$ZnH(PMP\cdot Pyu)^{1-} + H^+ \rightleftarrows ZnH_2(PMP\cdot Pyu)$	6.42(25°)	56
	$Zn(PMP\cdot Pyu)_2^{6-} + H^+ \rightleftarrows ZnH(PMP\cdot Pyu)_2^{5-}$	9.17(25°)	56
	$ZnH(PMP\cdot Pyu)_2^{5-} + H^+ \rightleftarrows ZnH_2(PMP\cdot Pyu)_2^{4-}$	8.14(25°)	56
	$ZnH_2(PMP\cdot Pyu)_2^{4-} + H^+ \rightleftarrows ZnH_3(PMP\cdot Pyu)_2^{3-}$	7.05(25°)	56
	$ZnH_3(PMP\cdot Pyu)_2^{3-} + H^+ \rightleftarrows ZnH_4(PMP\cdot Pyu)_2^{2-}$	5.87(25°)	56
L' = α-ketoglutarate			
L'' = α-alaninate	Zn^{2+}	β_{11} 5.97, β_{12} 10.00(25°)	29
L'' = L-glutamate	Zn^{2+}	β_{11} 5.64(25°)	29
L' = glyoxylate			
L'' = glycinate	Ni^{2+}	β_{11} 8.083, β_{12} 12.915, β_{22} 14.69(25°)	24
	Zn^{2+}	β_{11} 6.746, β_{12} 10.695, β_{22} 12.225(25°)	24
L'' = α-alaninate	Ni^{2+}	β_{11} 7.555, β_{12} 12.30, β_{22} 13.73(25°)	24
	Zn^{2+}	β_{11} 6.32, β_{12} 10.17, β_{22} 12.03(25°)	24

TABLE 2 (Continued)

	log equilibrium constant	Ref.
L' = glyoxylate (continued)		
L'' = α-aminoisobutyrate	Ni^{2+} β_{11} 7.36, β_{12} 12.35, β_{22} 14.17(25°)	24
	Zn^{2+} β_{11} 6.28, β_{22} 12.6(25°)	24
L'' = ethylenediamine	$Ni^{2+} + (glyox{\cdot}en)^{1-} \rightleftarrows Ni(glyox{\cdot}en)^{+}$ 7.10(25°)	26
	$Ni^{2+} + 2(glyox{\cdot}en)^{1-} \rightleftarrows Ni(glyox{\cdot}en)_2$ 12.2(25°)	26
	$Ni(glyox{\cdot}en)^{+} + H^{+} \rightleftarrows NiH(glyox{\cdot}en)^{2+}$ 5.05(25°)	26
	$Ni(glyox{\cdot}en)^{+} + glyox^{1-} \rightleftarrows Ni(Glyox_2{\cdot}en)$ 2.45(25°)	26
	$Zn^{2+} + (glyox{\cdot}en)^{1-} \rightleftarrows Zn(glyox{\cdot}en)^{1-}$ 5.28(25°)	26
	$Zn^{2+} + 2(glyox{\cdot}en)^{1-} \rightleftarrows Zn(glyox{\cdot}en)$ 9.05(25°)	26
	$Zn(glyox{\cdot}en)^{+} + H^{+} \rightleftarrows ZnH(glyox{\cdot}en)^{2+}$ 5.86(25°)	26
	$Zn(glyox{\cdot}en)^{+} + glyox^{1-} \rightleftarrows Zn(glyox_2{\cdot}en)$ 1.95(25°)	26
L' = salicylaldehyde anion, Sal^{1-}		
L'' = glycinate	Mn^{2+} β_{11} 7.26, β_{12} 9.15, β_{22} 13.04(25°)	30
	Ni^{2+} β_{11} 10.75, β_{12} 15.62, β_{22} 18.89(25°)	30
	Cu^{2+} β_{11} 16.15(25°)	30
	Zn^{2+} β_{11} 9.65, β_{12} 13.42, β_{22} 16.73(25°)	30
L'' = ethylamine	$Be^{2+} + (Sal{\cdot}Eta)^{1-} \rightleftarrows Be(Sal{\cdot}Eta)^{+}$ 10.4	35
	$Be^{2+} + 2(Sal{\cdot}Eta)^{1-} \rightleftarrows Be(Sal{\cdot}Eta)_2$ 18.3	35
	Zn^{2+} β_{11} 7.78, β_{22} 14.9(25°)	31
L'' = α-alaninate	Zn^{2+} β_{11} 9.56, β_{22} 16.23(25°)	31

Ligand	Equilibrium / constant	Ref.
L'' = β-alaninate	Zn^{2+} β_{11} 8.40, β_{22} 16.20(25°)	31
L'' = ethylenediamine	Zn^{2+} β_{11} 11.10, β_{12} 15.30(25°)	33
L^- = pyridoxal anion, PL^{1-}		
L'' = glycinate	Ni^{2+} β_{11} 10.30, β_{22} 19.84(25°)	39
	$Ni(PL{\cdot}Gly) + H^+ \rightleftarrows NiH(PL{\cdot}Gly)^+$ 7.18(25°)	39
	$Ni(PL{\cdot}Gly)_2^{2-} + H^+ \rightleftarrows NiH(PL{\cdot}Gly)_2^{1-}$ 7.48(25°)	39
	$NiH(PL{\cdot}Gly)_2^{1-} + H^+ \rightleftarrows NiH_2(PL{\cdot}Gly)$ 6.88(25°)	39
	$Ni(PL{\cdot}Gly)_2^{2-} + H^+ \rightleftarrows NiH(PL{\cdot}Gly)_2^{1-}$ 8.1	49
	$NiH(PL{\cdot}Gly)_2 + H^+ \rightleftarrows NiH_2(PL{\cdot}Gly)_2$ 7.3	49
	$Cu^{2+} + (PL{\cdot}Gly)^{2-} \rightleftarrows Cu(PL{\cdot}Gly)$ 14.4(25°)	(51)
	$Cu(PL{\cdot}Gly) + H^+ \rightleftarrows CuH(PL{\cdot}Gly)^+$ 6.05(25°)	(51)
	Zn^{2+} β_{11} 8.43, β_{22} 16.86(25°)	39
	$Zn(PL{\cdot}Gly) + H^+ \rightleftarrows ZnH(PL{\cdot}Gly)^{1+}$ 7.34(25°)	39
	$Zn(PL{\cdot}Gly)_2^{2-} + H^+ \rightleftarrows ZnH(PL{\cdot}Gly)_2^{1-}$ 7.64(25°)	39
	$ZnH(PL{\cdot}Gly)_2^{1-} + H^+ \rightleftarrows ZnH_2(PL{\cdot}Gly)_2$ 7.04(25°)	39
	$Zn^{2+} + (PL{\cdot}aa)^{2-} \rightleftarrows Zn(PL{\cdot}aa)$ β_1 7.4-8.1	51
	aa = glycinate and other amino acids	
	$Zn(PL{\cdot}Gly)^{2-} + H^+ \rightleftarrows ZnH(PL{\cdot}Gly)_2$ $K1_H = 8.1$	49
	$ZnH(PL{\cdot}Gly)_2^{1-} + H^+ \rightleftarrows ZnH_2(PL{\cdot}Gly)_2$ $K2_H = 7.3$	49
L'' = Valinate	$Mg^{2+} + (PL{\cdot}Val)^{2-} \rightleftarrows Mg(PL{\cdot}Val)$ <3.5(25°)	51

TABLE 2 (Continued)

	log equilibrium constant	Ref.
L^- = pyridoxal anion, PL^{1-} (continued)		
L" = Valinate (continued)	$Mn^{2+} + (PL{\cdot}Val)^{2-} \rightleftarrows Mn(PL{\cdot}Val)$ 5.0(25°)	51
	$Ni^{2+} + (PL{\cdot}Val)^{2-} \rightleftarrows Ni(PL{\cdot}Val)$ 10.8(25°)	51
	$Ni(PL{\cdot}Val) + H^+ \rightleftarrows NiH(PL{\cdot}Val)^+$ 6.7(25°)	51
	$Cu^{2+} + (PL{\cdot}Val)^{2-} \rightleftarrows Cu(PL{\cdot}Val)$ 14.5(25°)	51
	$Cu(PL{\cdot}Val) + H^+ \rightleftarrows CuH(PL{\cdot}Val)^+$ 5.6(25°)	
L" = α-alaninate	$Al(PL{\cdot}\alpha\text{-}Ala)^{1+} + H^+ \rightleftarrows AlH(PL{\cdot}\alpha\text{-}Ala)^{2+}$ 5.2(D_2O, 25°)	17
	$Zn(PL{\cdot}\alpha\text{-}Ala) + H^+ \rightleftarrows ZnH(PL{\cdot}\alpha\text{-}Ala)^+$ 6.6 (D_2O, 25°)	17
L' = pyridoxal 5'-phosphate, PLP^{3-}		
L" = glycinate	Zn^{2+} β_{11} 10.28, β_{12} 13.79, β_{22} 18.05(25°)	41
	$Zn(PLP{\cdot}Gly)^{2-} + H^+ \rightleftarrows ZnH(PLP{\cdot}Gly)^{1-}$ 7.45(25°)	41
	$ZnH(PLP{\cdot}Gly)^{1-} + H^+ \rightleftarrows ZnH_2(PLP{\cdot}Gly)$ 5.44(25°)	41
	$Zn(PLP{\cdot}Gly)_2^{6-} + H^+ \rightleftarrows ZnH(PLP{\cdot}Gly)_2^{5-}$ 8.32(25°)	41
	$ZnH(PLP{\cdot}Gly)_2^{5-} + H^+ \rightleftarrows ZnH_2(PLP{\cdot}Gly)_2^{4-}$ 6.94(25°)	41
L" = α-alaninate	Zn^{2+} β_{11} 10.26, β_{12} 13.82, β_{22} 17.08(25°)	41
	$Zn(PLP{\cdot}\alpha\text{-}Ala)^{2-} + H^+ \rightleftarrows ZnH(PLP{\cdot}\alpha\text{-}Ala)^{1-}$ 7.40(25°)	41
	$ZnH(PLP{\cdot}\alpha\text{-}Ala)^{1-} + H^+ \rightleftarrows ZnH_2(PLP{\cdot}\alpha\text{-}Ala)$ 5.45(25°)	41
	$Zn(PLP{\cdot}\alpha\text{-}Ala)_2^{6-} + H^+ \rightleftarrows ZnH(PLP{\cdot}\alpha\text{-}Ala)_2^{5-}$ 8.60(25°)	41
	$ZnH(PLP{\cdot}\alpha\text{-}Ala)_2^{5-} + H^+ \rightleftarrows ZnH_2(PLP{\cdot}\alpha\text{-}Ala)_2^{4-}$ 7.16(25°)	41

thorough characterization of the solution equilibria with the techniques then available. However, it was often found possible to simplify the equilibria by adjusting pH and concentrations so that only a few species need be considered, and quite useful results over limited concentration ranges were obtained. The more recent development and application of high-speed digital computers to this area now permits a wider range of conditions to be investigated and analyzed.

Eichhorn and Dawes [10] studied Schiff base formation in pyridoxal-alanine solutions containing Ni(II) or Cu(II). The spectra of the mixtures were quite different from those of the individual metal complexes of the components and showed that Schiff base complex formation does indeed occur. A continuous variation study demonstrated the formation of complexes in solution in which the metal ion-alanine-pyridoxal ratio is 1:2:2. Spectra of pyridoxamine-pyruvate-metal ion solutions were found to be distinctly different on first mixing the components, but on standing the spectra became identical to those of pyridoxal-alanine-metal ion solutions owing to transamination.

Christensen [46] isolated numerous metal chelates of pyridoxylidene amino acids in crystalline form. In particular, those of L-valine received special attention. An interesting paradox was observed. The crystalline 1:2:2 metal-PL^{1-}-Val^{1-} complexes of Mn(II), Fe(II), Fe(III), Ni(II), and Zn(II) and the 1:1:1 Cu(II) complex were all found to be electrically neutral, i.e., contained no counterions. This peculiar result was interpreted to indicate that the pyridine ring nitrogen atom of the pyridoxal moiety could be either protonated, or not, so as to preserve electroneutrality in the solid phase. This would require two pyridine protons for the Mn(II), Fe(II), Ni(II), and Zn(II) chelates, one for Fe(III), and none for Cu(II). IR spectral data confirmed these conclusions and clearly demonstrated the presence of the pyridinium group with the first set of divalent metal ions and its absence with the Cu(II) chelate. Thus, the compositions of the solids are

$MH_2(PL\cdot Val)_2$[M = Mn(II), Fe(II), Ni(II), Zn(II)], Fe(III)H(PL•Val), and Cu(PL•Val).

Titration of solutions of the dissolved metal chelates confirmed pyridine protonation and gave a pyridine protonation constant of about $10^{5.5}$ for the 1:1:1 Cu(II) complex:

$$Cu(PL\cdot Val) + H^+ \rightleftarrows CuH(PL\cdot Val)^+ \qquad (31)$$

The bis-Ni(II) chelate showed two protonation steps:

$$Ni(PL\cdot Val)_2^{2-} + H^+ \rightleftarrows NiH(PL\cdot Val)_2^{1-} \quad K_{1_H} \quad 10^{8.1} \qquad (32)$$

$$NiH(PL\cdot Val)_2^- + H^+ \rightleftarrows NiH(PL\cdot Val)_2 \quad K_{2_H} \quad 10^{7.3} \qquad (33)$$

The tendency for the Schiff bases to dissociate in acidic media proved to be a complication in this latter system. Dissociation created even more severe problems with the Mn(II) and Zn(II) chelates; nevertheless, evidence for pyridine protonation under pH conditions similar to those found for protonation of the Ni(II) complexes was still shown by these metal ions.

The protonation constant for the pyridine nitrogen atom is about $10^{8.66}$ for free PL^{1-} [47] and decreases to $10^{5.88}$ for $H(PL\cdot Val)^{1-}$ [48]. The basicity of the Schiff base pyridine nitrogen in the Ni(II)(PL•Val) chelate very closely resembles the former, while that of the Cu(II)(PL•Val) chelate is closer to that of the latter. Christensen [46] interpreted this striking sensitivity of the pyridine nitrogen atom to the nature of the metal ion as an indication that the metal ion-phenolate bond is largely ionic with Ni(II) and is more covalent with Cu(II).

In another paper Christensen [49] reported the results of a spectrophotometric study of the competition between excess valinate and $PL\cdot Val^{2-}$ for Cu(II). The data permitted the evaluation of the constant for the reaction:

$$Cu^{2+} + PL\cdot Val^{2-} \rightleftarrows Cu(PL\cdot Val) \quad \beta_{MSB} = 10^{14.4} \qquad (34)$$

Thus, the complex formed between Cu(II) and the Schiff base is even more stable than that formed between Cu(II) and a strongly complexing ligand such as nitrilotriacetate [50]:

$$Cu^{2+} + NTA^{3-} \rightleftarrows Cu(NTA)^{1-} \quad \beta_1 = 10^{12.96} \tag{35}$$

Davis et al. [51] also performed a spectrophotometric study of pyridoxylidenevaline metal complexes. They observed the formation of highly stable complexes which followed the Irving-Williams order Cu(II) >> Ni(II) > Zn(II) > Mn(II) > Mg(II). These authors also confirmed Christensen's observation that pyridine protonation of the Schiff base chelates occurs in solution. Values of the ring nitrogen protonation constant

$$M(PL{\cdot}Val) + H^{+} \rightleftarrows MH(PL{\cdot}Val)^{1+} \tag{36}$$

were reported to run from $10^{5.6}$ for Cu(II) to $10^{6.7}$ for Ni(II), in good agreement with Christensen's results.

The Bjerrum pH-titration technique for determining stability constants [52] provides direct information about the course of the hydrogen ion activity as a mixture containing ligands and a complexing metal ion is titrated with acid or base. In turn the hydrogen ion activity is a function of the concentrations of basic species in solution and provides a monitor of these concentrations. This method was adopted to the investigation of Schiff base equilibria in the presence and absence of metal ions by Leussing and Schultz [19]. Straightforward numerical and graphical methods have been developed and extensively employed for obtaining the complex equilibrium constants in binary systems [53], but the Schiff base systems involve ternary complexes (two ligands, one metal ion) and are not amenable to these approaches. Computer-based curve-fitting techniques were developed to analyze the data [18,19,30].

The potentiometric data yield values of i, j, and the formation constants β_{ij}, for reaction scheme (30) regardless of whether or not A and O interact or are independently bound in the complex. Thus, pertinent to the present topic, these β_{ij} represent the sum of the formation constants of Schiff base complex, carbinolamine complex, and conventional mixed complex for species having the same i and j. The formation constants for the mixed complex may be estimated with reasonable accuracy, in the absence of unusual steric or charge effects, from the formation constants of the corresponding binary complexes. In general, the mixed complexes have been found to make only small contributions to the experimentally determined β_{ij}. Thus, pertinent to Schiff base systems the equilibrium constant found for the formation of a species $M(A)_i(O)_j$ is essentially the same as that for the formation of the Schiff base complex, $M(A{\cdot}O)_i$-$(A)_{j-1}$, where A•O represents a Schiff base.

The experimentally determined composite formation constants for some pyruvate-glycinate-metal ion systems are presented in Table 2. While these values contain all the stability information, the magnitudes of the interactions involved are more clearly seen by comparing the stepwise constants in the manner shown in Figure 2. In Figure 2 an upward arrow represents in the acquisition of a glycinate ion by the metal ion and a downward arrow indicates the acquisition of a pyruvate ion. The number above each arrow is the log of the addition constant for the particular step.

The values given for the formation of binary complexes along the uppermost and lowermost arrows show the much weaker affinity of pyruvate for Ni(II) compared to glycinate. Also the values for both ligands show the characteristic decrease in stability which arises from binding and statistical factors as the degree of

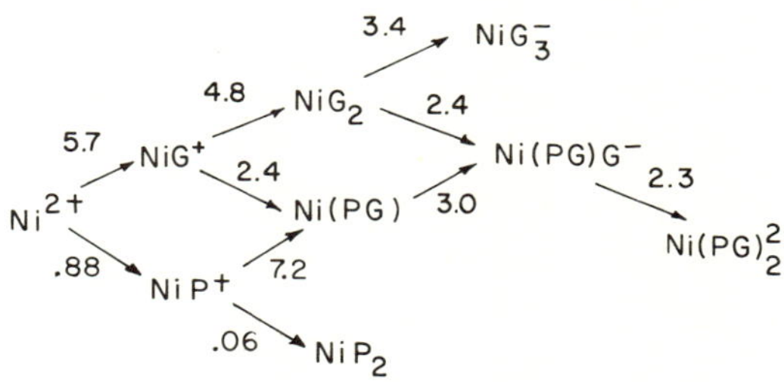

FIG. 2. The stepwise addition of pyruvate (P^-) and glycinate (G^-) to Ni(II). The number associated with each arrow is the log of the stepwise constant.

coordination increases. In contrast, when the ternary complexes are formed, it is seen that the presence of a ligand of one kind considerably enhances the addition of a ligand of the other kind, i.e., the presence of pyruvate in the metal ion coordination sphere enhances the coordination of glycinate, and vice versa. This enhancement arises from the presence of strong interligand attractive forces in the complexes. The addition of a second glycinate to Ni(Pyu•Gly) to give $Ni(Pyu)(Gly)_2^{1-}$ is seen to occur with a constant which is "normal" for glycinate binding, indicating that the second glycinate is independently bound in a species more appropriately formulated as $Ni(Pyu{\cdot}Gly)(Gly)^{1-}$. Pyruvate addition to this last species is once again considerably enhanced from interligand association.

Strong absorption in the UV region shows that the associated ligands are coordinated to the metal ion predominantly in the form of the Schiff base rather than as the carbinolamine.

An interesting and instructive comparison is shown in Table 3 for the stepwise coordination of the ligands N-pyruvideneglycinate, iminodiacetate (IDA^{2-}), and 1,10-phen to various metal ions. IDA^{2-} is analogous to $(Pyu{\cdot}Gly)^{2-}$ in that each is a tridentate ligand having one nitrogen donor atom and two carboxylate donor groups. 1,10-phen is a typical unsaturated ligand which forms complexes that show the effects of metal-ligand π back-bonding.

For the formation of the 1:1 complexes with various metal ions in the series Ca(II) to Zn(II), IDA^{2-} and $Pyu{\cdot}Gly^{2-}$ are seen in Table 3 to form complexes having remarkably similar stabilities. However, the behavior for the addition of the second ligand differs between these species. Compared to K_1 the K_2 values show a decrease with IDA^{2-} that is not observed with $(Pyu{\cdot}Gly^{2-})$. The decrease with increasing coordination is typical of the behavior of saturated ligands, but the unsaturated ligands, such as 1,10-phen, show stepwise constants which are relatively unchanged or may even show an increase as the number of coordinated ligands increases. Thus, $Pyu{\cdot}Gly^{2-}$ shows coordination behavior that is characteristic

TABLE 3

A Comparison of N-Pyruvideneglycinate, Iminodiacetate, and 1,10-Phenanthroline Complex Formation Constants

$$ML_{i-1} + L \overset{K_i}{\rightleftharpoons} ML_i$$

M\L		log equilibrium constant		
		IDA^{2-}	$Pyu{\cdot}Gly^{2-}$	1,10-phen
Ca^{2+}	K_1	2.59	3.3	--
Mn^{2+}	K_1	--	5.0	4.5
	K_2	--	4.0	4.15
Ni^{2+}	K_1	8.2	7.6	8.0
	K_2	6.3	7.2	8.0
Zn^{2+}	K_1	7.3	7.1	6.3
	K_2	5.3	6.3	5.6

of a tridentate ligand having one nitrogen and two oxygen donor atoms, but in which there is also a high degree of unsaturation.

The d-d electronic spectra of the Ni(II) complexes also provide valuable information as to the nature of the bonding in the adduct complexes [19]. The $^3A_{2g} \rightarrow {}^3T_{2g}$ and $^3A_{2g} \rightarrow {}^3T_{1g}$ transitions of octahedral $Ni(Pyu{\cdot}Gly)(H_2O)_3$ are found to be centered at 9660 cm^{-1} and 15,700 cm^{-1}. These band positions are very close to those observed for $Ni(Gly)(H_2O)_4^+$, 9660 cm^{-1} and 15,530 cm^{-1}, showing that the ligands in these two mono complexes have similar ligand field strengths. However, with the bis complexes the ligand field strength of $(Pyu{\cdot}Gly)^{2-}$ is somewhat greater than that of Gly^{1-}: in $Ni(Pyu{\cdot}Gly)_2^{2-}$, the first transition ($^3A_{2g} \rightarrow {}^3T_{2g}$) lies at 10,260 cm^{-1}, whereas with $Ni(Gly)_2$ it is found at 9900 cm^{-1}. In contrast, the second transitions in the bis complexes have similar values, 16,330 cm^{-1} and 16,230 cm^{-1}, respectively. This spectral behavior shows that in the bis complexes $(Pyu \cdot Gly)^{2-}$ also occupies a higher position in the nephelauxetic series [54] than does glycinate and this is consistent with a higher degree of unsaturation (and more metal-ligand π back-bonding) in the adduct.

Steric hindrance between groups which lie on either side of the $>$C=N- group of aliphatic Schiff bases is observed to have a larger effect on the stabilities of the metal ion complexes than occurs in the free Schiff bases. With the free Schiff bases rotation about this bond to reduce steric repulsions can readily occur, but coordination to a metal ion forces the substituents into cis positions where maximum interference exists. In Table 2 the stabilities of the N-pyruvidene-α-alaninato complexes are seen to be appreciably lower than those of the pyruvidene glycinato complexes even though the corresponding binary bisalaninato and glycinato complexes are seen to have similar stabilities [21]. Pyruvidene isoleucinate complexes are found to be even weaker than those of α-alaninate.

Replacing pyruvate by glyoxylate reduces steric repulsions in the Schiff base complexes and even α-aminoisobutyrate is found to form a Schiff base complex that is about as stable as that of glycinate [24]. The imidazolidines formed between glyoxylate and 1,2-diamines do not form particularly stable complexes owing to their bidentate character and the relatively low σ-donor properties of cyclic nitrogen [26]. These ligands have similar coordination behavior to proline, consistent with their lower basicity.

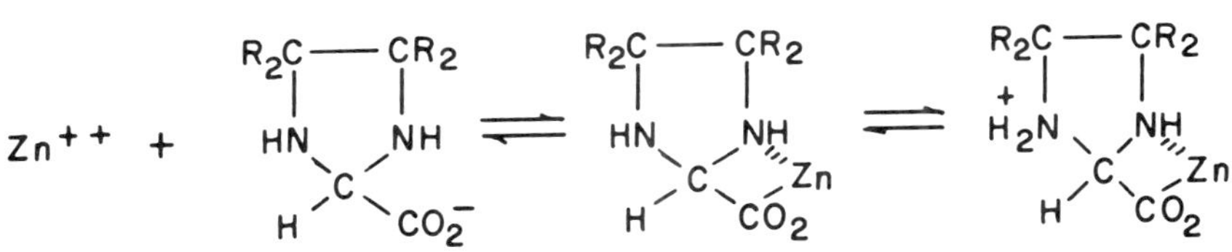

XXXII XXXIII

(37)

In the presence of Zn(II) or Ni(II) an excess of glyox$^-$ is able to open the ring of the en derivative to form a complexed tetradentate ligand Zn(en•glyox$_2$). The lack of UV absorbance indicates that this bisglyoxylate adduct exists as the carbinolamine Zn(O_2CCH(OH)NHCH_2CH_2NH(OH)CHCO_2) [26].

An interesting difference between the Schiff base and simple binary complexes is observed with respect to their tendencies to form 5- and 6-membered chelate rings [21]. Owing to a large degree to entropy effects, the stabilities of binary complexes having 6-membered chelate rings are appreciably lower than those having 5-membered rings, e.g., K_1 for Ni(β-Ala)$^{1+}$ is $10^{4.46}$ but for Ni(Gly)$^{1+}$ K_1 is $10^{5.65}$. This order of 6 < 5 is reversed in the pyruvidene-amino acid complexes; for example, β_{11} is $10^{8.34}$ for Ni(Pyu•β-Ala) compared to $10^{8.07}$ for Ni(Pyu-Gly). The bond angles about the unsaturated imine nitrogen are responsible for this reversal since they tend to direct the carboxylate group of the amino acid moiety away from the metal ion. The additional length of the 6-membered ring reduces the ring strain required to coordinate the carboxylate group to a metal ion coordination site. This does not seem to apply to tridentate Schiff bases which form two 6-membered chelate rings since N-salicylidene-β-alaninato zinc(II), as an example, is found to be appreciably less stable than N-salicylideneglycinato zinc(II) [31].

The configuration of the imine bond was found to have a significant influence on the coordination of N-αketoglutaridine α-alaninate relative to that of N-pyruvidene-α-glutamate [29]. These two Schiff bases are isomers, differing in the configuration of the C=N-CH group and, as reported above, the uncomplexed forms have about equal stabilities. However, the formation constant of Zn(Pyu•Glu)$^{1-}$ (XXXV) has been found to be about 10 times larger than that for Zn(α-Kg•Ala)$^{1-}$ (XXXIV). In Zn(Pyu•Glu)$^{1-}$ the propionate group, which is α to the C=N group, is oriented so that coordination to the metal ion is possible. In the isomeric Zn(αKg•Ala)$^{1-}$ complex the propionate group is directed away from the metal ion and is unable to form the third chelate ring.

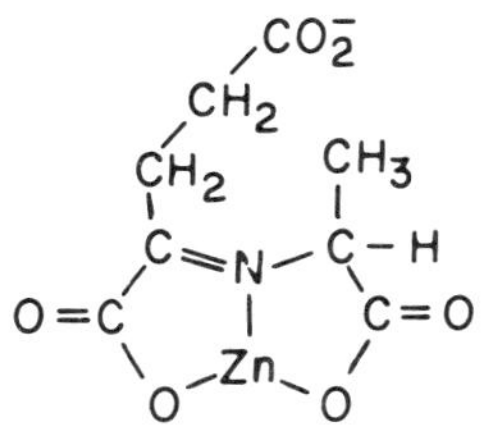

$Zn(\alpha KG\cdot ALA)^-$

XXXIV

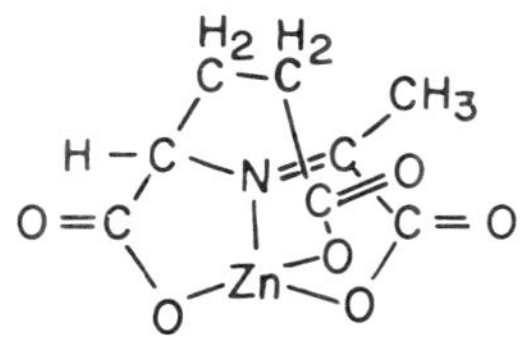

$Zn(PYU\cdot GLU)^-$

XXXV

Evidence for the formation of the isomeric mixed complexes $Zn(Pyu\cdot Glu)(Ala)^{2-}$ and $Zn(Pyu\cdot Ala)(Glu)^{2-}$ and the mixed Schiff base complex $Zn(Pyu\cdot Glu)(Pyu\cdot Ala)^{3-}$ was also obtained [29]. The stabilities of these more complicated species are consistent with predictions based on the stabilities of the simpler complexes from which they are derived.

The Schiff base-metal ion systems in which aromatic o-hydroxyaldehydes are employed present more complicated equilibrium situations than do the aliphatic systems: protonation of the hydroxy group of the free aldehyde cannot be neglected nor can protonation of uncomplexed Schiff base. Furthermore, the pyridoxal and pyridoxamine systems present an additional protonation site at the ring nitrogen atom, and the problem of a multiplicity of protonated species is further compounded with their phosphate derivatives. Nevertheless, the curve fitting procedures previously worked out to analyze potentiometric titration data proved to be readily applicable to these more complicated systems.

Salicylaldehyde-amino acid-metal ion mixtures form a series of complexes in solution which are qualitatively similar to those found for pyruvate and glyoxylate. Complexes were found having the compositions M(II)(Sal•Gly), $M(II)(Sal\cdot Gly)(Gly)^{1-}$, and $M(II)(Sal\cdot Gly)_2^{2-}$ [18,30]. Quantitatively, the salicylaldehyde Schiff bases have a considerably higher affinity for metal ions than do the aliphatic Schiff bases owing to the higher affinity of salicylaldehyde anion itself over pyruvate and glyoxylate for metal ions. In addition, the planar aromatic rings offer less steric interference to complex formation than does the methyl group of pyruvate. The Schiff base $(Sal{\cdot}\alpha\text{-}Ala)^{2-}$ forms complexes that are practically as stable as those formed by $(Sal{\cdot}Gly)^{2-}$ [31], but $(Sal{\cdot}\beta\text{-}Ala)^{2-}$ forms complexes that are substantially less stable owing to the presence of _two_ 6-membered chelate rings. The stabilities of the bis complexes are also high, probably from the influence of π back-bonding. Cu(II), which tends to be 4-coordinate, is an exception to this rule [30] and forms only a 1:1 complex having a high stability:

$$Cu^{2+} + (Sal{\cdot}Gly)^{2-} \rightleftarrows Cu(Sal{\cdot}Gly) \quad K_1 = 10^{15.67}$$

The Schiff base formed between ethylamine and salicylaldehyde is bidentate and forms much weaker complexes than the tridentate Schiff bases derived from the amino acids [31]. Because $Sal{\cdot}Eta^{1-}$ also has a highly basic imine group, a good proportion of the Schiff base in solution containing a coordinating metal may be present as uncomplexed H(Sal•Eta). Green and Sleet [35] investigated the formation of Be(II) complexes of N-salicylidene ethylamine. The species $Be(Sal{\cdot}Eta)^{1+}$ and $Be(Sal{\cdot}Eta)_2$ were found to have formation constants substantially higher than those reported for the corresponding Zn(II) complexes. Since Be(II), being a hard acid, is a better oxygen binder than nitrogen binder, the high affinity of Be(II) for the Schiff base may be a manifestation of strong metal ion-phenoxide bonding. The neutral bis Schiff base complex of Be(II) was found to be highly extractable from water into toluene.

Pyridoxal and pyridoxal 5'-phosphate in the presence of glycinate and Ni(II) or Zn(II) were found to form a series of Schiff base complexes having compositions analogous to those formed by the aliphatic carbonyl donors [39,41]. The metal ion complexes of $(PL{\cdot}Gly)^{2-}$ are appreciably weaker than those of $(Sal{\cdot}Gly)^{2-}$, the more so with Zn(II) than with Cu(II) and Ni(II). This decrease in affinity for metal ions arises in part from the lower basicity of the $(PL{\cdot}Gly)^{2-}$ imine group but the change is larger for the metal ions than for the proton. This suggests that metal ion-ligand π-bonding may also be involved and is appreciably greater with the Schiff base derived from salicylaldehyde than from pyridoxal.

The presence of the phosphate group causes complex stability to be restored owing to a strong electrostatic interaction between the negative PLP phosphate group and positively charged ions. This effect compensates for the weakening effect of the heterocyclic nitrogen so that the stability of the Schiff base complex, $Zn(PLP\cdot Gly)_2^{6-}$, is similar to that observed for the salicylaldehyde derivative. The high charge of the bis complex, $Zn(PLP\cdot Gly)_2^{6-}$, causes some destabilization in this species relative to the mono complex, however.

The results of the potentiometric titrations confirmed the findings of Christensen [46] and Metzler et al. [51] that pyridine nitrogen protonation occurs with the PL Schiff base complexes, and, in fact, the potentiometric technique permitted a more detailed assessment of this behavior. The pK_a values found for M(PL•Gly) are 7.18 for M equals Ni(II) and 7.34 for M equals Zn(II). The bis complexes $M(PL\cdot Gly)_2^{2-}$ yielded two values statistically distributed about that found for the corresponding mono complex

$$MH_2(PL\cdot Gly)_2 \rightleftarrows MH(PL\cdot Gly)_2^{1-} + H^+ \quad pK_{1a} = 6.9(Ni) \quad 7.0(Zn) \tag{38}$$

$$MH(PL\cdot Gly)_2^{1-} \rightleftarrows M(PL\cdot Gly)_2^{2-} + H^+ \quad pK_{2a} = 7.5(Ni) \quad 7.6(Zn) \tag{39}$$

These values contrast markedly with the pK_a (5.88) of the pyridinium group in the corresponding proton-stabilized Schiff base (XIX). Thus, the influence of these metal ions on the basicity of the ring nitrogen atom of the coordinated ligand is seen to be considerably less than that of an imine proton.

The value of β_{11} for the formation of Ni(PL•Gly) is about 2 log units greater than that of Zn(PL•Gly), but in spite of this the Ni(II) Schiff base complexes show pK_a values which are only a few tenths of a pK_a unit more acidic than those of the Zn complexes. Very likely the metal-phenolate bond is highly ionic with these two metal ions. A very strong metal-ligand interaction, such as occurs with Cu(II), causes the pyridinium pK_a to be lowered to about 5.6

[46]. In this case there is probably more transfer of charge from the aromatic ring to the metal ion. The shifts of the UV band positions when the pyridoxal Schiff base forms complexes confirms that Zn(II), at least, interacts more strongly with the imine group than with the phenolate group [30].

Similar results were observed for the protonation behavior of the pyridoxal 5'-phosphate Schiff base Zn(II) complexes. The pK_a of the pyridinium group of $ZnH(PLP{\cdot}Gly)^{1-}$ is 7.4 [41], close to that of the PL analog, but the high charge of the bis complex $Zn(PLP{\cdot}Gly)_2^{6-}$, brings about an increase in the pK_a for the ionization of the last proton; the pK_a values for the successive ionization steps of the two pyridinium groups of $ZnH_2(PLP{\cdot}Gly)_2^{4-}$ are 6.9 and 8.3 [41]. Correcting these values for the statistical factor which applies to the ionization of identical groups in a molecule gives values which are close to that of the monocomplex, 7.2 and 8.0.

The phosphate group of $ZnH(PLP{\cdot}Gly)^{1-}$ has a pK_a of 5.44 [41], which is close to that of the phosphate group in PLP itself. It was not possible to determine these values for phosphate protonation in the bis complex owing to its extensive dissociation at $pH < 6$, but it is likely that the values lie in the same range. Typical distribution curves for the Zn(II)-PLP-Gly system are presented in Figure 3.

NMR investigations of pyridoxal-alanine interactions in the presence of Zn(II) or Al(III) have been made by Gansow and Holm [17]. These authors confirmed the formation of 1:1 and 2:1 aldimine: metal ion complexes in D_2O. The Zn(II) complexes were found to exchange rapidly with free aldimine on the NMR time scale, but those of Al(III) were found to be essentially "inert," so that the spectra of the individual 1:1 and 2:1 complexes could be readily identified. As the solution pD was increased above 7 the effect of the deprotonation of the pyridine ring nitrogen on the chemical shifts of the Zn(II) aldimine complexes was observed. The Al(III) aldimines were found to undergo deprotonation at lower

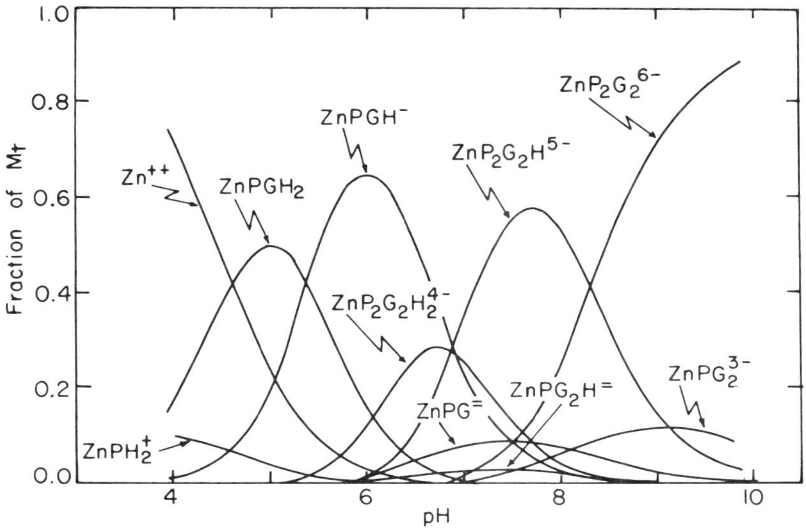

FIG. 3. The fraction of Zn(II) present in various forms as a function of pH in a Zn(II)-pyridoxal 5'-phosphate(P)-glycinate(G) system. 0.004 M Zn(II); 0.00995 M PLP; 0.0200 M Gly. (From Ref. 41, by permission of the publishers.)

pD (∿5.2), showing that the Al(III) aldimines are similar to those of Cu(II) in this respect. Using DL-alanine these authors also found evidence for the formation of isomeric bisaldimine Al(III) complexes that are not formed when only L-alanine is used. The diastereoisomers differ in the positions of the 2-methyl proton resonances of the pyridoxal moieties.

Abbott and Martell [16] have confirmed the observations of Gansow and Holm and explained the results on the basis that the 2-CH_3 group of one ligand is in the shielding region of the azomethine π system of the other ligand. Because of this the chemical shift of the 2-CH_3 protons is hypersensitive to small steric interactions with the asymmetric center of the amino acid portion of the other ligands. The shifts are so sensitive that "mixed" ligand complexes formed by using D,L-mixtures of amino acids are easily identifiable.

In Figure 4 are shown portions of the NMR spectra obtained by Abbott and Martell in the 2-CH_3 region for solutions of 0.10 M

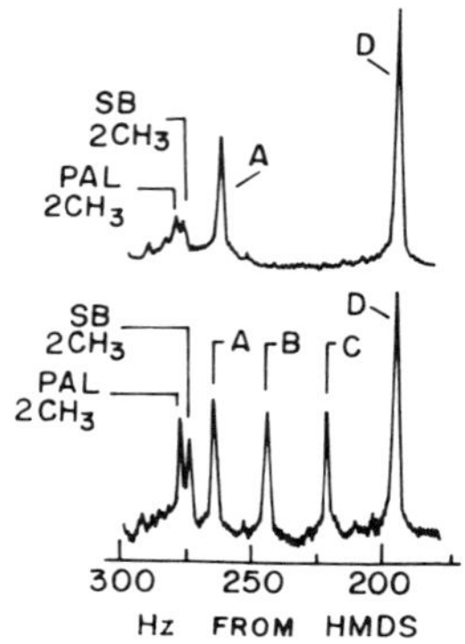

FIG. 4. 100-MHz NMR spectra. Top: 0.10 M L-valine, 0.10 M pyridoxal, 0.050 M Al(III), pD 8.9; Bottom, 0.10 M DL-valine, 0.10 M pyridoxal, 0.050 M Al(III), pD 9.7. See text for further discussion. (From Ref. 16, by permission of the publishers.)

pyridoxal, 0.050 M Al(III) at pD 8.9 for (i) 0.10 M L-valine and (ii) 0.10 M DL-valine. In solutions containing only the pure enantiomorph the spectrum of $Al(PL{\cdot}Val)_2^{1-}$ shows two resonances which are designated in Figure 4 as A and D. The proposed structures for these isomers are as shown below.

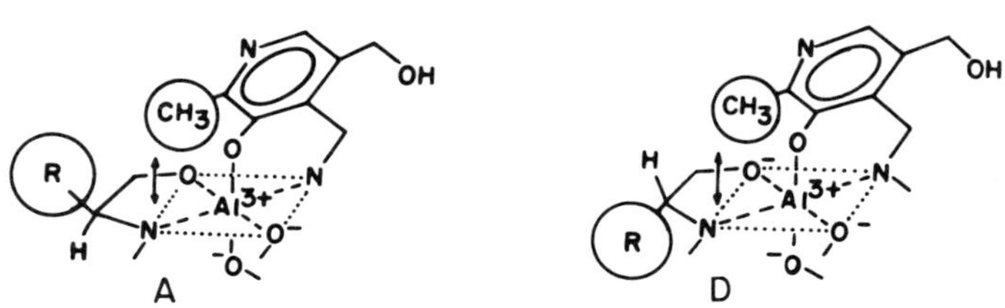

Resonance A: maximum steric repulsion: minimum shielding.
Resonance D: minimum steric repulsion: maximum shielding.

The low field resonance at A is assigned to the isomer in which both of the 2-CH_3 groups are crowded and in which the shielding is a minimum. Resonance D is assigned to the isomer in which shielding is a maximum and steric repulsion is a minimum. Consistent with these assignments resonance D is seen to arise from the most energetically favorable and, therefore, the most abundant of these isomers in solution. Resonances B and C that arise using the racemic mixtures are assigned to the crowded and uncrowded 2-CH_3 resonances of the identical D,L and L,D complexes.

In mixtures containing two different amino acids, the number of possible 2-CH_3 resonances increases. The four theoretical 2-CH_3 bands of pyridoxylidene glycinatopyridoxylidene norvalinato Al(III) were found to be clearly distinguishable. In the most complicated case involving two racemic Schiff bases 16 different 2-CH_3 resonances are possible. For the system pyridoxal, DL-valine, DL-norvaline Al(III), 12 resonances were found and were assigned to definite species.

The pK_a of the heterocyclic nitrogen of $Al(PL{\cdot}Val)_2^{1-}$ was found to be around 5.8 in reasonable agreement with the results of Gansow and Holm [17].

When comparing the stabilities of the Schiff base metal ion complexes with the stabilities of the corresponding binary complexes formed by the component ligands it becomes apparent that a correlation exists between the two sets of values, i.e., those factors which cause a given metal ion to form either a strong or weak complex with a Schiff base are determined to a large extent by those factors which cause the metal ion to form strong or weak complexes with the individual ligands comprising the Schiff base. In fact it was found possible to demonstrate that an almost quantitative relationship exists between the formation constants obtained for glycinato Schiff base complexes, β_{11}^*, and the stability constants for the formation of the corresponding mixed complexes in which the ligands are independently attached to the metal ion, β_{11}^* [18].

The mixed ligand formation constant is defined for the reaction:

$$M^{m+} + \rangle C{=}O + Gly^- \overset{\beta_{11}^*}{\rightleftharpoons} (\rangle C{=}O)\, M^{m+}(Gly^-) \tag{40}$$

Considering only statistical factors, it can be shown [18] that β^*_{11} is related to the individual stepwise addition constants for the binary systems according to equation (41):

$$\log \beta^*_{11} = \log (K_{1_{CO}} K_{2_{Gly}} + K_{2_{CO}} K_{1_{Gly}}) \qquad (41)$$

where the K_{iA} are defined for the reactions

$$M(A)_{i-1} + A \rightleftarrows M(A) \qquad (42)$$

A plot of the logarithms of the Schiff base formation constants, β_{11}, versus the logarithms of the corresponding β^*_{11} for a variety of metal ions and carbonyl donors is shown in Figure 5. The points, for which the equilibrium constants range over 10 orders of magnitude, are seen to fall along a straight line having a slope of 1.0. The intercept of 2.0 corresponds to the log of the equilibrium constant for fusing two coordinated bidentate ligands

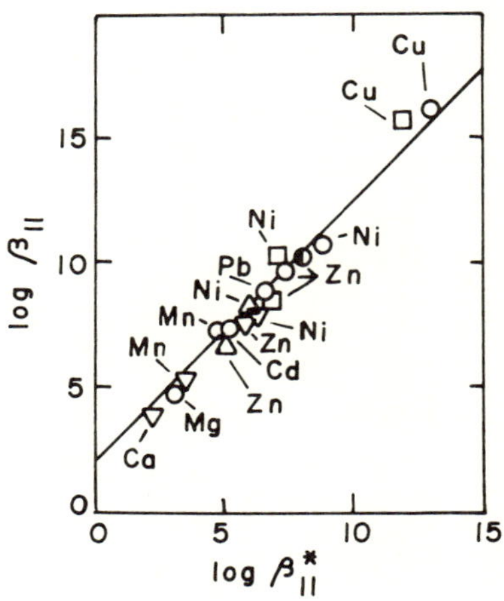

FIG. 5. The enhancement of stability on forming a glycinate Schiff base complex from the complex containing the independently bound ligands. (○) salicylaldehyde; (△) glyoxylic acid; (▽) pyruvic acid; (□) pyridoxal; (◐) pyridoxal 5'-phosphate. β_{11} is the Schiff base formation constant and β^*_{11} is the mixed complex formation constant. (From Ref. 18 by permission of the publishers.)

into one coordinated tridentate ligand [eq. (43)]. The indepen-

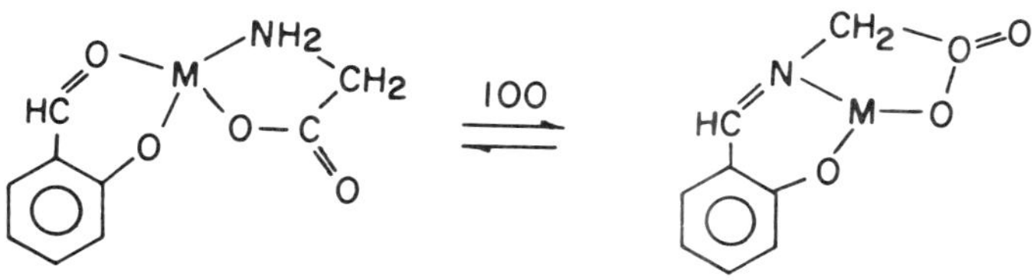 (43)

dence of this condensation constant on metal ion suggests that the reaction is chiefly entropy-controlled. Although glycinate was the only amino acid examined, presumably other amino acids which do not involve appreciably different steric or charge factors will behave similarly.

The fusion constant for the salicylaldehyde-ethylamine-Zn(II) system is about 10^3:

$$\text{(Sal)M(Eta)} \rightleftarrows \text{M(Sal}\cdot\text{Eta)} + H_2O \qquad (44)$$

This system, which involves the merging of one-coordinated bidentate chelate and one-coordinated monodentate ligand to give a one-coordinated bidentate ligand is qualitatively different from the glycinate reactions described above. The higher fusion constant probably results from the relatively low coordinating tendency of the monodentate ligand to a metal ion compared to a bidentate ligand, rather than from an unusually stable Schiff base complex. In other words, β_{11}^* is probably low with respect to the amino acid systems rather than β_{11} being unusually high.

Pyruvidinepyridoxamine 5'-phosphate, $PMP\cdot Pyu^{4-}$, reacts with Zn(II) to form highly stable 1:1 and 1:2 complexes:

$$Zn^{2+} + PMP^{3-} + Pyu^{-} \underset{}{\overset{\beta_{11}}{\rightleftharpoons}} Zn(PMP\cdot Pyu)^{2-} + H_2O \qquad (45)$$

$$Zn^{2+} + 2PMP^{3-} + 2Pyu^{-} \underset{}{\overset{\beta_{22}}{\rightleftharpoons}} Zn(PMP\cdot Pyu)_2^{6-} + 2H_2O \qquad (46)$$

The formation constants are one to 2 orders of magnitude greater than those found for the formation of the analogous $(Pyu \cdot Gly)^{2-}$-Zn(II) complexes. This is a manifestation in the Schiff base of the high affinity of PMP^{3-} for Zn(II) compared to glycinate and once again illustrates the influence of the stabilities of binary complexes on the Schiff base complexes derived from them.

Although highly stable Zn(II) complexes are formed by the ketimine $(PMP \cdot Pyu)^{4-}$, the complexes are weaker than those formed by the aldimine tautomer:

$$Zn^{2+} + PMP \cdot Pyu^{4-} \underset{}{\overset{\beta_{ZnSB} = 10^{7.4}}{\rightleftharpoons}} Zn(PMP \cdot Pyu)^{2-} \tag{47}$$

$$Zn^{2+} + PLP \cdot Ala^{4-} \underset{}{\overset{\beta_{ZnSB} = 10^{9.7}}{\rightleftharpoons}} Zn(PLP \cdot Ala)^{2-} \tag{48}$$

Thus, metal ion complexing is seen to be strongly influenced by the positions of the proton and double bond of the Schiff base >C=N-C-H group. There appear to be two reasons for the lower ketimine complex stability. For one, the lower ketimine basicity will be reflected in weaker Zn(II)-ketimine σ bonding than is found in the aldimine. For another, molecular models show that chelate ring strain and steric hindrance exists in the ketimine complex. The chelate ring formed by the pyridoxamine nitrogen, the phenolate oxygen, and the Zn(II) ion is considerably crowded, while coordination of the ring arising from the amino acid moiety to the metal ion causes strain. In contrast to this situation, the aldimine provides almost optimum geometry for planar terdentate coordination. As a result metal ion chelation considerably stabilizes the aldimine with respect to the ketimine shown in reaction (49).

The protonation scheme of the PMP ketimine complexes is similar to that described above for the aldimine complexes. The protonation constant for the reaction (50) coincides within experimental error with that found for pyridine protonation of pyridoxamine (51).

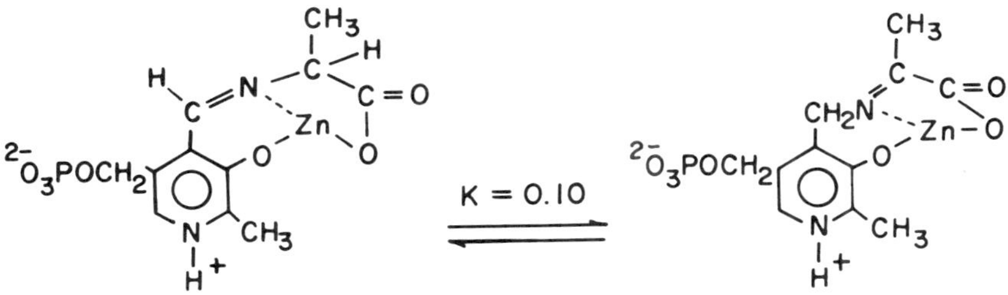

(49)

$$Zn(PMP\cdot Pyu)^{2-} + H^+ \underset{}{\overset{K_H = 10^{8.44}}{\rightleftharpoons}} Zn(PMP\cdot Pyu)H^{1-} \tag{50}$$

$$HPMP^{2-} + H^+ \overset{K_H = 10^{8.43}}{\rightleftharpoons} H_2PMP^{1-} \tag{51}$$

The bis complex $Zn(PMP\cdot Pyu)_2^{6-}$ is somewhat more basic with respect to uptake of the first proton:

$$Zn(PMP\cdot Pyu)_2^{6-} + H^+ \overset{K_{1_H} = 10^{9.17}}{\rightleftharpoons} Zn(PMP\cdot Pyu)_2H^{5-} \tag{52}$$

but the second proton is slightly more acidic:

$$Zn(PMP\cdot Pyu)_2H^{5-} + H^+ \overset{K_{2_H} = 10^{8.14}}{\rightleftharpoons} Zn(PMP\cdot Pyu)_2H_2^{4-} \tag{53}$$

Correcting for the statistical factor gives intrinsic stepwise protonation constants or $K_{1_H} = 10^{8.87}$ and $K_{2_H} = 10^{8.44}$, respectively. The latter value is practically identical to the pyridine protonation constant of $HPMP^{2-}$, while the former is only slightly higher owing to the high negative charge of $Zn(PMP\cdot P)_2^{6-}$. More direct evidence for pyridine protonation is obtained from the NMR studies on pyruvidenepyridoxamine by Gansow and Holm [17], who report pK_a of about 8.3 for $ZnD_2(PM\cdot Pyu)_2$ in D_2O. This value is in good agreement with that found potentiometrically for the PMP^{3-} system.

The basicity of the heterocyclic nitrogen atom in the Zn(II)-ketimine complex is greater than in the protonated uncomplexed ketimine. This same effect was observed with the aldimines but the difference is even greater between the ketimines because it is the phenol proton, as opposed to aldimine imminium proton, that is replaced by the Zn(II) ion. Owing to this situation the phenolate oxygen experiences a change from a highly covalent bond in the protonated ketimine to a highly ionic bond in the Zn(II) complex.

In addition to pyridine protonation the Zn(II)-(PMP•Pyu) complexes can be further protonated with constants characteristic of PMP phosphate protonation:

$$\mathrm{ZnH(PMP{\cdot}Pyu)^{1-} + H^+ \xrightleftharpoons{K_{2_H} = 10^{6.42}} ZnH_2(PMP{\cdot}Pyu)} \quad (54)$$

$$\mathrm{ZnH_2(PMP{\cdot}Pyu)_2^{4-} + H^+ \xrightleftharpoons{K'_{3_H} = 10^{7.05}} ZnH_3(PMP{\cdot}Pyu)_2^{3-}} \quad (55)$$

$$\mathrm{ZnH_3(PMP{\cdot}Pyu)_2^{3-} + H^+ \xrightleftharpoons{K'_{4_H} = 10^{5.87}} ZnH_4(PMP{\cdot}Pyu)_2^{2-}} \quad (56)$$

4. THE INFLUENCE OF METAL IONS ON THE RATES OF SCHIFF BASE FORMATION

Schiff base formation and hydrolysis rates in the absence of metal ions have been studied for many years, and extensive reviews have been published on this subject [55,56]. Relatively few studies have been performed on metal ion effects, however, and attention here will be concentrated on this area. Nevertheless, the general features of Schiff base formation reactions in the absence of metal ions will first be discussed for those systems pertinent to transamination.

Owing to their chelating properties, Schiff bases in general are stabilized by the presence of metal ions. Thus, metal ions inhibit hydrolysis rates regardless of whether or not they exhibit kinetic activity. When back reactions are inhibited, rates of formation may appear to be faster and metal ions will tend in many cases to show apparent kinetic activity. In addition, metal ions will complex with the reactants and reduce the concentrations of the free reactants in solution. Complexing will therefore tend to reduce the forward rate if the binary complexes so formed are unreactive. On the other hand, if the binary complexes are active this will tend to compensate for, if not override, the effect of the lower concentrations of the free species. Owing to this variety of parallel effects which must be sorted out, deriving a quantitative rate law for metal ion-dependent paths in Schiff base formation may be a complicated process. Careful work and a detailed knowledge of the equilibria and reaction rates in the binary and ternary systems is required.

Schiff base formation is generally conceded to take place in stepwise fashion with the formation of an intermediate carbinolamine [38,55,56]:

$$\text{-}\overset{\overset{\displaystyle O}{\|}}{C}\text{-} + H_2N\text{-} \underset{k_{-1}}{\overset{k_1}{\rightleftharpoons}} \text{-}\overset{HO}{\overset{|}{C}}\text{-}\overset{H}{\overset{|}{N}}\text{-} \underset{k_{-2}}{\overset{k_2}{\rightleftharpoons}} \text{-C=N-} + H_2O \tag{57}$$

With aromatic aldehydes in which the ortho position is occupied by a hydroxy group, with pyruvate and likely with other α-ketoacids, the intermediate carbinolamine is not observed because it rapidly decomposes through either a backward path to give reactants or a forward path to give products. In the absence of the ortho hydroxy group and with aliphatic aldehydes, the formation of the intermediate may be observed [55-58].

Application of the steady state approximation to those cases where an unstable intermediate is formed in low concentration yields the rate expression

$$\frac{-d[\text{-NH}_2]}{dt} = k_f[\text{-NH}_2][\text{>C=O}] - k_b[\text{>C=N-}] \tag{58}$$

where

$$k_f = \frac{k_1}{1 + (k_{-1}/k_2)} \qquad k_b = \frac{(k_{-1}k_{-2})/k_2}{1 + (k_{-1}/k_2)} = \frac{k_1/\beta_{SBO}}{1 + (k_{-1}/k_2)} \tag{59}$$

Thus, these cases are characterized by a simple second-order disappearance of reactants and the rate of formation of Schiff base is at all times equal to the rate of disappearance of carbonyl compound or amine.

In general, k_f is proton-dependent and conforms to an equation of the type [13,18,31,59]:

$$k_f = \sum_{i=0}^{n} k_i^H \; [H]^i \tag{60}$$

Mathematically it is not possible to distinguish between proton catalysis or reactivity arising from definite stable protonated species in which the proton is not directly involved in the reaction. This proton ambiguity arises in many situations and either one, or both, possibilities are likely. Furthermore, general acid-base catalysis may be present and introduce additional terms into the equation for k_f. Values of k_i^H which have been reported for some pertinent Schiff base formation reactions are given in Table 4A.

The solvent-catalyzed reactions of salicylaldehyde anion with various primary amines given in Table 4A show k_0^H values which decrease in the order EnH^+ > Eta > en > β-Ala^{1-} > Gly^- > α-Ala^{1-}. With the exception of α-Ala-, which shows steric inhibition by the α-side chain, the variations in these rate constants seem to be determined more by electrostatic factors than differences in amine basicity; the rate constant for the reaction of the negatively charged Sal^- ion is largest with the positively charged enH^+ ion and is the least with the smallest negatively charged ion, Gly^-. The basicities of these two amines lie in the reverse order, Gly^- > enH^+.

The paths giving rise to the k_1^H terms in the Sal rate laws probably arise from the reaction of neutral HSal with unprotonated amines. Since coulombic attraction and repulsion is thereby minimized, the values of k_1^H are expected to correlate with the basicity

TABLE 4

Rate Constants for Schiff Base Formation

A. Metal-Independent Paths

$$\frac{d[\gt C{=}N^{-R}]_{\Sigma}}{dt} = \sum_i k_i [H^+]^i [\gt C{=}O][RNH_2]$$

$\gt C{=}O$	RNH_2	k_0 (M^{-1} sec^{-1})	k_1 (M^{-2} sec^{-2})	k_2 (M^{-3} sec^{-1})	Ref.
Isobutyraldehyde	CH_3NH_2	53	27	--	58
	n-$PrNH_2$	50	--	--	58
	$CH_3O(CH_2)_3NH_2$	20	--	--	58
	$CH_3O(CH_2)_2NH_2$	5.2	∿23	--	58
	$(CH_3O)_2CHCH_2NH_2$	1.06	7.5	--	58
	$F_3CCH_2NH_2$	0.0049	0.29	--	58
Pyruvate	$^-O_2CCH_2NH_2$	0.15	9.2×10^7	--	61
	PMP^{3-}	--	3×10^8	2×10^{16}	44
Salicylaldehyde anion	$EtNH_2$	1.0	2.5×10^9	1.6×10^{15}	31
	$^-O_2CCH_2NH_2$	0.18	0.67×10^9	6.9×10^{15}	59
	$^-O_2CCHCH_3NH_2$	0.037	0.15×10^9	1.4×10^{15}	31
	$^-O_2CCH_2CH_2NH_2$	0.48	1.8×10^9	6.0×10^{15}	31
	$H_2NCH_2CH_2NH_2$	1.50	3.1×10^{10}	4.8×10^{18}	33
				$k_3 = 7.8 \times 10^{24}$ M^{-4} sec^{-1}	
3-Oxypyridine-4-aldehyde anion	$^-O_2CCH_2NH_2$	3.3	0.72×10^9	4.0×10^{15}	13
	$^-O_2CCH_2CH(CH_3)NH_2$	4.0	0.94×10^9	2.7×10^{15}	13
	$^-O_2CCH_2CH_2CH(CO_2^-)NH_2$	1.3	0.58×10^9	2.7×10^{15}	13

TABLE 4 (Continued)

B. Metal-Dependent Paths

$$\frac{d[\rangle C{=}N^{\diagup R}]_\Sigma}{dt} = \sum_i k_i [H^+]^i [\rangle C{=}O][RNH_2] + (k_0^M + k_1^M[H^+] + k_2^M[RNH_2])[M][\rangle C{=}O][RNH_2]$$

$\rangle C{=}O$	RNH_2	M	k_0^M $M^{-2}\ sec^{-1}$	k_1^M $M^{-3}\ sec^{-1}$	k_2^M $M^{-3}\ sec^{-1}$	kp_0[a] sec^{-1}	kp_1[a] $M^{-1}\ sec^{-1}$	kp_2[a] $M^{-2}\ sec^{-1}$	Ref.
Pyruvate	$^{-}O_2CCH_2NH_2$	Zn(II)	3.4×10^3	1.03×10^9	3.4×10^7	0.0058	2×10^3	0.32	18,61
	PMP^{3-}			evidence					
Salicyl-aldehyde anion	$EtNH_2$	Zn(II)	8.0×10^4	∿0	--	0.75	0	--	31,38
	$^{-}O_2CCH_2NH_2$	Mg(II)	18	--	--	0.015		--	18
		Mn(II)	6.95×10^2	5.2×10^9	--	0.015	1.1×10^5	--	18
		Co(II)	$<7 \times 10^2$	--	--	$<10^{-4}$			18
		Ni(II)	$<7 \times 10^2$	--	--	$<10^{-6}$			18
		Cu(II)	$<1.5 \times 10^6$	--	--	$<2 \times 10^{-7}$			18
		Zn(II)	5.9×10^4	3.85×10^{10}	1.2×10^9	0.0026	1.6×10^3	0.37	18
		Cd(II)	1.0×10^4	4.4×10^{11}	2.45×10^7	0.062	2.7×10^6	1.6	18
		Pb(II)	7.2×10^6	3.3×10^{13}	--	1.9	8.7×10^6	--	18
	α-Ala⁻	Zn(II)	7.2×10^3	4.5×10^9	--	6×10^{-4}	3.8×10^2	--	31,33
	β-Ala⁻	Zn(II)	6.5×10^4	10.4×10^9	--	0.026	4×10^3	--	31,33
	$H_2NCH_2CH_2NH_2$	Zn(II)	1.6×10^8	--	6.9×10^{11}	0.70	--	52	33

[a] $kp_0 = k_0^M/\beta_{11}^*$; $kp_1 = k_1^M/\beta_{11}^*$; $kp_2 = k_2^M/\beta_{12}^*$.

of the amine Eta > β-Ala^- > Gly^- > enH^+, and this is the actual order observed. Thus, both electrostatic interactions and amine basicity appear to influence the rates of Schiff base formation. The electrostatic term suggests the preequilibrium formation of an ion pair or encounter complex of the type involved in metal ion substitution reactions [60]. Pertinent to Schiff base formation this would involve a sequence such as

$$\mathrm{Sal^- + EnH^+ \underset{fast}{\overset{K_{os}}{\rightleftharpoons}} Sal^-\ldots EnH^+ \underset{slow}{\overset{k_{add}}{\rightleftharpoons}} adduct} \quad (61)$$

where the observed forward rate constant is the product, $K_{os} \cdot k_{add}$.

The rate constant for the reaction of 3-oxypyridine-4-aldehyde anion with glycinate has been found to be about 20 times larger than that found for the corresponding reaction between Sal^- and Gly^- [13]. This faster formation rate almost completely accounts for the greater stability, noted above, of the pyridine Schiff base over the Sal Schiff base. Nucleophilic attack at the carboxaldehyde carbon is promoted by the substitution of a more electronegative nitrogen atom for a ring $\geqslant$CH group. This substitution, however, has little influence on the dissociation rate.

Pyruvate ions react with glycinate at almost the same rate as does Sal^- [61].

$$\mathrm{Pyu^- + Gly^- \xrightarrow{k_o^H = 0.18\ M^{-1}\ sec^{-1}} (Pyu{\cdot}Gly)^{2-} + H_2O} \quad (62)$$

$$\mathrm{Sal^- + Gly^- \xrightarrow{k_o^H = 0.15\ M^{-1}\ sec^{-1}} (Sal{\cdot}Gly)^{2-} + H_2O} \quad (63)$$

The paths first order in H^+ also have similar rate constants, albeit the pyruvate reaction is significantly slower:

$$\mathrm{Pyu^- + Gly^- + H^+ \xrightarrow{k_1^H = 9.2 \times 10^7\ M^{-2}\ sec^{-1}} (Pyu{\cdot}Gly)^{2-} + H^+ + H_2O} \quad (64)$$

$$\mathrm{Sal^- + Gly^- + H^+ \xrightarrow{k_1^H = 6.7 \times 10^8\ M^{-2}\ sec^{-1}} (Sal{\cdot}Gly)^{2-} + H^+ + H_2O} \tag{65}$$

This difference may result in part from the higher affinity of Sal^- for protons, but the ratio of the rate constants is considerably greater than the ratio of the Pyu^{1-}/Sal^- protonation constants, suggesting that another factor, such as protonation of the intermediate carbinolamine, is more important.

Pyridoxamine 5'-phosphate has been found to react with pyruvate via the pathways [44]:

$$\mathrm{H^+ + PMP^{3-} + Pyu^{1-} \xrightarrow{k_1^H = 3 \times 10^8\ M^{-2}\ sec^{-1}} H(PMP{\cdot}Pyu)^{3-} + H_2O} \tag{66}$$

$$\mathrm{2H^+ + PMP^{3-} + Pyu^{1-} \xrightarrow{k_2^H = 2 \times 10^{16}\ M^{-3}\ sec^{-1}} H(PMP\ Pyu)^{3-} + H_2O + H^+} \tag{67}$$

$$\mathrm{3H^+ + PMP^{3-} + Pyu^{1-} \xrightarrow{k_3^H = 9 \times 10^{21}\ M^{-4}\ sec^{-1}} H(PMP{\cdot}Pyu)^{3-} + H_2O + 2H^+} \tag{68}$$

The rate constant for the first of these reactions is strikingly close to the value $9 \times 10^7\ M^{-2}\ sec^{-1}$ found for the reaction of pyruvate with glycinate via an analogous path. This suggests that the formation of the PMP ketimine does not differ significantly from the reaction in which simple amino acids react with pyruvate. The higher proton terms in the PMP rate law probably arise from paths involving stable protonated PMP species. The path yielding the k_3^H term, for instance, seems to originate simply from the reaction of phosphate-protonated PMP with pyruvate without appreciable acid catalysis being provided by this group [44].

Detailed studies of the influence of divalent metal ions on the reactions of glycinate with pyruvate or salicylaldehyde anion revealed rate laws of the form,

$$\frac{dSB}{dt} = (\Sigma_i^H[H^+] + k_o^M[M^{2+}] + k_1^M[M^{2+}][H^+] + k_2^M[M^{1+}][Gly^-])[{>}C{=}O][Gly^-] \quad (69)$$

where the k_i^H are identical to those which govern the rates in the absence of metal ions [18,59,61]. This equation shows that the metal-dependent terms arise from solvent-catalyzed and proton-catalyzed paths, as well as those which are second order in glycinate.

Rates in the Sal system were studied in the presence of Mg^{2+}, Mn^{2+}, Co^{2+}, Ni^{2+}, Cu^{2+}, Zn^{2+}, Cd^{2+} or Pb^{2+}, and some, but not all, of the metal ions were found to be kinetically active. The rate constants which have been determined are given in Table 4B.

Divalent metal ions which were found to be kinetically active in promoting the formation of N-salicylideneglycinate are, in order of decreasing k_o^M, $Pb^{2+} > Zn^{2+} > Cd^{2+} > Mn^{2+} > Mg^{2+}$. With the exception of an inversion in the order of Zn^{2+} and Cd^{2+}, the order is the same for the metal ion-dependent proton-catalyzed path, k_1^M. Divalent metal ions for which no activity was found are Co^{2+}, Ni^{2+}, and Cu^{2+}. These metal ions displayed no activity even though the reactions conditions were such that they formed both binary and Schiff base complexes in ample concentrations. This group of metal ions merely serve the function of trapping Schiff base that is formed along metal ion-independent pathways.

Farago and Matthews [62] investigated the effect of Cu(II) on the rate of formation of the Schiff base formed between pyridoxal phosphate and glutamate. At pH 4 and 25°C the rate was found to be first order in PLP^{3-} and first order in Glu^{2-} but zero order in Cu(II). Thus, as was found with the Sal^--Gly^- system, Cu(II) serves merely as a trap for Schiff base formed via metal-independent paths. McDonnell et al. [38] also found no evidence for Cu(II) kinetic activity in the Cu(II)-glycinate-5-sulfosalicylaldehyde system.

Metal ions often function as catalysts by polarizing reactive functional groups. With Schiff base formation this would facilitate the attack of the amine on the carbonyl carbon (see structure below).

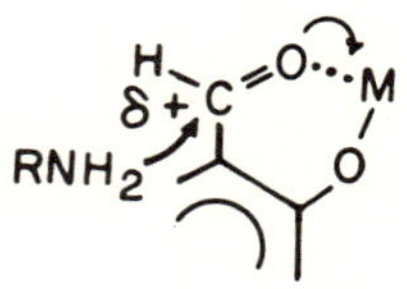

Such an effect does not seem to be very important in the case of salicylaldehyde and related compounds because the Sal^- complexes of those metal ions Cu^{2+} and Ni^{2+}, that interact strongly with Sal^-, show no activity. Activation of the amine through metal ion complexing can be ruled out since coordination of the amine electron pair to the metal ion can only lead to deactivation. Therefore, those paths which are first order in each of M^{2+}, Sal^-, and Gly^- suggest the preequilibrium formation of a reactive mixed complex. This step is followed by rate-limiting carbinolamine formation and dehydration within the corrdination sphere of the metal ion:

$$M^{2+} + Sal^- + Gly^- \underset{\text{fast}}{\overset{\beta^*_{11}}{\rightleftharpoons}} M(Sal)(Gly) \tag{70}$$

$$M(Sal)(Gly) \xrightarrow{k_p} M(Sal{\cdot}Gly) + H_2O \tag{71}$$

where β^*_{11} is the formation constant for the mixed complex, as previously defined, and k_p incorporates rate constants for the formation, dissociation, and dehydration of the low-level carbinolamine intermediate [18].

The pattern observed for metal ion activity lends support to this hypothesis. All of the divalent metal ions which have been studied are labile and undergo rapid ligand substitution reactions compared to the time scale of Schiff base formation, so the first step of the assumed sequence presents no difficulties. Because Cu^{2+}, Ni^{2+}, and Co^{2+} have β^*_{11} values which are comparable to, if not larger than, those of the other metal ions, their negligible activity must arise from a low value of k_p. The inactive metal ions differ from the others in that they possess ligand field stabilization energies arising from unfilled d orbitals. Attendent with this stabilization are well-defined coordination geometries and a high affinity for amino nitrogen. Both of these properties would tend to inhibit the ability of the amine nitrogen to associate with the carbonyl carbon. On the other hand, the large metal ions with low metal-ligand bond energies and those with no ligand field stabilization forces more readily permit the ligand mobility and the metal-amine bond breaking processes that are necessary for carbinolamine formation to occur within the primary or secondary coordination sphere of the metal ion. Incorporated in this mechanism are the views of Pearson and Lanier [63] who have shown in NMR studies that a very dynamic situation exists in complexes and partial dissociation of coordinated ligands readily occurs. The nitrogen-metal bond of an amino acid chelate is often broken while the metal-oxygen bond remains intact, and vice versa. Thus, the primary kinetic role of the metal ion in these reactions seems simply to be one of bringing the ligands together in a mixed complex without unduly restricting their ability to react further with each other. The term <u>promnastic</u> (Προμασθα--to be matchmaker) has been used to describe this effect [18].

One success of the promnastic model was to quantitatively account for the different rates of reaction of pyruvate and salicylaldehyde anion with glycinate [18]. Sal^{1-} forms stronger complexes with Zn^{2+} than Pyu^{1-} and therefore has a higher value of β_{11}^*. After correcting for β_{11}^*, the value of k_{p0} for Pyu^{1-} lies within a factor of about 2 of that for Sal^{1-}, while k_{p1} and k_{p2} are seen in Table 4B to be almost identical for the two systems. The ability of the promnastic model to explain these differences in the metal-dependent rates is strong support in its favor.

The variation of the amine donor on the k_0^M values for the Zn(II)-catalyzed rates of Sal Schiff base formation is seen in

Table 4B to have little influence with the exceptions of α-Ala$^-$, which presents steric hindrance, and ethylenediamine. The constant values with Eta, β-Ala$^-$, and Gly$^-$ and the high values for en can be explained by the opposing effects that structural differences have on β^*_{11} and k_{p0}. An increase in the metal ion-amine bond strength will increase the value of β^*_{11} but will simultaneously tend to reduce k_p, and vice versa. Therefore, the observed rate constants which are equal to the product β^*_{11} k_p will show less variation with amine structure than will β^*_{11} and k_p independently. After taking β^*_{11} into account, k_p values vary considerably with amine structure and are ordered Eta $\sim$ en > β-Ala^{1-} > Gly^{1-} > α-Ala^{1-}. Monodentate Eta is the most reactive, and it is this ligand which is least constrained when the metal-amine bond is broken in the mixed complex. The different reactivities of Zn(Sal)(en)$^+$ and Zn(Sal)(β-Ala) may indicate the existence of electrostatic effects, repulsion between the negative phenolate and the β-Ala^{1-} carboxylate oxygen atoms will tend to keep the two ligands separated in the Zn(II) coordination sphere and will reduce the ability for the addition of β-Ala$^-$ to Sal$^-$. Zn(Sal)(Gly) reacts at a still lower rate apparently because the shorter glycinate chain provides fewer opportunities for bringing the amine and carbonyl groups into proximity than occurs with β-alaninate. Thus, the influence of electrostatic forces is manifested in the reactions of these metal ion complexes just as they are in the formation of uncomplexed Schiff bases.

Earlier in this chapter the results of an NMR investigation conducted by Abbott and Martell [16] were discussed. These authors also made some interesting observations regarding the mechanism of formation of the isomeric bis (pyridoxylidene-DL valinato)Al(III) complexes. Figure 6 shows NMR spectra obtained at various times after mixing a pyridoxal, DL-valine, Al(III) solution at pH 10.2. A 1:1 Schiff base Al(III) complex forms rapidly and serves as a precursor for the formation of the isomeric bis complexes. The bis complexes that appear at first are seen to be the more sterically hindered high energy forms giving rise to the resonances designated

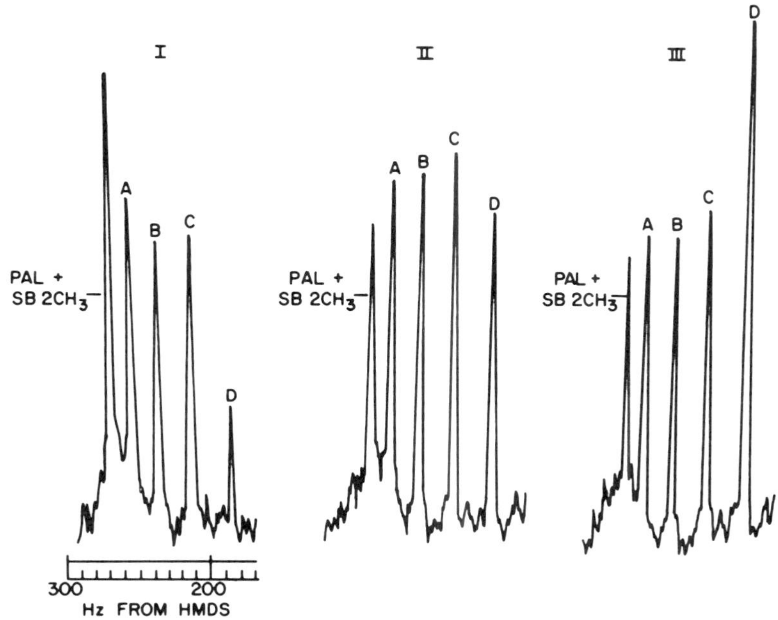

FIG. 6. The 100-MHz NMR spectrum of 0.10 DL-valine, 0.10 M pyridoxal, and 0.050 M Al(III) at pD 10.2. I, initially; II, at 20 min; III, at 1 hr. See text for further discussion. (From Ref. 16, by permission of the publishers.)

as A, B, and C. The most stable complex D, is formed in low yield in the initial stages of the reaction, but, as expected, increases in concentration as time goes on.

Abbott and Martell concluded that such a result can only be explained by the addition to the 1:1 complex of a Schiff base molecule that had been previously formed in the bulk of the solution. In such a case three situations arise, which, quoting the authors, are:

1. The attacking Schiff base is of the same absolute configuration as the coordinated Schiff base and the axial site attacked is on the same side of the equatorial plane as the alkyl group of the coordinated Schiff base. If this is the case, severe steric interaction will occur between the alkyl groups of the attacking Schiff base and the coordinated Schiff base.

2. The attacking Schiff base is of different absolute configuration from the coordinated Schiff base. If this is the case, there is no steric interaction of alkyl groups whether the alkyl group of the coordinated Schiff base is on the same or opposite side of the equatorial plane as the axial site being attacked.
3. The attacking Schiff base is of the same absolute configuration as the coordinated Schiff base, but attack occurs at the axial site on the other side of the equatorial plane of the 1:1 complex from its alkyl group. Here again, no steric effects are observed.

Situation 1 has the highest activation energy but is the path along which are formed the species giving rise to resonance D. Because the most stable complex has the highest activation energy for formation it is initially formed at a slower rate than its isomers. These results show that Al(III), owing either to the relative inertness of its ligand exchange processes, or to the fact that ligand mobility was blocked in these rate studies by the presence of a previously coordinated Schiff base, or for both reasons, behaves like Co(II), Ni(II), and Cu(II) and reacts with Schiff base formed along metal-independent paths but does not undergo a promnastic-type reaction.

5. TRANSAMINATION REACTIONS

Enzymic transamination processes are considered to proceed via a prototropic shift in a pyridoxal phosphate Schiff base:

$$\rangle C{=}O + H_2NCHRCO_2^- \underset{+H_2O}{\overset{-H_2O}{\rightleftarrows}} \rangle C{=}NCHRCO_2^- \rightleftarrows \rangle CHN{=}CRCO_2^- \underset{-H_2O}{\overset{+H_2O}{\rightleftarrows}} \rangle CHNH_2 + RCOCO_2^- \qquad (72)$$

The reaction is considered to proceed by way of a carbanion intermediate which is resonance-stabilized by the pyridinium group [reaction (73)]. In enzyme systems evidence for such an intermediate has been found through the appearance of an intense 500-nm absorption band, which is not attributable to other species [64]. The Schiff base formed from pyridoxal and diethylaminomalonate has

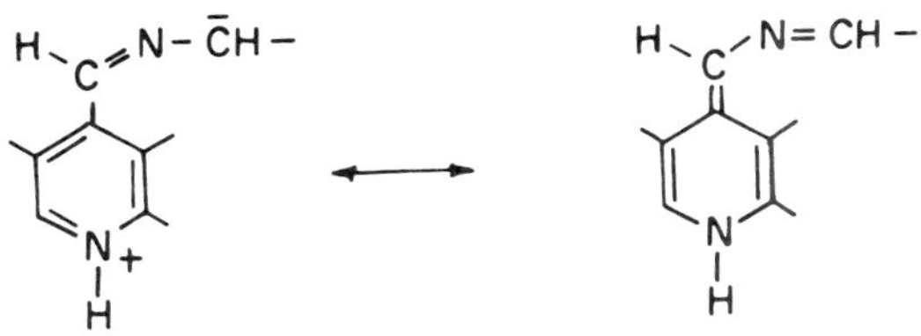

(73)

recently been reported to exist principally in the form of the 1,4-dihydropyridine tautomer [65].

Studies by Snell, Metzler, and coworkers [1-6,47,48,51] showed that transamination occurs on heating pyridoxal, or other appropriate aldehyde, with a suitable complexing metal ion to 100°C. Transamination is observed in the absence of a complexing metal ion but is accompanied by extensive decarboxylation. Braunstein and Shemyakin [9] and Metzler et al. [8] proposed that metal ion catalysis proceeds through the formation of Schiff base complexes. The metal ion was said to effect catalysis by stabilizing the Schiff base toward hydrolysis, maintaining planarity of the Schiff base π-electron system, and decreasing the electron density on the α-carbon of the amino acids [8].

Bonds other than the α-C-H bond can be activated, giving rise to elimination or activation of R, or decarboxylation [8]:

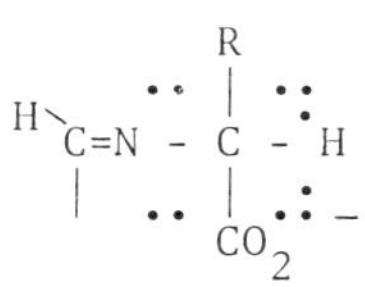

Dunathan [66,67] has shown that the conformation about the α-carbon atom has an important influence on the course of the reaction. The bond which is perpendicular to the plane of the conjugated π system is the one most likely to be activated. The orbital of the electron pair, which thereby becomes available as the bond is broken, has the proper geometry to interact with the Schiff base π orbitals. Structure XXXVI shows the conformation suitable for racemization and transamination, structure XXXVII is appropriate for loss, or activation, of R, and XXXVIII shows the conformation suitable for decarboxylation. Chelate formation by a metal ion tends to hold the carboxylate group in the plane of the pyridoxal ring and prevents conformation XXXVIII from being attained, thereby inhibiting decarboxylation.

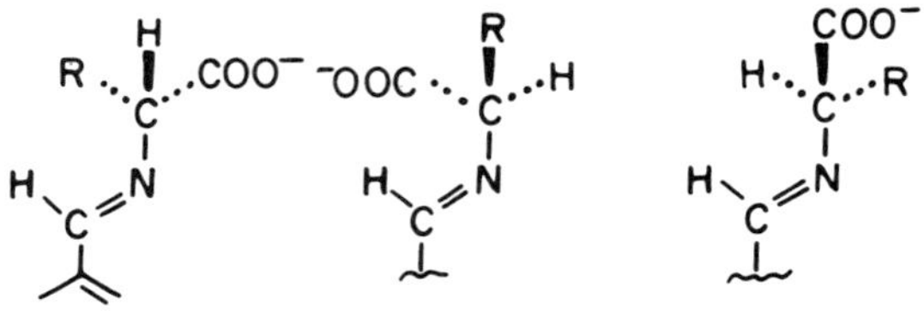

XXXVI XXXVII XXXVIII

Blake et al. [68] have examined the effect of deuteration of the α-carbon on the rate of the Al(III)-catalyzed transamination reaction of PLP and L-alanine. At 100°C, L-protioalanine was found to react with pyridoxal 2.4 times faster than L-deuterioalanine, but the Al(III) complex of the protio form was found to react 3.5 times faster than the deuterio complex. The isotope effects are consistent with mechanism in which the amino acid C-H (or C-D) bond is weakened in the activated complex and complexing by Al(III) further weakens this bond.

Matsumoto and Matsushima have recently shown [69] that an intensely colored species with an absorption maximum at 488 nm is formed as a transamination intermediate when Al(III) is added to preequilibrated "neutral" methanolic solutions of pyridoxamine and ethyl pyruvate. The absorption band arising from this species slowly increases after mixing the components, reaches maximum intensity in 15 min, and then slowly disappears. After 4 hr only the absorption band of the aldimine is observed. It is suggested that this intermediate is a carbanion which has been stabilized through coordination to Al(III).

Thanassi et al. [70] quantitatively studied metal ion-catalyzed rates of transamination of 3-hydroxypyridine-4-aldehyde by glutamic acid. Analysis of the pH-rate profile for the reaction in the absence of metal ions demonstrated that transamination occurred via the diprotonated Schiff base with a first-order rate constant of 1.4×10^{-4} sec^{-1} [see reaction (74)]. This reaction also showed imidazole catalysis with a second-order rate constant of 1.2×10^{-3} M^{-1} sec^{-1}.

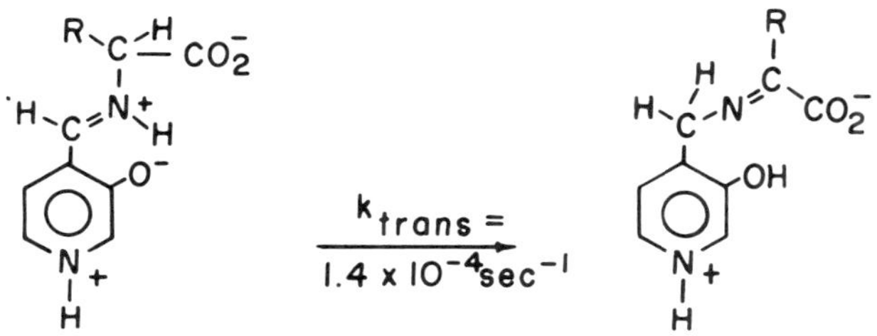

(74)

Apparent first-order rate constants for transamination in the presence of 0.1 M metal ion catalysts at pH 4.6 to 4.8 were determined and are given in Table 5 in order of decreasing metal ion

TABLE 5

Metal Ion Catalysis of Transamination of 3-Hydroxypyridine-4-aldehyde by Glutamic Acid[a] M^{n+} = 0.01 M

M	pH	$k_{obs} \times 10^5$ sec^{-1}
Mn^{2+}	4.67	22
Fe^{2+}	4.71	16
In^{3+}	4.60	10
Al^{3+}	4.68	9
Ni^{2+}	4.65	6
Co^{2+}	4.72	4
Mg^{2+}	4.74	2.7
Pb^{2+}	4.80	2.5
Ba^{2+}	4.84	2.3
None	4.74	2.3
Hg^{2+}	4.69	2.2
Zn^{2+}	4.64	2.0

[a]From Reference 70, with permission.

activity. Because the formation constants for the complexes present in these systems are not known it is not possible to obtain absolute rate constants from the data. Qualitatively, Mn(II) and Fe(II) are shown to be particularly effective catalysts under the conditions described while Zn(II) is seen to slightly supress the rate.

In contrast to this behavior observed with Zn(II), Matsushima and Martell [15], for methanolic pyridoxal alaninate, and Hopgood [71], for aqueous pyridoxal 5'-phosphate alaninate, have found Zn(II) to be a good catalyst for transamination. The conditions employed by Hopgood pH > 5.6 and large excess of Zn(II) (0.01 M), were such as to ensure the essentially quantitative formation of the Zn(II) aldimine complexes. Under these conditions it was found that the metal ion independent rates of transamination were negligibly slow compared to the Zn(II)-catalyzed path. It is likely that the very low Zn(II) activity observed by Thanassi et al. is due to the extensive dissociation of the aldimine complex under their experimental conditions. This is suggested by the distribution curves for the PLP-Ala-Zn(II) system in Figure 3 which show that extensive dissociation of the aldimine complexes occurs below pH 5. It is highly probable that a similar situation occurs with 3-hydroxypyridine-4-aldehyde.

Hopgood [71] studied the influence of the amino acid side chain on the rate of PLP transamination. The results presented in Table 6 show that transamination activity decreases in the order alanine > methionine > phenylalanine > α-amino-n-butyric acid > glycine > serine > isoleucine > threonine > valine. Electron-donation by the side chain to the α-carbon atom appears to be rate-enhancing, but very bulky side chains act to decrease the rate. A plot of log (k_f/k_{ala}) vs the Taft σ^* parameter for the side chain revealed that with small side chains the rate constant increases as the electron-donating ability of the side chain increases to a critical point with α-amino-n-butyric acid, where a downward break occurs. Thus, it appears that inductive electron donation by the side chain is a rate enhancing factor in addition to the effect of conjugative electron withdrawal by the PLP π system. The bulky side chains possibly hinder the approach of acid and base catalysts.

TABLE 6

Transamination Rates in the Systems Zn(II) (0.01 M)-Amino Acid (0.03 M)-Pyridoxal Phosphate (1×10^{-4} M)[a]

Amino acid	T = 25°C; pH 5.6		pH 7.0
	K_{app}	$k_{trans} \times 10^6$ sec^{-1}	$k_{trans} \times 10^6$ sec^{-1}
Glycine	0.79	0.31	0.24
Alanine	31.7	2.62	2.95
α-Amino-n-butyric acid	2.54	0.62	1.20
Phenylalanine	1.62	1.0	--
i-Leucine	--	--	∿0.05
Valine	--	--	0.02
Aspartic acid	11.5	9.2	2.28
Glutamic acid	2.25	9.7	--
Ornithine	--	--	3.5
Lysine	--	--	2.0
Serine	1.29	0.25	0.11
Threonine	--	--	0.025
Methianine	--	--	2.0
Asparagine	--	--	4.0
Glutamine	--	--	2.0

[a]From Reference 71, with permission.

Hopgood's results indicate that the Zn(II)-catalyzed prototropic shift in the PLP-alanine Schiff base occurs via <u>protonated</u> aldimine chelates and from his data it is possible to estimate an average rate constant of about 3.6×10^{-6} sec^{-1} for transamination via a typical path, such as,

$$\mathrm{ZnH(PLP\cdot Ala)^{-1}} \xrightarrow[k_b \sim 1 \times 10^{-5}\ \mathrm{sec}^{-1}]{k_f \sim 3.6 \times 10^{-6}\ \mathrm{sec}^{-1}} \mathrm{ZnH(PMP\cdot Pyu)^{-1}} \qquad (75)$$

The conversion of pyridoxamine and pyridoxamine-5'-phosphate ketimines to the aldimines proceeds at a somewhat faster pace than do the reverse reactions, and because of this pyridoxamine transamination rates have received more quantitative investigation than have the pyridoxal reactions. Banks et al. [45] report that transamination in the pyridoxamine-pyruvate system proceeds via monoprotonated N-pyruvidenepyridoxamine with a first-order rate constant of $9.5 \times 10^{-6}\ \text{sec}^{-1}$:

$$\text{H(PM}\cdot\text{Pyu)}^{1-} \xrightarrow{9.5 \times 10^{-6}\ \text{sec}^{-1}} \text{H(PL}\cdot\text{Ala)}^{1-} \tag{76}$$

Felty and Leussing [44] found two paths for the analogous pyridoxamine 5'-phosphate reaction:

$$\text{H(PMP}\cdot\text{Pyu)}^{3-} \xrightarrow{1.7 \times 10^{-6}\ \text{sec}^{-1}} \text{H(PLP}\cdot\text{Ala)}^{3-} \tag{77}$$

and

$$\text{H}^+ + \text{H(PMP}\cdot\text{Pyu)}^{3-} \xrightarrow{2 \times 10^{2}\ \text{M}^{-1}\ \text{sec}^{-1}} \text{H(PLP}\cdot\text{Ala)}^{3-} + \text{H} \tag{78}$$

The last path probably involves transamination via the pyridine-protonated species $H_2(\text{PMP Pyu})^{2-}$. Since the pK_a of this species has not been determined for certain, but is less than or equal to 7.0, the first-order rate constant for the reaction

$$\text{H}_2\text{(PMP}\cdot\text{Pyu)}^{2-} \rightarrow \text{H}_2\text{(PLP}\cdot\text{Ala)}^{2-} \tag{79}$$

can only be said to be equal to or greater than $2 \times 10^{-5}\ \text{sec}^{-1}$. Thus, the pyridine protonated Schiff base transaminates at a rate that is at least a factor of 10 greater than the unprotonated pyridine form.

A path analogous to this last one was not resolved by Banks et al. [45] in their pyridoxamine rate study and the higher rate constant reported by them for the monoprotonated Schiff base may arise either from the unresolved contribution of such a path, or from actual chemical differences between the PM and PMP molecules. Buffer catalysis was also observed in the pyridoxamine study. General acid-base catalysis was not observed for the pyridoxamine phosphate system, since PMP was the sole buffering agent.

Mg(II) and Ca(II) were found not to catalyze transamination

in the pyridoxamine system at pH 7.0, however, Cu(II) was qualitatively observed to exert a powerful catalytic effect [45]. The rate law for Cu(II) catalysis was not determined.

Catalytic activity of Zn(II) in a transamination system was first reported by Matsushima and Martell [15]. When pyridoxamine, α-ketoisovalerate and zinc acetate are brought together in methanol a two-step reaction is observed: formation of the ketimine chelate followed by isomerization of the ketimine to the aldimine. At low concentration levels, the transamination rates increase with increasing Zn(II) concentration, but at higher concentrations the rates level off to a limiting value. The first-order rate constant for pyridoxal formation was found to reach a limiting value of 1.86×10^{-4} sec^{-1}, 25°C in the presence of excess ketoacid and zinc acetate over pyridoxamine (1.0×10^{-4} M pyridoxamine, 1×10^{-3} M α-ketoisovalerate, 5.0×10^{-4} zinc acetate). This limit probably represents the rate constant for the isomerization of a complex such as ZnH_i(PM•KVal), where i is equal to 1 or 0. The presence of acetate in these experiments provides the opportunity for base catalysis so a direct comparison of this value with the lower rate constants reported below for PMP transamination in aqueous media may not be valid.

Another interesting and highly significant result of these studies was the observation that the addition of water to the methanolic solutions caused a marked decrease in the transamination rate. When the solvent contained 5% H_2O (v/v) the rate was observed to be less than one-third the value observed for solutions containing only 0.1% H_2O. This behavior may play an important role in determining the rate of enzyme reactions owing to the presence of nonpolar protein groups near the active site. It will be necessary to characterize the effect of solvent on the various equilibria and reaction steps involved in the entire reaction sequence before it is possible to properly understand this phenomenon, however.

The rate of the prototropic shift in the Zn(II) ketimine complexes was qualitatively examined in NMR studies performed by

Gansow and Holm [17] on pyridoxamine-pyruvate-zinc(II) mixtures. From pD 5 to 7, transamination was found to be so slow that about 30 min was required after mixing the components to detect aldimine species. Above pD 7, however, the rates were found to be faster and detectable levels of aldimine were formed within the time required to obtain the first spectra. Al(III) gave similar results: no significant transamination was observed below pD 6.5 after 3 hr, but above this value rates increase markedly with increasing pD.

Felty and Leussing [44] performed a detailed spectrophotometric study of the Zn(II)-catalyzed transamination of pyridoxamine 5'-phosphate-pyruvate mixtures in aqueous solutions at 25°C. It was found that metal-catalyzed transamination proceeded via the pyridine protonated ketimine complexes, $ZnH(PMP \cdot Pyu)^{1-}$, $ZnH(PMP)_2^{5-}$, and $ZnH_2(PMP \cdot Pyu)_2^{4-}$. Significantly, the phosphate-protonated complex $ZnH_2(PMP \cdot Pyu)$ appeared to be inactive, suggesting that internal base catalysis by unprotonated phosphate is involved in the transamination process. Molecular models show that in the ketimine complex the phosphate group of pyridoxamine can swing to a position well suited to aid the removal of the methylene proton that is involved in the tautomerization process (see structure below).

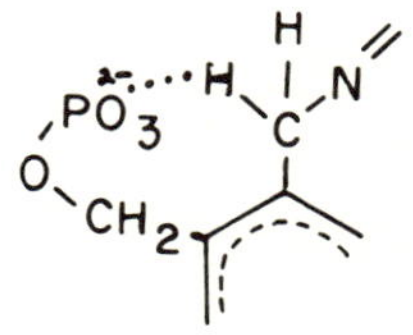

In addition to phosphate catalysis, catalysis by other basic species was observed to cause the rates of aldimine formation to markedly increase above pH 7. This increase in rate coincides with the appearance in solution of species having unprotonated pyridine groups, and was, therefore, attributed to pyridine base catalysis.

Average rate constants for reaction along typical paths were found to be

$$\mathrm{ZnH(PMP{\cdot}Pyu)^{1-}} \xrightarrow{2 \times 10^{-5}\ \mathrm{sec}^{-1}} \mathrm{ZnH(PLP{\cdot}Ala)^{1-}} \quad (80)$$

(internal phosphate catalysis)

$$\mathrm{HPMP^{2-}} + \mathrm{ZnH(PMP{\cdot}Pyu)^{1-}} \xrightarrow{0.012\ M^{-1}\ \mathrm{sec}^{-1}} \mathrm{HPMP^{2-}} + \mathrm{ZnH(PLP{\cdot}Ala)} \quad (81)$$

(pyridine group catalysis)

Although the value 2×10^{-5} sec^{-1} for the first reaction is about an order of magnitude greater than that found for first-order transamination via $\mathrm{H(PMP{\cdot}Pyu)^{3-}}$, it lies only at the lower limit set for the first-order rate constant for the reaction of $\mathrm{H_2(PMP{\cdot}Pyu)^{2-}}$. Thus, it appears that transamination activity is more to be associated with the presence of the pyridinium group than with any direct activating effect of the Zn(II) ion, on the Schiff base, e.g., through polarization. Because the basicity of the pyridine nitrogen atom is greater in a Zn(II) ketimine complex, such as $\mathrm{Zn(PMP{\cdot}Pyu)^{2-}}$, than it is in the protonated Schiff base $\mathrm{H(PMP{\cdot}Pyu)^{3-}}$ one immediate result of Zn(II) complexing is to produce Schiff base species in which the integrity of the pyridinium group is preserved to higher pH values than in its absence. Through this effect alone the presence of Zn(II) would serve to increase transamination rates in media at neutral pH and above. In addition, Zn(II) also promotes the reaction in the manner first suggested by Metzler et al. [8], that is, by increasing the concentration of Schiff base in solution through the formation of stable complexes.

The Schiff base complexes formed by Zn(II) and other metal ions also appear to be more susceptible to base catalysis than the uncomplexed ligands. This increased reactivity may result from the ability of the metal ion either to interact with the base and maintain it for an extended period in the vicinity of the coordinated Schiff base, that is, to exert a kind of promnastic effect, or to stabilize the planar carbanion intermediate. This last function may be more important with strongly polarizing metal ions such as Al(III) and Cu(II) than with Zn(II).

6. CONCLUSIONS

Studies of the effects of complexing metal ions on the formation of Schiff bases and their reactions have shown that metal ions are indeed able to promote reactions along paths that are analogous to enzyme-catalyzed reactions. An effective mode by which enzymes serve to increase reaction rates is to bring the reactants together in a preequilibrium step and thereby reduce the kinetic process from a second or higher order to first order. Metal ions have been found to promote Schiff base formation similarly. On the other hand, those metal ions which impose stringent coordination geometries on bound ligands are found to inhibit Schiff base formation and hydrolysis processes. The type of inhibition is also observed with enzymes, for example, when the Zn(II) ion of carboxypeptidase is replaced by more sterically demanding metal ions.

Factors which are important in enzymic transamination processes have also been found to be important in the metal ion-catalyzed reactions: protonation of the pyridoxal pyridine nitrogen atom, establishing the correct conformation about the Schiff base imine bond, maintaining the planarity of the π system and activating the tautomerizing proton through base catalysis. In connection with this last item, the data obtained for Zn(II)-catalyzed pyridoxamine 5'-phosphate transamination indicate that the phosphate group serves as a base catalyst. This has important implications regarding the enzymic reaction. Finally, the studies on the model systems show that the polarity of the solvent has a large effect on the reaction rates, and this probably is a factor which is important in the enzymic reactions.

Thus, it appears very likely that the metal ion reaction systems serve as good models for the enzyme systems. By suitably varying the nature of the metal ion and the experimental parameters it should be possible to increase considerably our understanding of the chemical factors which are involved in the more complicated biological processes.

ABBREVIATIONS AND SYMBOLS

ABA^{1-}: Aminobutyrate

Aib^{1-}: α-Aminoisobutyrate

Ala^{1-}: Alaninate

$BuNH_2$: Butylamine

Butal: Butyraldehyde

En(en): Ethylenediamine

en•$glyox^{1-}$: Adduct of en and $glyox^{-}$

Eta: Ethylamine

Glu^{2-}: Glutamate

Gly^{1-}: Glycine

$Glyox^{1-}$: Glyoxylate

IDA^{2-}: Iminodiacetate

α-Kg^{2-}: α-Ketoglutarate

$KVal^{1-}$: α-Ketoisovalerate

Leu^{1-}: Leucinate

Mea: Methylamine

NTA^{3-}: Nitrilotriacetate

$oxybenzal^{1-}$: Oxybenzaldehyde anion

3-oxypy-4-al^{1-}: 3-Oxypyridine-4-aldehyde anion

phen: 1,10-Phenanthroline

PL^{-}: Pyridoxal

PL•Gly^{2-}: Adduct of PL^{-} and Gly^{1-}

PLP^{3-}: Pyridoxal 5'-phosphate

PLP•Ala^{4-}: Adduct of PLP^{3-} and Ala^{1-}

PLP•Gly^{4-}: Adduct of PLP^{3-} and Gly^{1-}

PM^{1-}: Pyridoxamine

PM•$KVal^{2-}$: Adduct of pyridoxamine and $KVal^{1-}$

PMP^{3-}: Pyridoxamine 5'-phosphate

PMP•Pyu^{4-}: Adduct of PMP^{3-} and Pyu^{1-}

PM•Pyu^{2-}: Adduct of PM^{1-} and Pyu^{1-}

$PrNH_2$: Propylamine

Pyu^{1-}: Pyruvate

Pyu•Ala^{2-}: Adduct of Ala^{1-} and Pyu^{1-}

Pyu•Glu^{3-}: Adduct of Glu^{2-} and Pyu^{1-}

Pyu•Gly^{2-}: Adduct of Gly^{1-} and Pyu^{1-}

Sal^{1-}: Salicylaldehyde anion

Sal•Ala^{2-}: Adduct of Ala^{1-} and Sal^{1-}

Sal•Eta^{1-}: Adduct of Eta and Sal^{1-}

Sal•Gly^{2-}: Adduct of Gly^{1-} and Sal^{1-}

Sarc^{1-}: Sarcosinate

SSal^{1-}: Sulfosalicylaldehyde anion

tetrameen: C,C,C',C',-Tetramethylethylenediamine

tetrameen•glyox^{1-}: Adduct of tetrameen and glyox^{1-}

Val^{1-}: Valinate

REFERENCES

1. E. E. Snell, *J. Amer. Chem. Soc., 67,* 194 (1945).
2. D. E. Metzler and E. E. Snell, *J. Biol. Chem., 198,* 353 (1952).
3. D. E. Metzler and E. E. Snell, *J. Amer. Chem. Soc., 198,* 363 (1952).
4. D. E. Metzler, J. B. Longnecker, and E. E. Snell, *J. Amer. Chem. Soc., 76,* 639 (1954).
5. J. Olivard, D. E. Metzler, and E. E. Snell, *J. Biol. Chem., 199,* 669 (1952).
6. D. E. Metzler, J. Olivard, and E. E. Snell, *J. Amer. Chem. Soc., 76,* 644 (1954).
7. E. Werle and W. Koch, *Biochem. Z., 319,* 305 (1949).
8. D. E. Metzler, M. Ikawa, and E. E. Snell, *J. Amer. Chem. Soc., 76,* 648 (1954).
9. A. E. Braunstein and M. M. Shemyakin, *Biokhimiya, 18,* 393 (1953).
10. G. L. Eichhorn and J. W. Dawes, *J. Amer. Chem. Soc., 76,* 5663 (1954).
11. H. Mix, *Z. Physiol. Chem., 315,* 1 (1959).
12. P. Zuman, *Coll. Czech. Chem. Commun., 15,* 839 (1950).
13. T. C. French, D. S. Auld, and T. C. Bruice, *Biochemistry, 4,* 77(1965).
14. D. Heinert and A. E. Martell, *J. Amer. Chem. Soc., 85,* 183, 188 (1963).
15. Y. Matsushima and A. E. Martell, *J. Amer. Chem. Soc., 89,* 1322 (1967).

16. E. H. Abbott and A. E. Martell, *J. Amer. Chem. Soc., 92,* 5845 (1970).

17. O. A. Gansow and R. H. Holm, *J. Amer. Chem. Soc., 91,* 573, 5984 (1969).

18. D. Hopgood and D. L. Leussing, *J. Amer. Chem. Soc., 91,* 3740 (1969).

19. D. L. Leussing and D. C. Schultz, *J. Amer. Chem. Soc., 86,* 4846 (1964).

20. J. Hine, J. C. Craig, Jr., J. G. Underwood, and F. A. Via, *J. Amer. Chem. Soc., 92,* 5194 (1970).

21. D. L. Leussing and E. Hanna, *J. Amer. Chem. Soc., 88,* 693 (1966).

22. J. Hine and C. Y. Yeh, *J. Amer. Chem. Soc., 89,* 2669 (1967).

23. J. Hine, C. Y. Yeh, and F. C. Schmalsteig, *J. Org. Chem., 35,* 340 (1970).

24. D. L. Leussing and E. Hanna, *J. Amer. Chem. Soc., 88,* 696 (1966).

25. J. Kuta and P. Valenta, *Coll. Czech. Chem. Commun., 28,* 1593 (1963).

26. A. Hilton and D. L. Leussing, *J. Amer. Chem. Soc., 93,* 6831 (1971).

27. D. D. Perrin, *J. Chem. Soc.,* 290 (1959).

28. H. A. Krebs, *Biochem. J., 54,* 82 (1953).

29. H. Scheidegger, W. Felty, and D. L. Leussing, *J. Amer. Chem. Soc., 92,* 808 (1970).

30. D. L. Leussing and K. S. Bai, *Anal. Chem., 40,* 575 (1968).

31. B. E. Leach and D. L. Leussing, *J. Amer. Chem. Soc., 93,* 3377 (1971).

32. H. N. Christensen, *J. Amer. Chem. Soc., 80,* 99 (1958).

33. R. S. McQuate and D. L. Leussing, *J. Amer. Chem. Soc., 97,* 5117 (1975).

34. R. W. Green and E. L. LeMesurier, *Aust. J. Chem., 19,* 229 (1966).

35. R. W. Green and R. J. Sleet, *Aust. J. Chem., 19,* 233, 2101 (1966).

36. K. K. Chatterjee and B. E. Douglas, *Spectrochim. Acta, 21,* 1625 (1965).

37. W. Bruyneel, J. J. Charette, and E. DeHoffmann, *J. Amer. Chem. Soc., 88,* 3808 (1966).

38. C. V. McDonnell, Jr., M. S. Michailidis, and R. B. Martin, *J. Phys. Chem., 74,* 26 (1970).

39. D. L. Leussing and N. Huq, *Anal. Chem., 38,* 1388 (1966).

40. J. R. Maley and T. C. Bruice, *J. Amer. Chem. Soc., 90,* 2843 (1968).

41. W. L. Felty, C. G. Ekstrom, and D. L. Leussing, *J. Amer. Chem. Soc., 92,* 3006 (1970).

42. F. J. Anderson and A. E. Martell, *J. Amer. Chem. Soc., 86,* 715 (1964).

43. V. R. Williams and J. Neilands, *Arch. Biochem. Biophys., 53,* 56 (1954).

44. W. Felty and D. L. Leussing, *J. Inorg. Nuc. Chem.,* in press.

45. B. E. C. Banks, A. A. Diamantis, and C. A. Vernon, *J. Chem. Soc., 1961,* 1698, 4235.

46. H. N. Christensen, *J. Amer. Chem. Soc., 79,* 4073 (1957).

47. D. E. Metzler and E. E. Snell, *J. Amer. Chem. Soc., 77,* 2431 (1955).

48. D. E. Metzler, *J. Amer. Chem. Soc., 79,* 485 (1957).

49. H. N. Christensen, *J. Amer. Chem. Soc., 80,* 2305 (1958).

50. G. Schwarzenbach and R. Gut, *Helv. Chim. Acta, 36,* 1589 (1956).

51. L. Davis, F. Roddy, and D. E. Metzler, *J. Amer. Chem. Soc., 83,* 127 (1961).

52. J. Bjerrum, "Metal Ammine Formation in Aqueous Solution," P. Haase and Son, Copenhagen, 1941.

53. F. J. C. Rossotti and H. Rossotti, "The Determinates of Stability Constants," McGraw-Hill, New York, 1961.

54. C. K. Jorgensen, "Absorption Spectra and Chemical Bonding in Complexes," Pergamon Press, Oxford, 1962.

55. T. C. Bruice and S. J. Benkovic "Bioorganic Mechanisms," Vol. 2, Benjamin, New York, 1966, Chap. 8.

56. W. P. Jencks, "Catalysis in Chemistry and Enzymology," McGraw-Hill, New York, 1969).

57. T. C. French and T. C. Bruice, *Biochemistry, 3,* 1589 (1964).

58. J. Hine and F. A. Via, *J. Amer. Chem. Soc., 94,* 190 (1972).

59. K. S. Bai and D. L. Leussing, *J. Amer. Chem. Soc., 89,* 6126 (1967).

60. G. G. Hammes and J. I. Steinfeld, *J. Amer. Chem. Soc., 84,* 4639 (1962).

61. D. L. Leussing and L. Anderson, *J. Amer. Chem. Soc., 91,* 4698 (1969).

62. M. E. Farago and T. Matthews, *J. Chem. Soc. (A), 1969,* 609.

63. R. G. Pearson and R. D. Lanier, *J. Chem. Soc. (A), 86,* 765 (1964).

64. W. T. Jenkins, *J. Biol. Chem., 239,* 1742 (1964).

65. E. H. Abbott and M. A. Bobrik, *Biochemistry, 13,* 840 (1973).

66. H. C. Dunathan, *Proc. Natl. Acad. Sci., 55,* 712 (1966).

67. H. C. Dunathan, L. Davis, P. G. Kury, and M. Kaplan, *Biochemistry, 7,* 4532 (1968).

68. M. I. Blake, F. P. Siegel, J. J. Katz, and M. Kilpatrick, *J. Amer. Chem. Soc., 85,* 294 (1963).

69. S. Matsumoto and Y. Matsushima, *J. Amer. Chem. Soc., 94,* 7211 (1972).

70. J. W. Thanassi, A. R. Butler, and T. C. Bruice, *Biochemistry, 4,* 1463 (1965).

71. D. Hopgood, *J. Chem. Soc. (Dalton), 1972,* 482.

72. R. W. Green and M. J. Rogerson, *Aust. J. Chem., 21,* 2427 (1968).

73. F. Olivo, C. S. Rossi, and N. Siliprandi, "Chemical and Biological Aseects of Pyridoxal Catalysis" (E. E. Snell, P. M. Fasella, A. Braunstein, and A. Rossi Fanelli, eds.), Pergamon, 1963, p. 91.

74. K. Nagano and D. E. Metzler, *J. Amer. Chem. Soc., 89,* 2891 (1967).

75. Y. Matsuo, *J. Amer. Chem. Soc., 79,* 2011 (1957).

Chapter 2

THE ROLE OF DIVALENT METAL IONS IN PHOSPHORYL AND NUCLEOTIDYL TRANSFER

Barry S. Cooperman
Chemistry Department
University of Pennsylvania
Philadelphia, Pennsylvania

1. INTRODUCTION

Phosphoryl and nucleotidyl transfer enzymes are extremely important and widespread in biology. They have in common the catalysis of nucleophilic reactions of phosphorus esters and the general requirement for divalent metal ions, particularly Mg^{2+}, for activity. This requirement has stimulated considerable interest as to the possible catalytic roles of divalent metal ions in these reactions. A common approach to this problem has been to study metal ion effects on nonenzymic reactions which are chemically similar to the enzymic reactions. In this review we will seek to interpret the results of these studies in terms of several possible roles for divalent metal ions and in terms of possible structures of transition states in the presence of metal ion. We also will discuss the mechanism of phosphoryl transfer in creatine kinase and pyruvate kinase. For these enzymes the geometries of the enzyme-metal ion-substrate complexes have been sufficiently well characterized that meaningful suggestions may be made as to the role of metal ion in the enzyme-catalyzed reaction.

This review emphasizes work done in the last five years. The reader is referred to other sources for reviews of older work [1], or more general consideration of nucleophilic reactions at phosphorus [2,3], and of more general treatments of enzymic phosphoryl and nucleotidyl transfer [4,5,6].

2. MECHANISMS OF PHOSPHORYL TRANSFER

2.1. General Remarks

The group of reactions under the heading of phosphoryl transfer reactions share the common feature that during the course of the reaction PO_3^-, metaphosphate, is transferred from one nucleophilic atom to another. In biological systems the most important nucleophilic atoms are oxygen and nitrogen, which participate in phosphoryl transfer via reactions (1)-(3). S → O transfer reactions are also of some possible importance, but these will not be discussed separately because of the great similarities between thio and oxy esters of phosphoric acid [2,7]. Biochemical examples of reactions (1)-(3) are listed in Table 1.

$$(\text{O} \rightarrow \text{transfer})\ \text{RO-PO}_3^{2-} + \text{R'O(H)} \rightleftarrows \text{R'OPO}_3^{2-} + \text{RO(H)} \quad (1)$$

$$(\text{N} \rightarrow \text{N transfer})\ \text{R}_3\overset{+}{\text{N}}\text{PO}_3^{2-} + \text{R}_3'\text{N} \rightleftarrows \text{R}_3'\overset{+}{\text{N}}\text{PO}_3^{2-} + \text{R}_3\text{N} \quad (2)$$

$$\begin{pmatrix}\text{O} \rightarrow \text{N} \\ \text{N} \rightarrow \text{O}\end{pmatrix}\text{transfer}\ \ \text{ROPO}_3^{2-} + \text{R}_3'\text{N} \rightleftarrows \text{R}_3'\overset{+}{\text{N}}\text{PO}_3^{2-} + \text{RO(H)} \quad (3)$$

A central issue in the detailed examination of the mechanism of phosphoryl transfer is the extent to which the reaction can be viewed as proceeding via a unimolecular, metaphosphate-type elimination mechanism [$S_N{}^1$(P)] [reactions (4)]:

$$\begin{aligned} &\text{XP}(=\text{O})(\text{O}^-)_2 \xrightarrow{\text{slow}} \text{PO}_3^- + \text{X(H)} \\ &\text{Y(H)} + \text{PO}_3^- \xrightarrow{\text{fast}} \text{YPO}_3^{2-} \end{aligned} \quad (4)$$

as opposed to a bimolecular nucleophilic displacement reaction (5) [S_N^2(P)]:

$$\text{Y(H)} + \text{XPO}_3^{2-} \rightarrow \left[\begin{matrix}\text{O}\ \ \text{O} \\ \diagdown\!\diagup \\ \text{Y-P-X} \\ | \\ \text{O}\end{matrix}\right] \rightarrow \text{YPO}_3^{2-} + \text{X(H)} \quad (5)$$

where the 5-coordinate bipyramidal phosphorus represents either a high-energy intermediate or the transition state. As several excellent reviews have appeared fairly recently on this question [2, 3,12,13], we will here present only those aspects which are of particular importance to understanding metal ion effects. Briefly, phosphoryl transfer can best be described as being a loosely coupled bimolecular reaction, in which, in the transition state "there is a

TABLE 1

Examples of Enzyme-Catalyzed Phosphoryl Transfer

Transfer type	Phosphoryl donor	Nucleophile	Enzyme	Ref.
O → O	ATP	Glucose	Glucokinase	4
	ATP	Acetate	Acetate kinase	4
	ATP	AMP	Adenylate kinase	4
	Phenyl phosphate	Uridine	Nucleoside phosphotransferase	8
O → N, N → O	ATP	Creatine, arginine	Creatine kinase, arginine kinase	4
	Inorganic pyrophosphate	Histidine side chain	Glucose 6-phosphatase	9
	Phosphoenol pyruvate	Histidine side chain	Glucose 6-phosphatase	9
	Phosphoenol pyruvate	Histidine side chain	Phosphotransferase	10
	ATP	Histidine side chain	Succinyl coenzyme A synthetase	11
N → N	Phosphoramidate	Histidine side chain	Glucose 6-phosphatase	9

small degree of bond formation by the attacking nucleophile, a large amount of breaking of the bond to the leaving group, and the principal driving force for the reaction comes from electron donation from the negative charges on the oxygen atoms to expel the leaving group." [14]. It is thus a borderline reaction between $S_N1(P)$ and $S_N2(P)$. The best recent evidence for this picture comes from studies of amine attack on phosphoramidates [15-17] [eq. (2)], hydrolysis of phosphate monoesters [18] [eq. (1)], and amine attack on phosphate monoesters and hydrolysis of phosphoramidates [12,19, 20] [eq. (3)]. The principle findings are: (i) the reactions involving amine attack are first order in amine concentration over wide concentration ranges, which is strong evidence for a bimolecular reaction; (ii) in each case the rate constant for the reaction is strongly dependent on the conjugate acidity of the leaving group, with Brønsted β's between 1.0 and 1.2, which supports the notion of a largely broken P-X bond in the transition state, and (iii) for both equations (2) and (3) there is a very weak dependence of rate constant on basicity of the attacking amine, with Brønsted β's between 0.0 and 0.25, indicating formation of a very weak P-N bond in the transition state. It is the presence of this weak bond which constitutes the principal difference between this description of phosphoryl transfer and a pure $S_N1(P)$ mechanism.

The mechanism of phosphoryl dianion hydrolysis (or other solvolysis) is much less certain, because of the unavailability of data examining the dependency of such reactions on the concentration or basicity of the nucleophile. Given the similarities between what is known for such reactions and reactions (2) and (3), it could be argued that solvolyses should proceed with considerable bimolecular character, since the transition state for reaction with a weak nucleophile such as water, should occur further along the reaction coordinate than for a strong nucleophile, such as an amine [22]. On the other hand, if (4) and (5) are viewed as competing routes for phosphoryl transfer, then it is clear that an S_N1 (P) mechanism would be favored by weak nucleophiles, especially for phosphate esters with good leaving groups. Thus it may be, as recent evidence suggests [21,136,137], that the hydrolysis of, for example, nitrophenyl phosphate, proceeds essentially via an S_N^1 (P) mechanism, although a definitive conclusion would require further experimentation.

2.2. pH Effects

Phosphomonoesters $ROPO_3H_2$, are dibasic acids. pK_{a_1} is approximately 2 and pK_{a_2} varies between 5.5 and 7.0, depending on the nature of R. Thus, in general, at physiological pH two species have to be considered, $ROPO_3H^-$ and $ROPO_3^{2-}$. Studies on the pH dependence of hydrolysis rate have shown that for simple alkyl phosphates, the monoprotonated ester is more reactive than the dianion. However, since both species hydrolyze much more rapidly than the corresponding diester monoanion $(RO)_2PO_2^-$, this greater reactivity is clearly not due to the decrease in negative charge at the phosphoryl group. Rather it is believed that the increased reactivity of the monoprotonated species is due to a transfer of the proton from the thermodynamically preferred site (a) to the

thermodynamically less preferred site (b), thus changing the poor leaving group RO^- into the good leaving group ROH. It is still an open question whether transfer occurs via a water molecule in the transition state or via equilibrium prior to the transition state. Either way, the concept of a proton transfer is supported by studies of the effect in variation of the conjugate acidity of ROH on pH rate profile, which demonstrate that for very good leaving groups (specifically, pK_a of ROH less than 5.5) the dianions hydrolyze more rapidly than the monoprotonated species [18].

For phosphoramidates it has been established that the species

$R_2NPO_3^{2-}$ is totally unreactive toward hydrolysis [23,24] so that in the pH region near neutrality the only species of interest is the monoanion $[R_2NPO_3H]^-$. This species can exist in two tautomeric forms, 1 and 2, and it has been shown conclusively that 1 is the

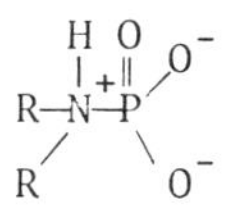

1

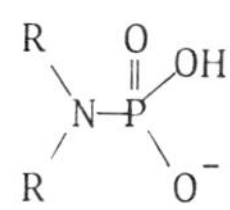

2

reactive species [12,16,23]. The overall equilibrium between 1 and 2 is shifted toward 1 as the basicity of the parent amine is raised. For the biologically important phosphoramidates phosphorylimidazole, phosphoramidate, phosphoryl lysine, phosphoarginine, and phosphocreatine, 1 should predominate. This zwitterionic form is well set up for a metaphosphate-type elimination, and is similar, in this sense, to the activated species postulated in $ROPO_3H^-$ hydrolysis.

3. MECHANISMS OF NUCLEOTIDYL TRANSFER

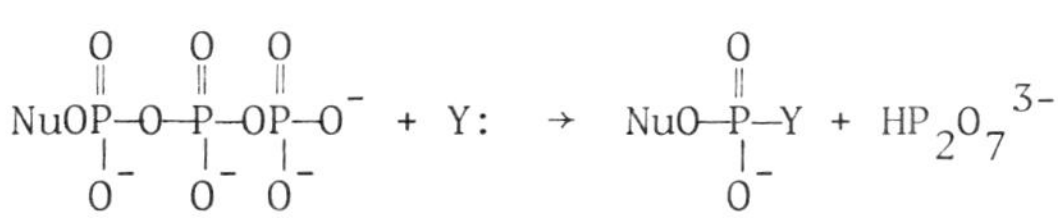

(6)

Nu = nucleoside residue

Nucleotidyl transfer reactions as depicted in reaction (6) involve nucleophilic attack at the α-P of the triphosphate chain, with inorganic pyrophosphate as the leaving group, and constitute one example of a more general class of reactions, i.e., nucleophilic attack on phosphodiester monoanions. These reactions differ from phosphoryl transfer in that they are much more S_N2(P)-like, the driving force for metaphosphate-type elimination being much less for the diesterified, monoanionic α-P position than for the dianionic monoesters discussed above. These conclusions are also based on Brønsted values. For amine and water attack on phosphodiester monoanions, the leaving group β varied between -0.9 and -1.0, while the β for attacking amines is 0.3 to 0.4. In contrast to phosphomonoester dianions, oxyanions can nucleophilically attack phosphodiester monoanions and the β for this attack is also approximetely 0.3. Thus, by comparison with phosphomonoester dianions, the rate of nucleophilic attack shows stronger dependence on the basicity of the nucleophile and less pronounced, though still strong, dependence on the conjugate acidity of the leaving group [12,13].

Of the three types of phosphate ester, monoester dianion, diester monoanion, and triester, the diester monoanions are by far the least reactive, as shown by the data in Table 2. The greater reactivity of monoester dianions can be understood on the basis of the contribution of the metaphosphate-type mechanism. That of the triesters may be explained by increased partial positive character at phosphorus as well as decreased charge-charge and/or charge-dipole repulsions between the phosphoryl group and incoming nucleophile.

An exception to the low reactivity of diesters is found in the case of 5-membered ring cyclic phosphates, such as ethylene phosphate, <u>3</u>, an analog of 2',3'-cyclic AMP, which undergoes P-O cleavage on hydrolysis some 10^7 times faster than the corresponding alicyclic diester, dimethyl phosphate, <u>4</u>. This finding has been

TABLE 2

Rate Constants for the $S_N2(P)$ Reactions of Various 2,4-Dinitrophenyl Phosphate Esters [13]

Nucleophile	$ArOPO_3^{2-}$	ArO–P(=O)(OMe)–O$^-$	ArO–P(=O)< (cyclic –O–(CH$_2$)$_3$–O–)
H_2O	1.9×10^{-4}	3.22×10^{-7}	8.35×10^{-6}
Acetate		7.4×10^{-6}	1.48×10^{-2}
Hydroxide		2.8×10^{-2}	64
Pyridine	3.90×10^{-2}	6.15×10^{-3}	1.24×10^{-2}
Hydroxylamine	6.48×10^{-2}	5.20×10^{-2}	10.7
Fluoride		1.92×10^{-3}	9.36

reasonably explained on the following basis [25]. In going from an acyclic diester to a high-energy trigonal bipyramidal intermediate (which is probably close to but not at the transition state) the angle α must be compressed from a normal angle of 110° in the 4-coordinate ground state to 90° in the 5-coordinate transition state.* In ethylene phosphate, the corresponding endocyclic O-P-O angle α' is compressed to 98° in the ground state because of constraints imposed by the ring, and the activation energy to attain the trigonal bipyramid is correspondingly reduced. From this argument it is to be expected that six-membered ring cyclic phosphates, in which an endocyclic O-P-O angle of 110° is attainable without inducing strain, would not be hydrolyzed rapidly and this has been found to be the case [26]. This probably accounts for the kinetic stability of the biologically important six-membered ring cyclic phosphate, 3',5'-cyclic AMP (which is, however, thermodynamically quite unstable toward hydrolysis [27]).

4. POSSIBILITIES FOR METAL ION PARTICIPATION

4.1. Discrete Roles

In this section we will list possible discrete roles for metal ion catalysis of both phosphoryl transfer and nucleophilic attack at phosphodiester monoanions. Similar listings have been made elsewhere [5,28,29].

4.1.1. Charge Shielding

In the previous section the point was made that anionic nucleophiles were prevented by charge repulsion from attacking dianionic phosphomonoesters and were slowed in attacking monoanionic diesters. By shielding these charges, metal ions could overcome this charge barrier. Catalysis from this effect should be more important for dianionic than for monoanionic species.

---

*It is well established that the leaving group must occupy an axial position in the trigonal bipyramid [25].

4.1.2. Coordination to Leaving Group

Since the rates of both phosphoryl transfer and nucleophilic attack on diesters show strong dependence on the pK_a of the conjugate acid of the leaving group, increase of acid strength of the leaving group through metal ion coordination should be strongly rate-enhancing in both cases.

4.1.3. Strain Induction

Metal ion binding in a bidentate mode to a single phosphoryl group, 5, could compress the O-P-O angle, α'', in the ground state,

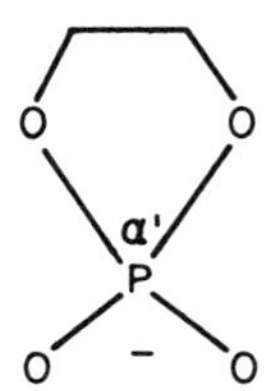

3

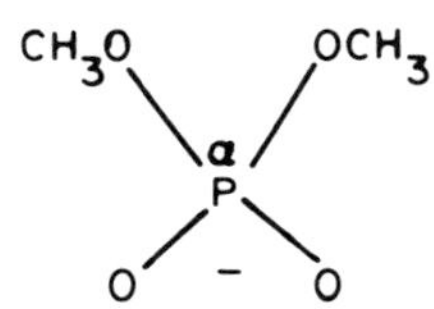

4

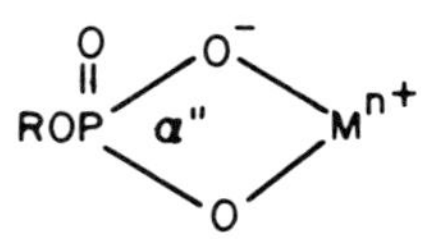

5

and reduce the activation energy necessary for attainment of the trigonal bipyramidal intermediate (or transition state) in phosphoryl transfer, in analogy to what has been discussed in the case of ethylene phosphate. This role is probably not as important for diesters, since their lower negative charge density reduces the likelihood of bidentate coordination.

4.1.4. Template

By forming a ternary complex, phosphate ester-metal ion-nucleophile, metal ions can transform bimolecular into unimolecular reactions and give rate accelerations in this manner, as has been shown for many other systems [30].

4.1.5. Activation of Nucleophile

Metal ions can increase the nucleophilicity of a bound nucleophile by lowering its pK_a, e.g., converting a bound neutral water into a bound hydroxide. In this example, the nucleophilicity of bound hydroxide would be expected to be lower than free hydroxide, but might be greater than free water. Activation of the nucleophile might be expected to be of somewhat greater importance for diester reactions, since these are more S_N2(P)-like than the monoester reactions.

4.1.6. Polarization or Electron Sink

Metal ions have been found to strongly accelerate reactions of carbonyl compounds (e.g., amino acid ester hydrolysis [31], β-keto acid decarboxylation [32]), and these effects have been reasonably attributed to the ability of the metal ion to withdraw electron charge and so polarize the carbonyl bond, thus increasing the positive character of the carbon. By contrast, the rates of both amine and water attack on simple phosphate monoesters and monoamides are hardly affected by addition of metal ions at concentrations sufficient to form the 1:1 $X\text{-}PO_3^{2-}M^{n+}$ complex [7,16,17,33,34], which is strong evidence against a polarization or electron sink effect. This lack of large catalytic effect is perhaps best explained by the high degree of metaphosphate character in phosphoryl transfer, since any effect of electron withdrawal on the positive character of phosphorus would be offset by a corresponding loss in driving force for leaving-group expulsion. Additionally, it has been argued, based on x-ray studies and vibrational spectra, that metal ion coordination does not change the charge density at phosphorus, since the removal of electron density from the oxygen bound to metal ion is compensated by increased π bonding between the oxygens and phosphorus [29].

For nucleophilic reactions at diesters, which have a lesser degree of metaphosphate character, polarization or electron-sink effects could be of some importance on the following grounds.

Triesters are much more reactive than diesters (Table 2), so that even though divalent metal ions do not polarize charge as well as protons or alkyl groups [29,35], even a partial effect in this direction could lead to appreciable catalysis.

4.2. Transition State Formalism

In the preceding discussion we listed the plausible catalytic roles for metal ions separately, and in a single direction of reaction. An alternative way to describe metal ion effects is through use of a formalism from transition state theory which has been previously applied to enzymic [36,37] and acid-base catalysis [38]. Thus (i) catalysis results when metal ion binds more tightly to the transition state for reaction than to the initial state of reactants (ii) inhibition results when metal ion binds less tightly to the transition state than to the initial state; and (iii) catalysis in one direction of reaction and inhibition in the opposite direction result when metal ion binds to the transition state more tightly than to the initial state of reactants in one direction of reaction and less tightly than to the initial state of reactants in the opposite direction of reaction. Condition (iii) must, of course, be accompanied by a change in the overall position of equilibrium and this underlines the point that metal ion is, strictly speaking, not a true catalyst when it forms part of the reactants in one or both directions of reaction. Shifts in the overall position of equilibrium may also accompany conditions (i) and (ii), although they will be minor if metal ion interactions with the reactants in both directions of reaction are similar.

Binding constants for many divalent metal ion-phosphate ester complexes have been determined, and a fair amount is known about the structure of these complexes in solution [39-41]. The structure of metal ion complexes of the transition state for phosphoryl transfer and nucleophile attack at phosphodiesters is discussed below.

4.3. Structure of Metal Ion Complexes of the Transition State

Our discussion of these structures is based on two simplifying assumptions. First, the transition state structure for both phosphoryl transfer and nucleophilic attack at phosphodiesters is a trigonal bipyramid with both entering nucleophile and leaving group in the apical positions. Second, this structure is not radically altered by the presence of metal ion, with the exception that we explicitly consider the possibility that metal ion binding may lead to a loss of ionizable protons. The first assumption appears to be quite valid both for phosphoryl transfer and for nucleophilic attack on simple acyclic diesters [12]. However, for the unusually reactive five-membered ring cyclic diesters previously discussed, it is (i) probable that the trigonal bipyramidal structure represents a metastable intermediate rather than the transition state, and (ii) clear that there is no requirement for a coapical arrangement of entering and leaving groups. We shall return to this point below. The second assumption is, in principle, arbitrary, since in the presence of metal ion reaction could proceed via a different pathway. However, it does allow a minimum estimate of metal ion effectiveness to be made, since it is an important consequence of transition state theory that when this assumption is incorrect, the observed rate increase in the presence of metal ion is less than the full catalytic potential of metal ion [37].

In the trigonal bipyramidal structure there are three sites with which metal ion can interact--the PO_3(R'') moiety, and the leaving and entering groups--leading to the four possible structures, (a)-(d), shown in Table 3. In these structures the parenthesis around R'' signifies that R'' may (nucleophilic attack on phosphodiesters) or may not (phosphoryl transfer) be present, and similarly for (H). In structure (a) we draw one of the M^{n+}-O bonds as a dashed line to indicate that both monodentate and bidentate coordination to the phosphoryl group are possible. Structures b' and c' are for bifunctional nucleophiles, such as HPO_4^{2-} (6) or

TABLE 3

Metal Ion Binding in the Transition State
and Possible Catalytic Roles

	One Site		Two Sites	Three Sites
Structure	(a)	(b)	(c)	(d)
	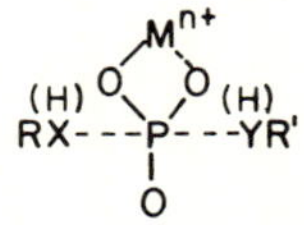	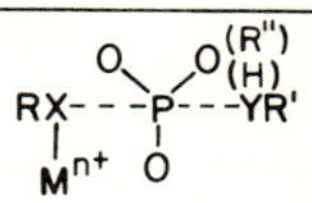	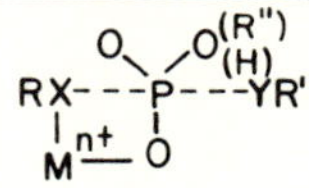	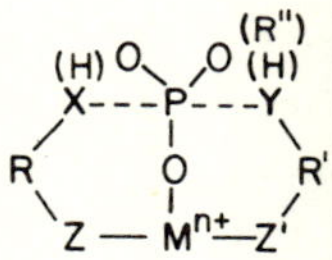
		(b')	(c')	
		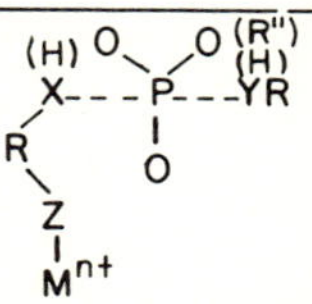	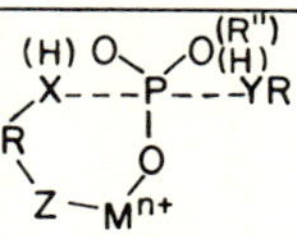	
forward	i, iii, vi	v	i, iv–vi	i, ii, iv–vi
reverse	i, iii, vi	ii	i, ii, vi	i, ii, iv–vi

Catalytic roles

i. charge shielding
ii. leaving-group coordination
iii. strain induction
iv. template
v. activation of nucleophile
vi. polarization

$CH_3CO_2^-$. For more acidic metal ions, such as Zn^{2+} or Fe^{3+}, the positions marked R and Z might also refer to the OH of a water molecule, such that a metal ion-bound hydroxy group could serve as a general base catalyst in the forward direction, (RX is nucleophile) and a metal ion-bound water molecule could serve as a general acid catalyst in the reverse direction (R'Y is nucleophile) (7).

The possible catalytic roles for metal ion in the forward and reverse directions are listed underneath each of these structures, but they could equally well be discussed in terms of the transition state formalism. For example, the catalytic role of metal ion in providing a template for reaction corresponds to the formation of a bi- or tridentate ligand in the transition state which binds metal

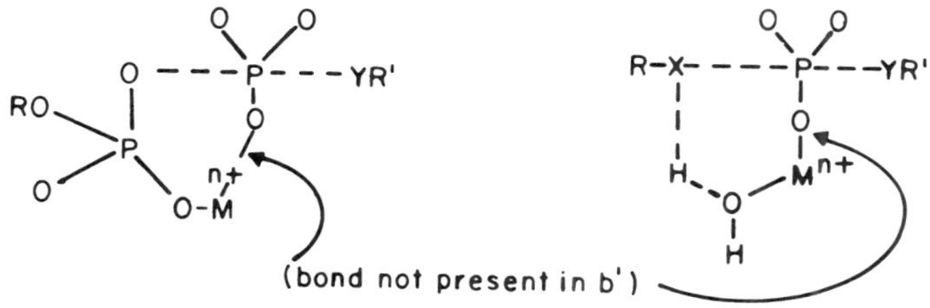

6 7

ion more tightly than do the reactants. Similarly, promotion of a nucleophile through a pK_a lowering corresponds to a more favorable electrostatic interaction between metal ion and the transition state as compared to metal ion and reactants, etc.

In structure (a), although bidentate coordination could be of quite general importance as a catalytic role (iii) for metal ion, catalysis via monodentate coordination is probably limited to nucleophilic attack by an oxygen anion, since rate studies indicate there to be no large repulsive charge-dipole interaction in attack by neutral nucleophiles on $PO_3(R'')$ (Table 2). Structures (b) and (b') can lead to catalysis via the roles indicated if the pK_a is higher than the pH at which the rate is being measured. However, if the pK_a is lower than this pH, one would predict metal ion inhibition in the forward direction of reaction.

The two-site coordination structures in (c) and (c') are probably of more general importance, since they allow metal ion to have several catalytic roles. In these structures only one of the incoming and leaving groups is coordinated to metal ion, and a choice can be made as to which site of coordination leads to more effective catalysis. Thus, for O → O transfer, coordination to the more basic oxygen would give the greater catalytic effect, on the basis of the correlation of rates of phosphate ester hydrolysis with conjugate acidity of the leaving group. N → O transfer via structure (c) should proceed with O coordinated preferentially, since direct coordination to N would destroy its nucleophilicity. Indirect coordination to the amine is possible in structure (c') but coordination to oxygen would have the greater catalytic effect, given the greater lability of P-N as opposed to P-O bonds.

Structure (d) is a special case in which both entering and leaving groups can be coordinated to metal ion simultaneously in the metal ion. Similar structures, with two four-membered rings, or with one six- and one four-membered ring, are unlikely because of ring strain.

We now return to the question of the transition state structure for cyclic diesters. Although entering and leaving groups need not be coapical, it has been shown that nucleophiles enter and leave from the apical positions [25]. Thus, the transition state structure might resemble 8 and it is clear that structures (a)-(c') (Table 3) could be appropriately modified for 8. Only the three-site coordinated structure (d) would be inapplicable.

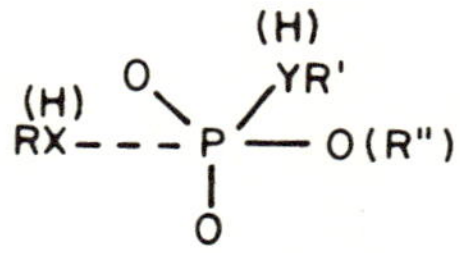

8

5. MODEL SYSTEM STUDIES FOR METAL ION CATALYSIS

A large number of studies have been made on metal ion nucleophilic reactions of phosphoryl compounds and a list of these studies is given in Table 4. Solutions containing metal ion, phosphoryl compound, and nucleophile can have a fairly complex chemistry, involving (i) several possible complexes with different stoichiometries, e.g., metal ion:phosphoryl compound 1:1, 1:2, 2:1, etc., (ii) for multidentate ligands, such as ATP, different metal ion binding modes; and (iii) complex pH dependence due to ionization of the phosphoryl group, the nucleophile, and the water bound to metal ion.

It can be a formidable undertaking to arrive at that point which necessarily precedes intelligent formulation of mechanism, that is, a knowledge of the stoichiometry, state of protonation,

TABLE 4

Bibliography of Metal Ion Effects on Phosphoryl Transfer and on Nucleophilic Attack on Phosphodiesters

Reaction	Ref.
Hydrolysis of simple phosphate esters and phosphoramides	
Phenyl salicylyl phosphate	42
Phenyl lactyl phosphate	42
Phosphoenolpyruvate	43
Acetyl phosphate	33,34,44
Carbamyl phosphate	45
Salicyl phosphate	46,47
AMP	48
Pyridyl methyl phosphate	49,50
2-[4(5)-Imidazolyl]phenyl phosphate	51
Methyl phosphate	52
Pyridine carbaldoxyimyl phosphate	53
Phosphorylimidazole	17
Phosphoramidate	16
Phenyl glycinyl phosphoramidate	42b
Benzyl phosphoroguanidate	54
Cysteamine S-phosphate	7
Hydrolysis of nucleoside di- and triphosphates	48,55-64
Hydrolysis of other organic di- and triphosphates	
Methyl triphosphate	56
Phenyl propyl triphosphate	58
P^1,P^2-Diphenyl pyrophosphate	65
P^1,P^2-Dithymidinyl pyrophosphate	66
3,6-Diaza-7-(2-hydroxyphenyl)heptylphosphonyl phosphate and -pyrophosphate	67
3,6,9-Triaza-10-(4-methyl-2-pyridyl)decylphosphonyl phosphate and -pyrophosphate	67

TABLE 4 (Continued)

Reaction		Ref.
Hydrolysis of inorganic pyrophosphate, oligophosphates, and polyphosphates		68-70
Hydrolysis of RNA		71-74
Hydrolysis of phosphate esters in the presence of metal ion gels, hydroxides, and other precipitates		1,75-82
Transfer to nucleophiles other than water		
Phosphoryl compound	Nucleophile	
ATP (γ-P)	Phosphate	58,84,85
ATP (γ-P)	Acetate	86,87
ATP (γ-P)	Ca^{2+} apatite	81
ATP (γ-P)	1,10-Phenanthroline-2-carbinol	88
ATP (α-P)	Internal-3'-OH	77-78
Phenyl triphosphate (γ-P)	Phosphate	58
P^1,P^2-Diphenyl pyrophosphate	Cyclodextrin	29
Phosphorylimidazole	Pyridine aldoxime anion	23
Phosphorylimidazole	Various amines	17
Benzyl phosphoroguanidate	Phosphates	54
8-Quinolyl phosphate	Benzyl alcohol	89

and dominant ground state structure of the catalytically active species in solution. The great majority of studies cited in Table 4 do not approach this level of sophistication and merely record metal ion effects which allow of no detailed mechanistic interpretation. Most of these studies will not be discussed further in this article. However, there are a small number of model systems which (i) have been well designed so as to simplify interpretation of results, or (ii) have been extensively studied kinetically, or (iii) structurally, or (iv) show very large rate enhancements by metal ion, or some combination of (i)-(iv), for which it is worthwhile to advance mechanistic proposals. In this section the discussion of these systems will be organized around structures (a)-(d) of Table 3.

Good evidence has been presented for strong metal ion catalysis of hydroxide ion attack on both acetyl phosphate and carbamyl phosphate [33,34,44,45]. However, in the former case the reaction h s been shown to proceed exclusively via C-O cleavage [34] and given the similarity of the two compounds, this is probably true for the latter case as well. We will not discuss these studies further, since we are here concerned with nucleophilic attack at phosphorus.

5.1. Models for Structure (a) (Table 3)

5.1.1. Co^{3+} Complexes of Methyl Phosphate

Farrell et al. [52], noting that PO_4^{3-} forms a bidentate four-membered ring with Co^{3+}, reasoned that such coordination would activate the phosphorus toward nucleophilic attack by strain induction (Section 4.1.3). Accordingly, they measured the rate of $CH_3OPO_3^{2-}$ hydrolysis when bound to triethylene tetramine-Co^{3+}, which has two coordination sites available for binding, and to $Co(NH_3)_5^{3+}$, which has a single site. They found the trien-Co^{3+} complex hydrolyzes 130 times faster than $CH_3OPO_3H^-$ [which in turn hydrolyzes faster than $CH_3OPO_3^{2-}$ (Section 2.2)], while the $Co(NH_3)_5^{3+}$ complex was at least about 10^2 slower.

However, the precise explanation of this strong catalytic effect is uncertain, because structural studies showed that methyl phosphate was bound through one oxygen rather than two in the trien-Co^{3+} complex, with water occupying the sixth position. Thus, the observed rate constant observed could be due either to (i) reaction funneling through a small amount of complex with bidentate coordination so that the observed catalysis would be only a fraction of the specific catalysis for the bidentate complex, or (ii) nucleophilic attack by bound water, with Co^{3+} acting as a template [i.e., transition state structure (c)].

5.2. Models for Structures (b) and (b') (Table 3)

5.2.1. 2-(4(5)-Imidazolyl)phenyl Phosphate (9)

Benkovic and Dunikoski [51] have shown that the Cu^{2+} complex of 9 hydrolyzes at least 10^4 times faster than 9 itself. Although the structure of the complex was not established, the weak calculated binding constant (4×10^2 M^{-1}) suggest that it is predominantly monodentate or bidentate with an unfavorable eight-membered ring 10. The authors suggest that the transition state

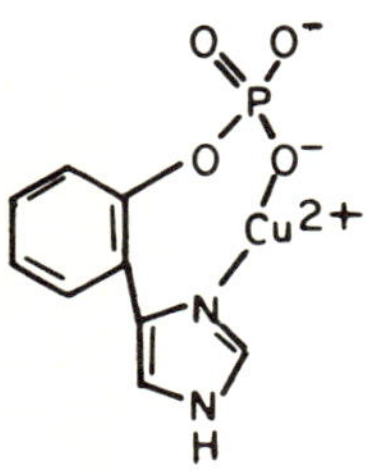

10

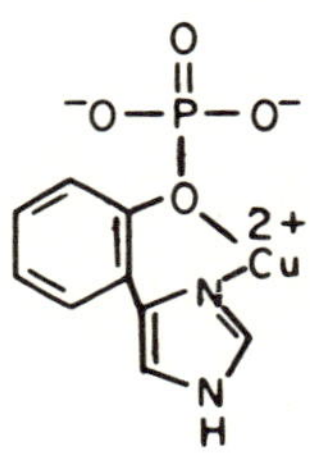

11

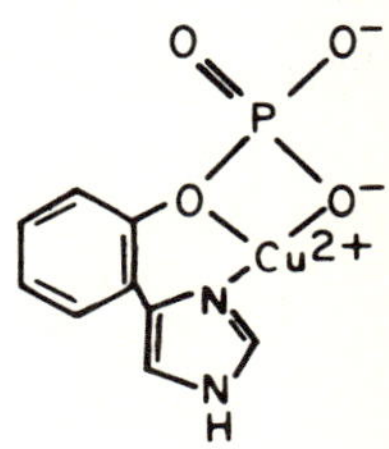

12

for the reaction may be represented either by 11 or 12 and in either case Cu^{2+} would serve to lower the pK_a of the 2-(4(5)-imidazolyl)phenyl leaving group. One unexplained result is the failure to observe significant catalysis by other divalent transition metal ions. For the proposed transition state, one could expect other ions to be effective catalysts, although, given the strong acidity of Cu^{2+} (Table 6), perhaps to a smaller extent than Cu^{2+}. This apparent anomaly may be due to many factors (phosphate vs imidazole as primary coordination site, hydroxy complexes, polynuclear complexes) (see below) and for the present the mechanism must be regarded as tentative.

5.2.2. 2-Pyridyl Methyl Phosphate

In a related study, Murakami and Takagi [49] have studied Cu^{2+} effects on 2-pyridyl methyl phosphate hydrolysis and have found an 18-fold acceleration in the complex. They consider the dominant structure of the Cu^{2+} complex in solution to be a bidentate complex

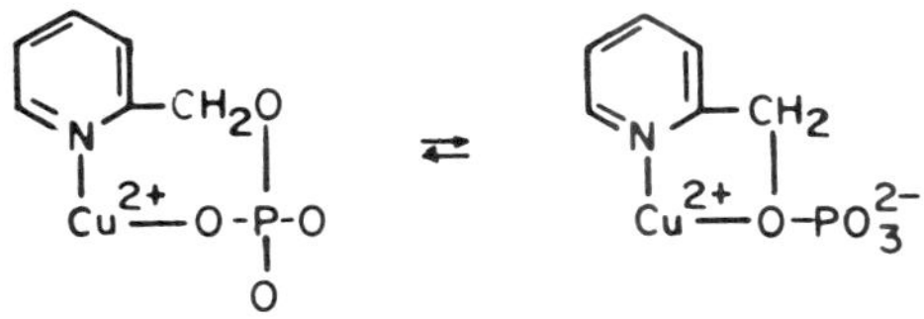

13 14

with a seven-membered ring, 13, with reaction proceeding via the five-membered ring transition state, 14. The smaller rate factor seen in this study as compared to the previous one might reflect a less favorable equilibrium between the five- and seven-membered ring chelates.

5.2.3. Cysteamine S-Phosphate

Benkovic and Miller [7] have demonstrated a very large catalytic effect of Cu^{2+} on cysteamine S-phosphate hydrolysis. Although they were unable to obtain a saturated rate with respect to Cu^{2+}, a minimal rate acceleration for the hydrolysis of the Cu^{2+} cysteamine S-phosphate complex of approximately 10^3 with respect to $\overset{+}{N}H_3CH_2CH_2$-$SPO_3H^-$ and 10^4 with respect to $\overset{+}{N}H_3CH_2CH_2SPO_3^{2-}$ is indicated. By contrast, Zn^{2+}, Co^{2+}, Ni^{2+}, Mn^{2+}, Ca^{2+}, Mg^{2+}, and Al^{3+} have minimal effects on rate while Fe^{3+} is strongly inhibitory. The authors attributed Cu^{2+} catalysis to the formation of 15, and Fe^{3+} inhibition to complexation solely with the phosphoryl oxygens, with con-

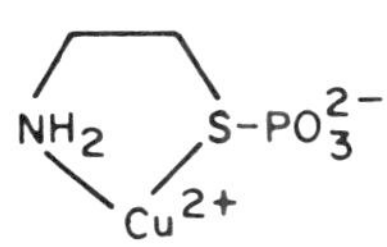

15

sequent interference with electron donation to phosphorus which provides the driving force for the spontaneous hydrolysis.

5.2.4. Phosphoenol Pyruvate

Benkovic and Schray [43] have investigated the catalytic effect of metal ions on phosphoenol pyruvate hydrolysis. Rate accelerations by Cu^{2+}, Al^{3+}, Fe^{3+}, Mg^{2+}, Mn^{2+}, Ni^{2+}, Co^{2+}, Ca^{2+}, and

Zn^{2+} were all quite modest, not exceeding tenfold. Hg^{2+}, on the other hand, was strongly catalytic, giving a rate acceleration of greater than 10^6, with about 10% of hydrolysis (above pH 4) proceeding via P-O cleavage. This effect is attributed to an addition reaction of Hg^{2+} to the double bond of phosphoenol pyruvate, with a corresponding increase in the conjugate acidity of the enolate leaving group. Although no structural evidence is presented, this interpretation of the Hg^{2+} effect is reasonable on the basis of the known reactivity of Hg^{2+} toward double bonds and the metal ion specificity of the catalytic effect.

5.2.5. Pyrophosphate Hydrolysis at Terminal Position

There is both direct [83a] and indirect [83b] evidence that the hydrolysis of unsymmetrical diesters of pyrophosphate 16 proceeds at a very much faster rate (10^5 to 10^6 times faster) than

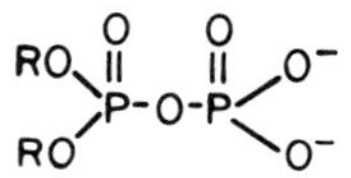

16

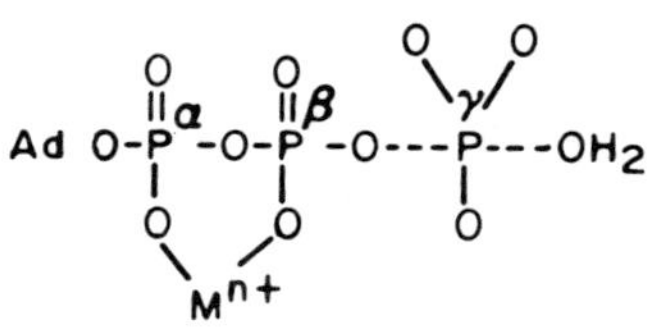

17

that of the corresponding monoesters, and this is as expected, since the pK_a of $(RO)_2PO_2H$ is about 2, while that of $ROPO_3H^-$ is about 6.5. This has led to speculation [2,55,56,67] that metal ions could catalyze phosphoryl transfer from, for example, ATP, by coordinating to the β-phosphoryl group and neutralizing its charge, while leaving the terminal position unbound, thus converting ATP into a pseudo unsymmetrical diester, 17. This mode of binding is unfavorable, since electrostatic interaction should lead to binding to the γ position, and this expectation is borne out by ^{31}P NMR data showing that in M^{2+}-ATP complexes, the dominant species in solution have metal ion binding to both β and γ positions, and for some metal ions to the α position as well [90,91]. Therefore, testing of the model requires putting a constraint on metal ion binding. To do this, Cooperman [67] synthesized a number of multidentate ligands containing pyrophosphate linkages, with the hope that the presence of the additional ligands would draw metal ion off the terminal phosphate, as shown in structures 18-21. This

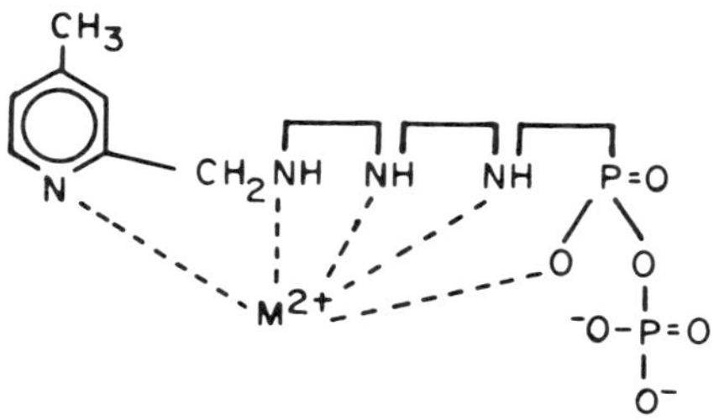

18

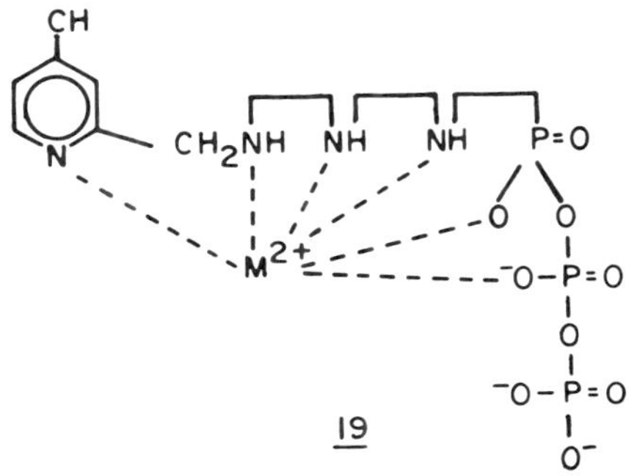

19

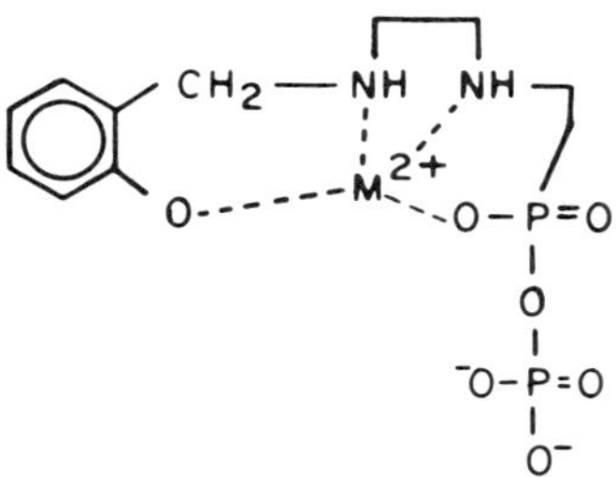

20

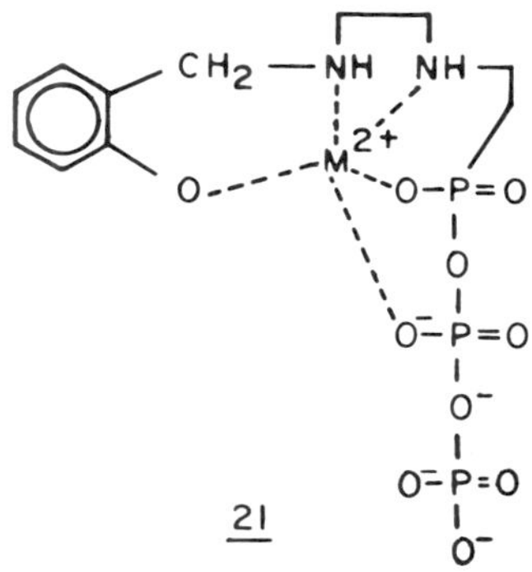

21

experiment failed, as none of the metal ions tested gave rate enhancements significantly larger than that found with ATP which has no such directional effect, even though ^{31}P NMR studies indicated that for some complexes the desired binding mode was substantially obtained. Despite this result, recent work on the mechanism of creatine kinase [92] suggests that in fact on the surface of the enzyme Mn^{2+} is bound to the β and not to the γ position of ATP (Section 6.1). In view of these disparate results, it is useful to examine the question of what is the maximum potential metal ion activation of phosphoryl transfer from ATP to water via a transition state such as 17, using a Brønsted β value of -1.23

for the dependence rate of phosphomonoester dianion hydrolysis on leaving-group conjugate acidity. The two leaving groups are ADP^{3-} in the uncatalyzed case and $[ADPM^{2+}]^{-1}$ in the case of metal ion catalysis.* The pK_a's of $ADPH^{2-}$ and MADPH are compared in Table 5, and the anticipated relative rates are calculated. The noteworthy result is that, even with the assumption of optimal binding, the maximum total rate acceleration possible with those metal ions most commonly associated with phosphoryl transfer enzymes, i.e., Mg^{2+}, Ca^{2+}, Mn^{2+}, and Zn^{2+}, varies between 10^2 and 10^3, which, though

TABLE 5

pK_a's and Calculated Relative Hydrolysis Rate Constants for M^{2+}-ATP Complexes of Structure 17

M^{2+}	pK_{MADPH}	$k_{cat_{rel}}$
none	6.4	1.0
Mg^{2+}	4.9	70
Ca^{2+}	5.2	30
Mn^{2+}	4.3	400
Co^{2+}	4.3	400
Ni^{2+}	4.2	500
Cu^{2+}	3.1	11,500
Zn^{2+}	4.2	500

[a]Calculated from data in References 93 and 94.

*The assumption here is that the rates of pyrophosphate hydrolysis show the same dependence on leaving group conjugate acidity as do both acyl and aryl monoesters [18]. This assumption is supported by Miller's results [83a], although it is true that the absolute rates of pyrophosphate hydrolysis are about two orders of magnitude slower than expected on the basis of the aryl and acyl monoester rates. [83c].

significant, is much less than the acceleration seen in unsymmetrical pyrophosphate diesters. With such a relatively small optimal rate factor to work with, the lack of significant metal ion catalysis in the model system may be explained by suboptimal binding and/or by reduced acidity of the metal ion in the multidentate complexes.

5.2.6. Pyrophosphate Hydrolysis-Internal Position

Tetas and Lowenstein [55] have measured metal ion effects on AMP formation from ATP via reaction (7).

$$\text{ATP} \xrightarrow{Mn^{2+}} \text{AMP} + \text{PPi} \tag{7}$$

This reaction most likely proceeds by attack of water or hydroxide at the α-P, and thus represents a nucleotidyl transfer process. At pH 9 and $[Mn^{2+}]$ = [ATP] = 0.02 M, rates of AMP formation were found to be between 3- and 60-fold greater than in the absence of metal ion, and for all of the metal ions tested the rate enhancement for this process was much greater than for the corresponding phosphoryl transfer reaction giving ADP and Pi formation. Although the data are insufficient for construction of a detailed model, a plausible role for metal ion is coordinate to the β and γ phosphoryl positions, 22, which would lower the pK_a of the leaving group, inorganic pyrophosphate. As discussed above, this type of coordination has

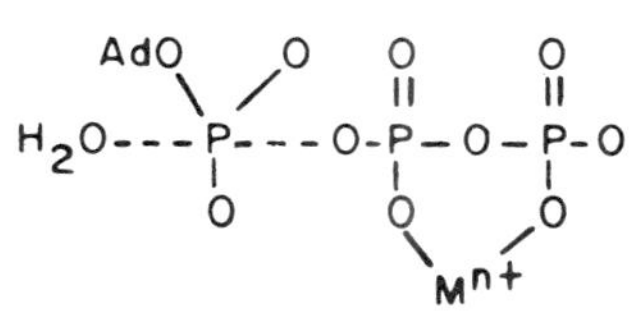

22

been shown to be the most favorable one in solution. It is also possible that metal ion coordinates to the α-P, thus reducing the charge barrier to hydroxide attack, but it is worth pointing out that Ca^{2+} and Zn^{2+}, which do not coordinate the α-P (at least not by NMR evidence [90]) are very effective in catalyzing AMP formation.

5.3. Models for Structures (c) and (c') (Table 3)

5.3.1. RNA

Metal ions have been shown to catalyze RNA depolymerization via formation of 2',3' cyclic phosphates. Pb^{2+} has been found to be most effective by far [71-74]. For example, at a concentration of 1 mM Pb^{2+} and pH 6.5 depolymerization is half over in about 20 min, whereas in the absence of Pb^{2+} the rate is very slow. From the published data, it is possible to estimate that Pb^{2+} catalyzes the reaction by at least a factor of 10^4. Zn^{2+} is also effective, giving a rate enhancement of 10^2. Other metal ions also catalyze the reaction to some extent. A plausible structure for the transition state is shown in 23. It is interesting that metal ion

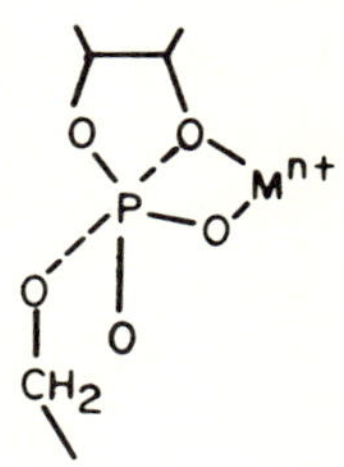

23

catalysis is sensitive to the nature of the neighboring base, and is much slower in regions of high secondary structure, so that the rate of the depolymerization reaction can be used as a probe of nucleic acid structure.

5.3.2. Pyridine Aldoxime Anion Attack on Phosphoryl Imidazole

Lloyd and Cooperman [23] demonstrated a rapid phosphoryl transfer from the zwitterionic species phosphoryl imidazole monoanion, N-methyl phosphoryl imidazole monoanion, and phosphoramidate monoanion to Zn^{2+}-pyridine aldoxime anion 24 and showed further that transfer occurs within the ternary complex of phosphoryl donor with 24. Phosphoryl imidazole and phosphoramidate were chosen as donors because of their chemical similarity to known or postulated phosphoryl enzyme intermediates, and Zn^{2+}-pyridine aldoxime anion was chosen as a nucleophile because in this complex metal ion was sterically prevented from direct coordination to the nucleophilic oxygen anion. In the absence of Zn^{2+}, it was shown that pyridine aldoxime anion does not attack N-methyl phosphoryl imidazole monoanion at a measurable rate, which is expected on the basis of the charge repulsion effect discussed earlier. The rates of phosphoryl transfer in the complex were from 1 to 4 x 10^3 times faster than simple hydrolysis, which shows that if the charge barrier can be overcome strong oxyanion nucleophiles will atack phosphoryl dianions much more rapidly than the weaker nucleophile, water.

The transition state for the reaction is thought to be a modified (having a five- rather than six-membered ring) type of structure (c') (25). Of the possible catalytic roles for metal ion in this structure, charge shielding is probably the most important, since no rapid N → N phosphoryl transfer could be demonstrated between the Zn^{2+} complex of pyridine carboxaldehyde hydrazone (26) and phosphoryl imidazole [53].

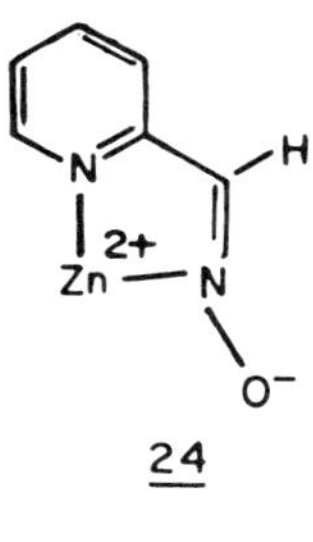

24

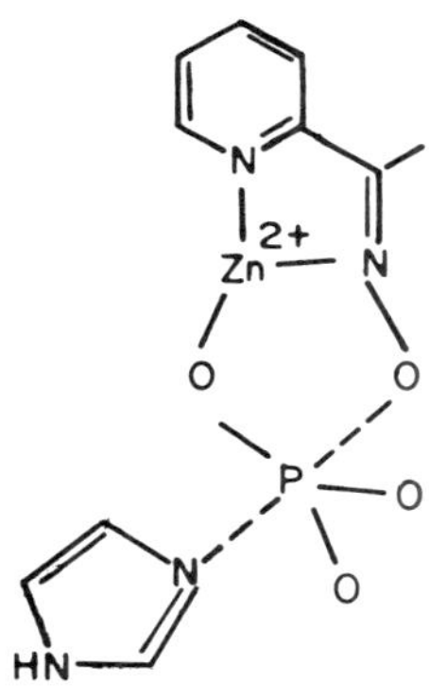

25

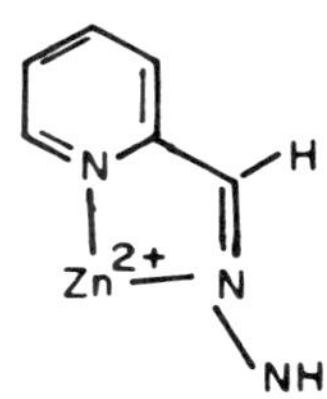

26

5.3.3. Pyridine Carbaldoximyl Phosphate Hydrolysis

Recent work has shown that Zn^{2+} catalyzes a very rapid hydrolysis of pyridine carbaldoximyl phosphate [53], and it is reasonable to assume that the reaction proceeds via a transition state, 27

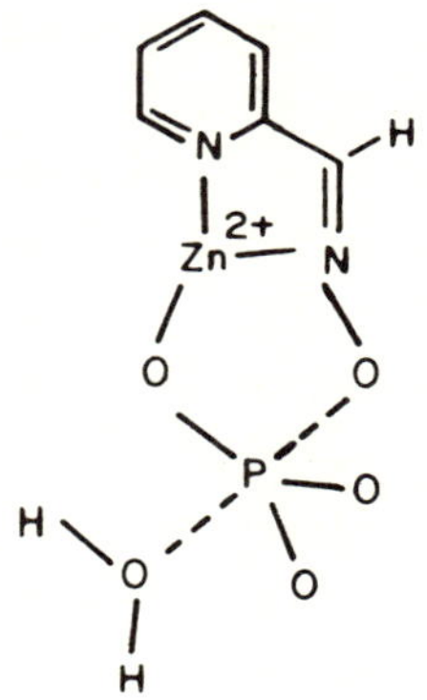

27

which is essentially the microscopic reverse of 25 with water substituting for imidazole. Of the possible catalytic roles for Zn^{2+}, leaving-group activation is clearly the major one. The pK_a of Zn^{2+}-PCAH is 6.0, and the measured rate for Zn^{2+}-PCAP hydrolysis corresponds closely to what is predicted for a leaving group of pK_a 6.0 [18]. Since the pK_a for PCAH is 10.0, Zn^{2+} catalysis may amount to as much as a factor of 10^5.

5.4. Models for Structure (d) (Table 3)

5.4.1. Phosphoryl Transfer from ATP to Inorganic Phosphate

Lowenstein has found that in the presence of divalent metal ion ATP will phosphorylate Pi via reaction (8) [84,85]:

$$ATP^{4-} + HPO_4^{2-} \longrightarrow ADP^{3-} + HP_2O_7^{3-} \quad (8)$$

The rate of the transphosphorylation reaction shows strong metal ion specificity as indicated in Table 6. For the Mn^{2+}-catalyzed reaction, the pH rate profile shows a maximum at pH 9.0. Miller

TABLE 6

Divalent Metal Ion-Catalyzed Phosphorylation of Pi by ATP

M^{2+}	Relative rate of phosphorylation[a]	Ionic radius (M^{2+})	pK_a of M^{2+}-H_2O[b]
Mn^{2+}	1.00	0.80	10.6
Ca^{2+}	2.60[c]	0.94	>13[d]
Cd^{2+}	0.65	0.92	11.6
Zn^{2+}	0.23	0.69	8.7
Co^{2+}	0.16	0.70	8.9
Fe^{2+}	0.12	0.76	7.2[e]
Cu^{2+}	0.08	0.92	6.8
Ni^{2+}	0.06	0.68	9.4
Be^{2+}	0.06	0.30	5.7[f]
Mg^{2+}	0.05	0.65	12.8

[a]Reference 84

[b]Unless otherwise indicated values are determined at 30°C, 0.1 N KCl [95a].

[c]Precipitate present.

[d]Estimated from values in Reference 40, pp. 41,42.

[e]25°C, 0.5 N KCl [96].

[f]20°C, 0.1 N KNO_3 [97].

and Westheimer [58] have obtained similar results studying the Ca^{2+}-catalyzed transphosphorylation using both ATP and phenylpropyl triphosphate, a synthetic triphosphate, as phosphoryl donors. Miller [100] suggests that the active nucleophile is PO_4^{3-}, which would explain the descending limb of the pH rate curve below pH 9 (pK_a HPO_4^{2-} is 12). This suggestion is supported by the failure to find transphosphorylation with AMP^{2-} as the phosphoryl acceptor. The drop in rate at high pH is complicated by precipitation problems, but Miller suggests it might be due to formation of metal ion-hydroxy complexes. This explanation is attractive since it accounts for the relative ineffectiveness of the more acidic metal ions in the transphosphorylation reaction (Table 6). For these ions hydrolysis could occur well below pH 9, at pH values where there would be little PO_4^{3-}. As there is evidence that Ba^{2+} and Sr^{2+} are also effective catalysts, the ineffectiveness of Mg^{2+} is most reasonably explained by its small ionic radius.

With respect to stoichiometry, there is clear evidence that the ATP_2Mn^{2+} complex is unreactive, which is reasonable. When ATP is varied at fixed Mn^{2+} concentration (0.05 M) it is found that the rate reaches a maximum at $[ATP]/[Mn^{2+}] = 0.7$, and this has been taken as evidence that the reactive complex has a stoichiometry of Mn_2ATP. This complex may be reactive, but it cannot account for the total observed reaction velocity, since at $[ATP] = [Mn^{2+}]$ the velocity is 71% that of the maximal velocity, whereas the Mn_2ATP concentration should have fallen approximately seven-fold [95]. It therefore seems reasonable to assume that the 1:1 complex is more reactive than the 2:1 complex. The variation with rate as a function of phosphate concentration at fixed MnATP was also studied and, although solubility problems prevented measurement over a wide concentration range, it would not be inconsistent with formation of a reactive ternary complex Pi-Mn^{2+}-ATP.

This discussion supports a transition state structure 28 similar to that proposed by Lowenstein, where an important role of divalent metal ion is to shield the negative charges on both phosphoryl donor and acceptor. Mn^{2+} is depicted as being bound to all

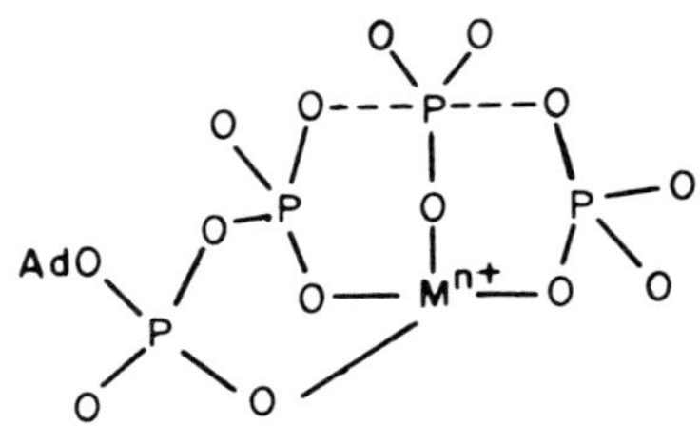

28

three phosphoryl positions in ATP because of NMR evidence that this is the case in the binary Mn-ATP complex. A decrease in rate on hydroxide formation could be due to an additional negative charge destabilizing the complex and/or a role for bound H_2O as a general acid catalyst for the leaving group, ADP^{3-}. The small ionic radius of Mg^{2+} may prevent it from forming a ternary complex having the correct orientation for reaction, thus explaining its poor catalytic ability. The failure of AMP^{2-} to act as phosphoryl acceptor may be due to the masking of its nucleophilicity on binding metal ion. PO_4^{3-}, having greater charge density, could be complexed yet still retain nucleophilicity.

Both Lowenstein and Miller were reluctant to calculate specific rate constants for the ATP-Pi transphosphorylation because of the complexity of the system. However, given the recent demonstrations of phosphoryl transfer in ternary complexes [23,88] it is reasonable to attempt a calculation of the rate constant for transfer, based on Lowenstein's data, in the ATP-Mn^{2+}-Pi complex, which comes out to be approximately 2×10^{-4} M^{-1} min^{-1} at pH 9 and 30°C. ATP hydrolysis is slow, and is usually measured at much higher temperatures. An extrapolation of rate constant values in the literature gives a value of 2×10^{-6} min^{-1} at pH 9 and 30°C [48,55,56,58] which is probably correct within a factor of three. So here, as with the transfer reaction <u>25</u>, metal ion allows anion attack on a phosphoryl dianion within a ternary complex at a rate significantly faster than the hydrolysis rate.

In related studies, it has been shown that divalent metal ions catalyze acetohydroxamate formation in solutions of acetic acid, hydroxylamine, and ATP [86,87]. It was proposed that the reaction proceeded though metal ion catalysis of ATP phosphorylation of acetate, presumably via a transition state similar to <u>28</u>, and that the resulting product, acetyl phosphate, was trapped by hydroxylamine to give acetohydroxamate. However, more recently metal ions have been shown to catalyze hydroxamate formation even in the absence of ATP [98], so that the proposed acetyl phosphate intermediate cannot be considered to have been unambiguously demonstrated.

5.4.2. Zn^{2+}-1,10-Phenanthroline-2-carbinol Attack on ATP

Sigman et al. [88] have recently shown that the phosphorylation of 1,10-phenanthroline-2-carbinol by ATP has an absolute Zn^{2+} requirement and proceeds via a ternary complex. The kinetics have not been fully worked out and a complete mechanism remains to be established. It has been suggested that in the Zn^{2+} complex the carbinol pK_a is lowered to 7.5, and that the transition state structure resembles 29. An alternate structure also considered by the authors 30 seems less likely, since there charge repulsion is unrelieved.

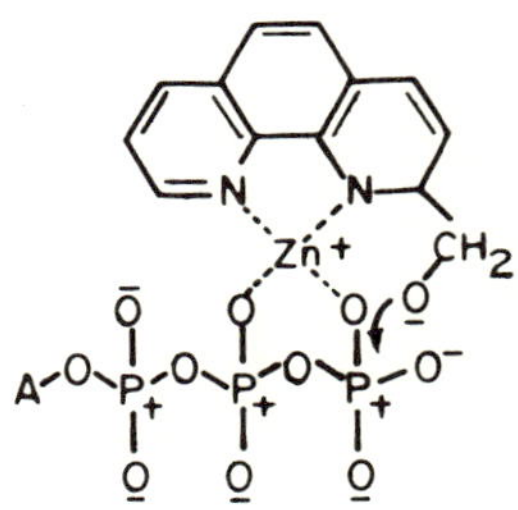

29

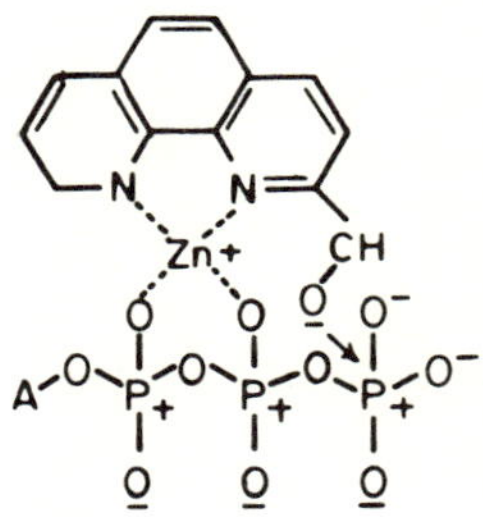

30

5.5. Hydroxy Complexes, Polynuclear Complexes, and Heterogeneous Catalysts

5.5.1. ATP + Cu^{2+}

Since the initial discovery by Tetas and Lowenstein of a very rapid Cu^{2+} catalysis of ATP hydrolysis to produce ADP and Pi [55], this reaction has been the subject of a number of studies [56,58, 59]. Spiro et al. [59] have examined both its concentration and pH dependence and concluded that (i) of the major species in solution, $CuATP^{2-}$, $CuATPH^{-}$, $CuATP(OH)^{3-}$, $CuATP(OH)_2^{4-}$, and $[CuATP(OH)]_2^{6-}$ [99], only the last named is of importance catalytically; and (ii) that the reaction is specific acid-catalyzed, i.e., that the rate law is $d(Pi)/dt = k[CuATP(OH)_2^{-6}]\ [H_3O^+]$, or its kinetic equivalent. No rationale was offered for why the dimeric species is more reactive than one of the monomeric species. Miller [100] took much the same approach, but also considered species with a Cu:ATP stoichiometric ratio of 2:1 and concluded that either the 2:1 or the 2:2 complex could be catalytically active. Two strongly acidic Cu^{2+} ions bound to a single ATP chain might effectively neutralize the γ-phosphoryl position, thus promoting both water and hydroxide ion attack, and accounting for the observed enhanced hydrolysis rates. Since the hydrolyses of both methyl triphosphate and phenylpropyl triphosphate are not markedly accelerated by Cu^{2+}, it seems reasonable to conclude that the adenosine moiety of ATP is participating in some way in the Cu^{2+}-catalyzed hydrolysis reaction. There is good evidence for Cu^{2+} binding to the purine (most likely to N-7) [56a,90,91] and it has been suggested that this binding could lead to a catalysis of hydrolysis via formation of a complex in which the β-phosphate is bound to Cu^{2+} but the γ-phosphate is not (similar to 17-21) [56]. However, this suggestion leaves unexplained the stoichiometry of the catalytically active species and at present the detailed role of the adenosine moiety remains unelucidated.

5.5.2. Other Examples

It has been known for some time that polynuclear hydroxy metal ion gels and metal ion hydroxide precipitates are very effective catalysts of phosphate ester hydrolysis [1] and references to these studies are given in Table 4. Because of the heterogeneity of these systems, little is known of these reactions in a mechanistic sense. One can speculate that phosphate esters could bind to metal ions at the surfaces of these gels and be subject to rapid hydrolysis via the mechanisms proposed for Cu^{2+} catalysis of ATP hydrolysis, but hard evidence is lacking. A very interesting example of heterogeneous phosphoryl transfer comes from the work of Krane and Glimcher [81] who showed that nucleotide di- and triphosphates bind to calcium apatite crystals and transfer their terminal phosphoryl groups to a crystal surface phosphate group, thus forming inorganic pyrophosphate. Although this reaction resembles the Mn^{2+}-catalyzed ATP-Pi transphosphorylation which takes place in solution, it proceeds considerably more rapidly. In this respect it parallels ATP hydrolysis, which is catalyzed to a much greater extent by metal ion hydroxide precipitates and gels than by the simple ions in solution [78].

An interesting example of a nucleotidyl transfer is the formation of 3',5'-AMP from ATP in the presence of divalent metal ion hydroxides [$Ba(OH)_2$ is the most efficient] at pH's greater than 11 [77]. This reaction has an absolute requirement for divalent ion, and the requirement of high pH probably reflects the necessity for ionization of the 3'-hydroxyl before reaction can occur. Metal ion may activate the pyrophosphate leaving group, as in reaction (7).

6. STUDIES ON PHOSPHORYL TRANSFER ENZYMES

In Section 5 we discussed results of kinetic studies for metal ion catalysis of phosphoryl transfer in terms of the structures presented in Table 3. This was possible, despite the frequent lack of direct evidence from either spectroscopic or crystallographic methods as to the structures of the complexes in solution, because the relative simplicity of the model compounds studied limited the number of structural possibilities which it was necessary to consider. For enzymic studies, such a simplification is not possible, since an activating metal ion need not be bound at all closely to the substrate. Thus, although there is a vast literature on phosphoryl transfer enzymes, which has been ably reviewed quite recently [4,5], we will here confine ourselves to a discussion of the two enzymes for which there is direct knowledge of the disposition of metal ion and substrate with respect to each other on the surface of the enzyme, creatine kinase and pyruvate kinase.*

*For *E. coli* alkaline phosphatase, both covalently and noncovalently bound phosphate have been found to perturb both optical spectra (Co^{2+} enzyme [101,102]) and ESR spectra (Mn^{2+} and Cu^{2+} enzymes [103-105]), but it is not clear that these perturbations are due to a direct interaction of phosphate with metal ion.

6.1. Muscle Creatine Kinase

Creatine kinase catalyzes reaction (9). Kinetic, isotope ex-

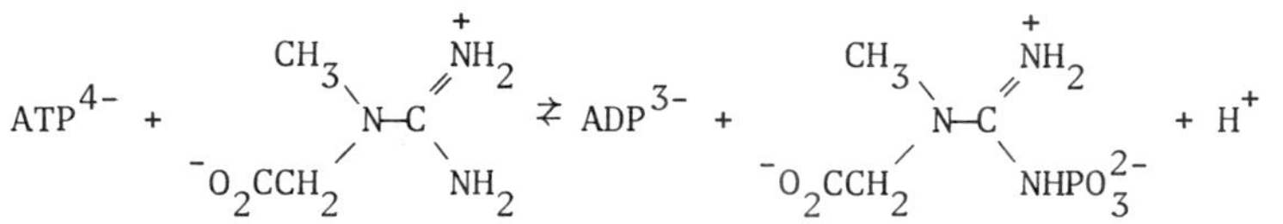
(9)

change and binding studies [106-109] have shown that phosphoryl group transfer is the rate determining step in the enzyme-catalyzed reaction and that the active substrates are the nucleotide-divalent metal ion complexes and uncomplexed creatine and phosphocreatine. Although maximal turnover rates are obtained with Mg^{2+}, the enzyme is active with Mn^{2+} as a cofactor, and Cohn and her coworkers [92, 110] have been able to use this paramagnetic ion as a structural probe. Their studies, and those of Milner-White and Watts [111], have led to the mechanistic proposal for creatine kinase (31) [112], which corresponds to structure (b') of Table 3. The salient features of this model are that Mn^{2+} is bound to the pyrophosphate chain of ADP, but is not bound either to the phosphoryl group undergoing transfer or to creatine. The evidence for the model is as follows.

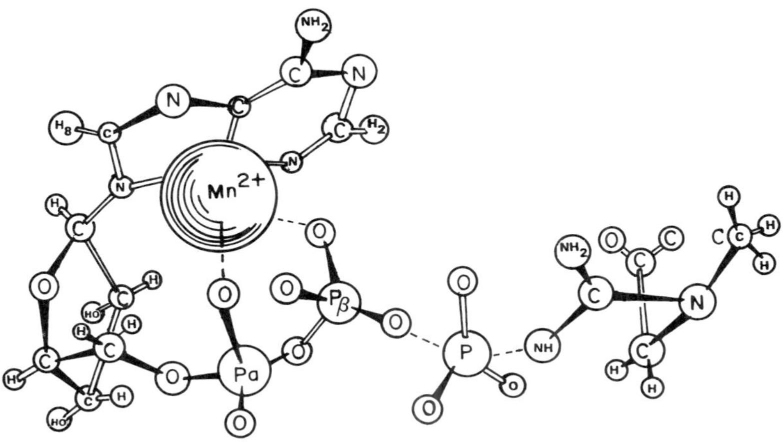

31

Mn^{2+} binds tightly to the enzyme only in the presence of nucleotides; furthermore the Mn^{2+} ESR spectrum of MnADP and MnATP are only slightly changed on formation of the ternary complexes with enzyme. These observations suggest that the Mn^{2+} environment in the ternary complex is similar to what it is in the binary complex with the nucleotides, i.e., bound to the pyrophosphate chain. Measurement of Mn^{2+} effects on the proton relaxation rates [$1T_1$ and $1/T_2$] [113] of the methyl and methylene protons of creatine in both the abortive [enzyme-ADP-Mn^{2+}-creatine] and reactive [enzyme-ATP-Mn^{2+}-creatine] quaternary complexes allowed distance calculations to be made. For both complexes the Mn^{2+}-CH_3 (10.3 ± 0.5 Å) and Mn^{2+}-CH_2 (9.8 ± 0.5 Å) distances were the same. These distances are too long to permit direct binding of Mn^{2+} to creatine. Furthermore, if one assumes (<u>31</u>) that the phosphoryl group undergoing transfer is shuttling back and forth between ADP and creatine without ever being very far from either, then the calculated distances are too large to permit Mn^{2+}-γ-phosphoryl group binding. However, if the ground-state structure of the reaction quaternary complex is not as close to the transition state for transfer as shown in <u>31</u>, then it is entirely consistent with the data obtained to allow the γ-phosphoryl group of ATP to bind to Mn^{2+}. Unfortunately it has not yet been possible with this system to study Mn^{2+} effects on ^{31}P relaxation rates, which should permit one to distinguish between these possibilities. Evidence of a different kind favoring the structure proposed in <u>31</u> comes from studies of enzymic inhibition by anions [111]. Nitrate was found to be more tightly bound than either acetate, arsenate, sulfate, or any of the halides. It was also especially effective in protecting enzyme against inactivation by iodoacetamide, but this protection was only manifested in the presence of metal ion-nucleotide complex and creatine. This led to the suggestion that nitrate was binding to the abortive quaternary complex as a transition state analog [37] for the metaphosphate ion, PO_3^-. The unique role of nitrate as an anionic ligand for the enzyme has also been demonstrated recently by magnetic resonance experiments [110].

We have previously argued (Sections 4.3 and 5.2.5.) that for a transition state such as 31, metal ion (especially Mg^{2+}) is unlikely to accelerate the reaction in the direction of phosphocreatine formation by more than 1 or 2 orders of magnitude, and that in the direction of ATP formation metal ion could well be inhibitory. Therefore, if 31 is correct, the direct catalytic role of metal ion in phosphoryl transfer is probably a minor one [although, given the enzyme's requirement of metal ion for activity, it may be absolutely crucial in an indirect sense, such as for proper nucleotide positioning at the active site]. It seems reasonable to assume that the positively charged residue(s) at the active site, which presumably form(s) the anionic binding site, also serve to relieve the charge repulsion between the anionic nucleophile ADP and the dianionic phosphoryl group of phosphocreatine. In this connection, it is interesting that both lysine [114,115] and histidine [116] have been implicated as essential residues at the active site.

6.2. Muscle Pyruvate Kinase

Pyruvate kinase catalyzes reaction (10):

$$ATP^{4-} + CH_3\overset{\overset{O}{\|}}{C}CO_2^- \rightleftarrows ADP^{3-} + CH_2{=}\overset{\overset{OPO_3^{2-}}{|}}{C}{-}CO_2^- + H^+ \qquad (10)$$

As with creatine kinase, Mn^{2+} can be used as the essential divalent metal ion cofactor and there is evidence that phosphoryl transfer is the rate-determining step [4]. By contrast with creatine kinase, pyruvate kinase binds Mn^{2+} in a binary complex as well as in a ternary complex with nucleotide [117]. Moreover, NMR studies have provided good evidence that Mn^{2+} coordinates directly to the phosphoryl group undergoing transfer, but not to the carboxylate group [118,119]. Mildvan [5] has proposed the mechanism shown in 32 for the enzyme, which corresponds to structure (c') of Table 3. Although this mechanism is attractive and consistent with all experimental data, it must be regarded as tentative, since the binding of

Mn^{2+} to the β-phosphoryl group on the surface of the enzyme has not yet been demonstrated.

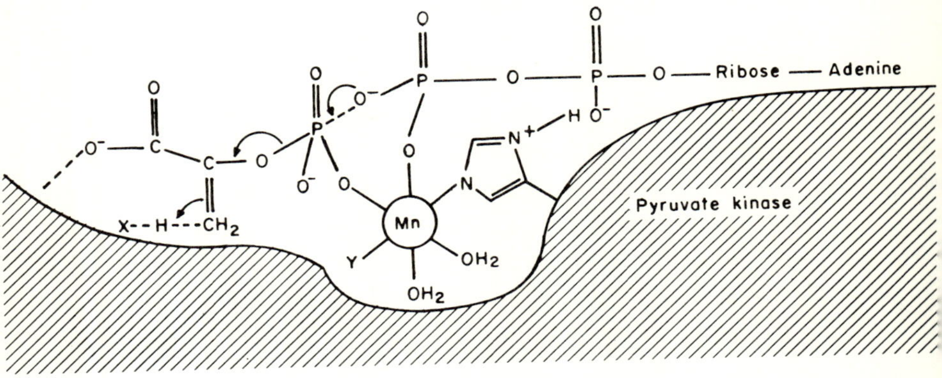

32

7. SUMMARY

The model system studies discussed in this review provide clear evidence for the versatility of metal ions in catalyzing nucleophilic attack on phosphoryl compounds through one-, two-, or three-site coordination. Of the six plausible, independent catalytic effects listed in Section 4, charge shielding, leaving-group activation, activation of nucleophile and template seem well demonstrated, strain induction is likely but not clearly demonstrated, while polarization has neither been demonstrated nor is it expected to be of major importance for phosphoryl transfer.

In some cases metal ion effects have been found to be quite large, with catalytic rate factors up to 10^5 cited. Given the very general requirement of phosphoryl and nucleotidyl transfer enzymes for divalent metal ions, it is reasonable to assume that at least some of the transition state structures will be relevant to enzyme-catalyzed reactions, and in our discussions of creatine kinase and pyruvate kinase we have presented evidence for two of them, (b') and (c') (Table 3). However, it should be remembered that with the possible exception of activation via bidentate coordination, amino acid residues at an enzymic active site can provide much the same kind of catalytic function described for divalent metal ion. We have already discussed the possibility of positively charged residues serving to shield charge in the creatine kinase reaction. Similarly, protonated histidine, aspartate, glutamate, lysine, and tyrosine could function as general acid catalysts to activate leaving groups, while the conjugate base forms of these residues could act as general bases to activate nucleophiles. Finally, the site itself could constitute a template by providing binding subsites for both phosphoryl (or nucleotidyl) donor and acceptor. Thus one can envisage a very great variety of mechanisms for enzymic phosphoryl and nucleotidyl transfer, ranging from transition state structures such as (d) (Table 3) where metal ion is multifunctional, to structures such as (a) or (b) (Table 3) where its role is quite limited, to structures where metal ion has no role at all. In fact, enzymes catalyzing phosphoryl transfer are known which seem not to have an absolute requirement for divalent metal ion [120-125].

ACKNOWLEDGMENTS

The author is indebted to Drs. Albert Mildvan, John Leigh, and Edward Thornton for helpful comments during the preparation of this manuscript.

ADDENDUM

This brief addendum covers some of the more important work reported since the writing of the body of the chapter was finished in May, 1973.

Two recently completed studies have attempted, with some success, to describe metal-ion catalysis of phosphoryl transfer at a much more detailed level than has been done previously. Steffens et al. [126] have found that phosphodiesters which show strong intramolecular catalysis of hydrolysis by neighboring carboxylate anions, such as lactic acid-O-phenyl phosphate (I), 3-hydroxybutyric acid-O-phenyl phosphate (II), and salicyclic acid-O-phenyl phosphate (III) are subject to a large further catalysis by added divalent metal ions. For the hydrolysis of I, they demonstrated a correlation between the ionic radius and catalytic effectiveness of added metal ion, suggesting the importance of bidentate coordination in the transition state. Measurement of the Zn^{2+}-catalyzed rates of hydrolysis of a series of substituted phenyl esters of III allowed a determination of a leaving group Brønsted β of 0.7, much less than the value of 1.17 found for the reaction in the absence of Zn^{2+}. These results, taken together, lead to the conclusion that the transition-state structure has metal ion bound in a bidentate manner either to two phosphoryl oxygens [structure (a), Table 3] or to a phosphoryl oxygen and the oxygen of the phenol leaving group [structure (c), Table 3]. Unfortunately most of the rate data reported were obtained at metal ion concentrations which are low compared to the anticipated metal-ion-phosphate diester binding constants (which, with one exception, the Mn^{2+}-I complex, were not measured). This means that the comparisons of observed rates reported are really comparisons of some undetermined combination of microscopic rate and equilibrium constants, introducing some uncertainty into the correlations made. An NMR study of the Mn^{2+}-I complex showed, surprisingly, that Mn^{2+} was weakly bound to the carboxylate and not at all to the phosphate, so that both of the postulated transition state structures are very different from the actual reactant state structure. Hsu and Cooperman [53], in a continuation of earlier work [23], determined the Brønsted leaving group β for Zn^{2+}-catalyzed phosphoryl transfer from a series of phosphoramidates to pyridine carbaldoxime anion, via the ternary complex, phosphoramidate-Zn^{2+}-pyridine carbaldoxime, and the Brønsted β for Zn^{2+} catalysis of nucleophilic attack on pyridine carbaldoximyl phosphate, using a series of substituted pyridines. For both reactions they found that the Brønsted β's are close to those for similar nonmetal ion-catalyzed reactions. They concluded that the transition state for the Zn^{2+}-catalyzed reaction is little distorted from that for the noncatalyzed reaction and that the reason that Zn^{2+} is an effective catalyst in both directions of reaction (i.e., a true catalyst) is that it has a specific favorable interaction with the transition state, in a structure of Type (c') (Table 3).

In other work, Ikenaga and Inoue [127] have extended the earlier studies of Eichhorn and his coworkers [74] through a very thorough investigation of divalent metal ion effects on the hydrolysis of a large number of dinucleotides. They have confirmed the earlier finding that the presence of a terminal 3'-phosphate in a dinucleotide significantly enhances ($\sim 10^2$ X) the catalytic effect of Zn^{2+} on dinucleotide hydrolysis. A similar though less marked effect was found with Cu^{2+}, but not with Co^{2+} and Ni^{2+}. A detailed study of Zn^{2+} effects on ApUp hydrolysis showed the necessity of a terminal phosphate dianion for reactivity and provided some evidence that the 1:1 Zn^{2+}-ApUp complex is the active species. Woltermann et al. [128] have shown that in the presence of hydrogen peroxide or MnO_4^-, VO^{2+} catalyzes a rapid hydrolysis of ATP. Although little is known about the mechanism, it is clear that under the experimental conditions VO^{2+} binds to ATP. The dependence of the rate of ATP hydrolysis on H_2O_2 concentration is taken as evidence that the oxidation of VO^{2+} to V(V) is the important factor in the observed catalysis. Suzuki et al. [129] have measured the effects of a series of Co^{3+}-polyamine complexes on ATP hydrolysis. Both phosphate and pyrophosphate release are catalyzed by complexes with two available cis-binding sites, while a planar complex and complexes with five and six amine ligands are ineffective. The most effective catalyst, Co^{3+}-diethylenetriamine, enhances ATP hydrolysis 150-fold at pH 4. Murakami and Sunamoto [130] have shown that both Cu^{2+} and Th^{4+}, but not Ni^{2+}, are effective catalysts for 8-quinolyl phosphate hydrolysis. Catalysis by Cu^{2+} is thought to proceed via a transition-state structure (c) (Table 3), analogous to that previously discussed (see Section 5.2.2) for Cu^{2+} catalysis of pyridyl-2-methyl phosphate. Th^{4+}, but not Cu^{2+}, is also an effective catalyst for pyridine-3-methyl phosphate hydrolysis, and this result is interpreted as showing that catalysis is due to direct coordination to phosphate, i.e., transition state structure (a) (Table 3). It has been found recently that Cu^{2+} and other metal ions catalyze phosphoryl transfer from 8-quinolyl phosphate to nucleophiles other than water, and two Japanese groups have demonstrated the practicability of using the Cu^{2+} complex as a phosphorylating agent for alcohols, nucleosides, and phosphates [131,132]. Schwartz [133] and Handschuh et al. [134] have recently reported on prebiotic models of phosphoryl transfer in heterogeneous systems. Finally, Benkovic and Schray [135] have published an excellent general review of phosphoryl transfer that is both clear and comprehensive.

REFERENCES

1. E. Bamann and H. Trapmann, *Advan. Enzymol., 21,* 169 (1959).
2. T. C. Bruice and S. J. Benkovic, "Bioorganic Mechanisms," Vol. 2, W. A. Benjamin, New York, 1966, Chapters 5, 6, 7.
3. J. R. Cox and O. B. Ramsay, *Chem. Reviews, 64,* 317 (1964).
4. J. F. Morrison and E. Heyde, *Annu. Rev. Biochem., 41,* 29 (1972).
5. A. S. Mildvan, *Enzymes, 2,* 446 (1970).
6. J. Imsande and P. Handler, *Enzymes, 5,* 281 (1961).
7. S. J. Benkovic and E. M. Miller, *Bioinorg. Chem., 1,* 107 (1972).
8. E. F. Brunngraber and E. Chargaff, *J. Biol. Chem., 245,* 4825 (1970).
9. R. C. Nordlie, *Enzymes, 4,* 543 (1971).
10. W. Kundig and S. Roseman, *J. Biol. Chem., 246,* 1393, 1407 (1971).
11. R. F. Ramaley, W. A. Bridger, R. W. Moyer, and P. D. Boyer, *J. Biol. Chem., 242,* 4287 (1967)
12. S. J. Benkovic and E. J. Sampson, *J. Amer. Chem. Soc., 93,* 4009 (1971).
13. A. J. Kirby and M. Younas, *J. Chem. Soc. (B),* 1165 (1970).
14. W. P. Jencks, "Catalysis in Chemistry and Enzymology," McGraw-Hill, New York, 1969, pp. 82-83.
15. G. W. Jameson and J. M. Lawlor, *J. Chem. Soc. (B),* 53 (1970).
16. W. P. Jencks and M. Gilchrist, *J. Amer. Chem. Soc., 87,* 3199 (1965).

17. G. J. Lloyd, C. M. Hsu, and B. S. Cooperman, *J. Amer. Chem. Soc., 93,* 4889 (1971).

18. A. J. Kirby and A. G. Varvoglis, *J. Amer. Chem. Soc., 89,* 415 (1967).

19. A. J. Kirby and W. P. Jencks, *J. Amer. Chem. Soc., 87,* 3209 (1965).

20. A. J. Kirby and A. G. Varvoglis, *J. Chem. Soc. (B),* 135 (1970).

21. D. G. Gorenstein, *J. Amer. Chem. Soc., 94,* 2523 (1972).

22. W. P. Jencks, "Catalysis in Chemistry and Enzymology," McGraw-Hill, New York, 1969, p. 511.

23. G. J. Lloyd and B. S. Cooperman, *J. Amer. Chem. Soc., 93,* 4883 (1971).

24. J. D. Chanley and E. Feageson, *J. Amer. Chem. Soc., 85,* 1181 (1963).

25. F. H. Westheimer, *Accounts Chem. Res., 1,* 70 (1968).

26. A. Eberhard and F. H. Westheimer, *J. Amer. Chem. Soc., 87,* 253 (1965).

27. P. Greengard, S. A. Rudolph, and J. M. Sturtevant, *J. Biol. Chem., 244,* 4798 (1969).

28. W. P. Jencks, "Catalysis in Chemistry and Enzymology," McGraw-Hill, New York, 1969, p. 112.

29. T. G. Spiro, in "Inorganic Biochemistry" (G. L. Eichhorn, ed.), Elsevier Amsterdam, 1971.

30. T. C. Bruice and S. J. Benkovic, "Bioorganic Mechanisms," Vol. I, W. A. Benjamin, New York, 1966, pp. 125ff.

31. R. B. Martin, *J. Amer. Chem. Soc., 87,* 2501 (1965).

32. S. Seltzer, G. Hamilton, and F. H. Westheimer, *J. Amer. Chem. Soc., 81,* 4018 (1959).

33. C. H. Oestreich and M. M. Jones, *Biochemistry, 6,* 1515 (1967).

34. J. Klinman and D. Samuel, *Biochemistry, 10,* 2126 (1971).

35. M. M. Jones, "Ligand Reactivity and Catalysis," Academic Press, New York, 1968, Chap. 3.

36. L. Pauling, *Chem. Eng. News, 24,* 1375 (1946).

37. R. Wolfenden, *Accounts Chem. Res., 5,* 10 (1972).

38. J. L. Kurz, *Accounts Chem. Res., 5,* 1 (1972).

39. R. Phillips, S.J. *Chem. Rev., 66,* 501 (1966).

40. L. G. Sillen and A. E. Martell, "Stability Constants of Metal Complexes. Special Publication #17," The Chemical Society, London, 1964.

41. A. A. Giallo, I. L. Hansen, H. Z. Sable, and T. J. Swift, *J. Biol. Chem., 247,* 5913 (1972), and references therein.

42. J. J. Steffens, E. J. Sampson, I. J. Siewers, and S. J. Benkovic, *J. Amer. Chem. Soc., 95,* 936 (1973).

43. S. J. Benkovic and K. J. Schray, *Biochemistry, 7,* 4097 (1968).

44. P. J. Briggs, D. P. N. Satchell, and G. F. White, *J. Chem. Soc. (B),* 1008 (1970).

45. C. H. Oestreich and M. M. Jones, *Biochemistry, 5,* 3151 (1966).

46. R. Hofstetter, Y. Murakami, G. Mont, and A. E. Martell, *J. Amer. Chem. Soc., 84,* 3041 (1962).

47. Y. Murakami and A. E. Martell, *J. Phys. Chem., 67,* 582 (1963).

48. L. B. Nanninga, *J. Phys. Chem., 61,* 1144 (1957).

49. Y. Murakami and M. Takagi, *J. Amer. Chem. Soc., 91,* 5130 (1969).

50. Y. Murakami, J. Sunamoto, and H. Ishizu, *Bull. Chem. Soc. Japan, 45,* 590 (1972).

51. S. J. Benkovic and L. K. Dunikoski, *J. Amer. Chem. Soc., 93,* 1526 (1971).

52. F. J. Farrell, W. Kjellstrom, and T. G. Spiro, *Science, 164,* 320 (1969).

53. C. M. Hsu and B. S. Cooperman, submitted for publication.

54. V. M. Clark, A. R. Todd Lord, and S. G. Warren, *Biochem. Z., 338,* 591 (1963).

55. M. Tetas and J. M. Lowenstein, *Biochemistry, 2,* 350 (1963).

56. P. W. Schneider and H. Brintzinger, *Helv. Chim. Acta, 47,* 1717 (1964).

56a. P. W. Schneider, H. Brintzinger, and H. Erlenmeyer, *Helv. Chim. Acta, 47,* 992 (1964).

57. H. Möll, P. W. Schneider, and H. Brintzinger, *Helv. Chim. Acta, 47,* 18 (1964).

58. D. L. Miller and F. H. Westheimer, *J. Amer. Chem. Soc., 88,* 1514 (1966).

59. T. G. Spiro, W. A. Kjellstrom, M. Zeydel, and R. A. Butow, *Biochemistry, 7,* 859 (1968).

60. M. Ikehara and H. Uno, *Biochim. Biophys. Acta, 85,* 512 (1964).

61. C. Liébecq and M. Jacquemotte-Louis, *Bull. Soc. Chem. Biol., 40,* 67 (1958).

62. C. Liébecq and M. Jacquemotte-Louis, *Bull. Soc. Chem. Biol., 40,* 759 (1958).

63. C. Liébecq, *Bull. Soc. Chem. Biol., 41,* 1181 (1959).

64. J. Blum and E. Felauer, *Arch. Biochem. Biophys.*, *81*, 285 (1959).

65. N. Hennrich and F. Cramer, *J. Amer. Chem. Soc.*, *87*, 1121 (1965).

66. G. Rutherford and A. R. Morgan, *Can. J. Biochem.*, *50*, 287 (1972).

67. B. S. Cooperman, *Biochemistry*, *8*, 5005 (1969).

68. Y. Moriguchi, *Bull. Chem. Soc. Japan*, *39*, 2656 (1966).

69. J. van Steveninck, *Biochemistry*, *5*, 1998 (1966).

70. J. M. Rainey, M. M. Jones, and W. L. Lockhart, *J. Inorg. Nucl. Chem.*, *26*, 1415 (1964).

71. K. Dimroth, H. Witzel, W. Hülsen, and H. Mirbach, *Ann. Chem.*, *620*, 94 (1959).

72. T. Lindahl, *J. Biol. Chem.*, *242*, 1970 (1967).

73. W. R. Farkas, *Biochim. Biophys. Acta*, *155*, 401 (1960).

74. R. Butzow and G. L. Eichhorn, *Biochemistry*, *10*, 2019 (1971).

75. E. Cherbuliez, J. P. Leber, and P. Stuck, *Helv. Chim. Acta*, *36*, 537 (1953).

76. W. W. Butcher and F. H. Westheimer, *J. Amer. Chem. Soc.*, *77*, 2420 (1955).

77. D. Lipkin, R. Markham, and W. H. Cook, *J. Amer. Chem. Soc.*, *81*, 6075 (1959).

78. G. Torralba, Master's Thesis, Washington University, St. Louis, Mo. (1960).

79. F. Millick and E. Hayes, *J. Amer. Chem. Soc.*, *86*, 2914 (1964).

80. R. Hofstetter and A. E. Martell, *J. Amer. Chem. Soc.*, *81*, 4461 (1959).

81. S. M. Krane and M. J. Glimcher, *J. Biol. Chem.*, *237*, 2991 (1962).

82. M. J. Selwyn, *Nature*, *219*, 490 (1968).

83a. D. L. Miller and T. Ukena, *J. Amer. Chem. Soc.*, *91*, 3050 (1969).

83b. D. Samuel and B. Silver, *J. Chem. Soc.*, 4321 (1961).

83c. S. J. Benkovic and R. C. Hevey, *J. Amer. Chem. Soc.*, *92*, 4971 (1970).

84. J. M. Lowenstein, *Biochem. J.*, *70*, 222 (1958).

85. J. M. Lowenstein, *Nature*, *187*, 570 (1960).

86. J. M. Lowenstein, *Biochim. Biophys. Acta*, *28*, 206 (1958).

87. J. M. Lowenstein and M. N. Schatz, *J. Biol. Chem.*, *236*, 305 (1961).

88. D. S. Sigman, G. M. Wahl, and D. J. Creighton, *Biochemistry, 11,* 2236 (1972).

89. K. Nagasawa and H. Yoshidome, *Chem. Pharm. Bull., 20,* 1840 (1972).

90. M. Cohn and T. R. Hughes, *J. Biol. Chem., 237,* 176 (1962).

91. H. Sternlicht, R. G. Shulman, and L. W. Anderson, *J. Chem. Phys., 43,* 3123 (1965).

92. M. Cohn, J. S. Leigh, Jr., and G. H. Reed, *Cold Spring Harbor Quant. Biol., 36,* 533 (1971).

93. M. M. Tacqui Khan and A. E. Martell, *J. Phys. Chem., 66,* 10 (1962).

94. M. M. Tacqui Khan and A. E. Martell, *J. Amer. Chem. Soc., 84,* 3037 (1962).

95. V. Handschin and H. Brintziger, *Helv. Chim. Acta, 45,* 1037 (1962).

95a. S. Chabarek, R. C. Courtney, and A. E. Martell, *J. Amer. Chem. Soc., 74,* 8057 (1952).

96. K. H. Gayer and L. Woontner, *J. Amer. Chem. Soc., 78,* 3944 (1956).

97. G. Schwarzenbach, *Pure Appl. Chem., 5,* 377 (1962).

98. J. M. Lawlor, *Chem. Commun.,* 404 (1967).

99. M. M. Tacqui Khan and A. E. Martell, *J. Amer. Chem. Soc., 88,* 668 (1966).

100. D. L. Miller, Ph.D. Thesis, Harvard University, Cambridge, Mass., 1965.

101. M. L. Applebury and J. E. Coleman, *J. Biol. Chem., 244,* 709 (1969).

102. R. T. Simpson and B. L. Vallee, *Biochemistry, 7,* 4344 (1968).

103. H. Csopak and K. E. Falk, *FEBS (Fed. Eur. Biochem. Soc.) Lett., 7,* 147 (1970).

104. C. Lazdunski, C. Petitclerc, D. Chappelet, F. Letterrier, P. Douzou, and M. Lazdunski, *Biochem. Biophys. Res. Commun., 40,* 589 (1970).

105. C. Lazdunski, D. Chappelet, C. Petitclerc, F. Letterrier, P. Douzou, and M. Lazdunski, *Eur. J. Biochem., 17,* 239 (1970).

106. J. F. Morrison and E. James, *Biochem. J., 97,* 37 (1965).

107. J. F. Morrison and W. W. Cleland, *J. Biol. Chem., 241,* 673 (1966).

108. J. F. Morrison and A. White, *Eur. J. Biochem., 3,* 145 (1967).

109. M. Cohn, *Quart. Rev. Biophys., 3,* 1 (1970).

110. G. H. Reed and M. Cohn, *J. Biol. Chem., 247,* 3073 (1972).

111. E. J. Milner-White and D. C. Watts, *Biochem. J., 122,* 727 (1971).

112. J. S. Leigh, Jr., Ph.D. Thesis, University of Pennsylvania, Philadelphia, 1971.

113. A. S. Mildvan and M. Cohn, *Advan. Enzymol., 33,* 1 (1970).

114. T. A. Mahowald, *Fed. Proc., 19,* 46 (1960).

115. R. Kassab, C. Roustan, and L. Pradel, *Biochim. Biophys. Acta, 167,* 308 (1968).

116. L. Pradel and R. Kassalb, *Biochim. Biophys. Acta, 167,* 317 (1968).

117. A. S. Mildvan and M. Cohn, *J. Biol. Chem., 241,* 1178 (1966).

118. T. Nowak and A. S. Mildvan, *Biochemistry, 11,* 2819 (1972).

119. A. S. Mildvan, J. S. Leigh, and M. Cohn, *Biochemistry, 6,* 1805 (1967).

120. M. J. Dowler and H. I. Nakada, *J. Biol. Chem., 243,* 1434 (1968).

121. E. F. Brunngraber and E. Chargaff, *J. Biol. Chem., 242,* 4834 (1967).

122. M. R. Stetten and D. Rounbehler, *J. Biol. Chem., 243,* 1823 (1968).

123. M. E. Holzer, K. D. Johnson, and R. A. Smith, *Biochim. Biophys. Acta, 122,* 232 (1966).

124. O. Walinder, O. Zetterqvist, and L. Engström, *J. Biol. Chem., 244,* 1060 (1969).

125. V. P. Hollander, *Enzymes, 4,* 449 (1971).

126. J. J. Steffens, I. J. Siewers, and S. J. Benkovic, *Biochemistry, 14,* 2431 (1975).

127. H. Ikenaga and Y. Inoue, *Biochemistry, 13,* 577 (1974).

128. G. M. Woltermann, R. A. Scott, and G. P. Haight, Jr., *J. Amer. Chem. Soc., 96,* 7569 (1974).

129. S. Suzuki, S. Kimura, T. Higashimaya, and A. Nakahara, *Bioinorg. Chem., 3,* 183 (1974).

130. Y. Murakami and J. Sunamoto, *Bull. Chem. Soc. Japan, 44,* 1827 (1971).

131. K. Nagasawa and H. Yoshidome, *Chem. Pharm. Bull., 21,* 2438 (1973).

132. H. Takaku and Y. Shimida, *Chem. Pharm. Bull., 22,* 1743 (1974).

133. A. W. Schwartz, *BioSystems, 5,* 119 (1973).

134. G. J. Handschuh, R. Lohrmann, and L. E. Orgel, *J. Mol. Evol., 2,* 251 (1973).

135. S. J. Benkovic and K. J. Schray, *Enzymes, 8,* 201 (1973).

136. J. Rebek and F. Gaviña, *J. Amer. Chem. Soc., 97,* 1591 (1975).

137. C. H. Clapp and F. W. Westheimer, *J. Amer. Chem. Soc., 96,* 6710 (1974).

Chapter 3

METAL ION-CATALYZED DECARBOXYLATIONS OF BIOLOGICAL INTEREST

R. W. Hay
Department of Chemistry,
University of Stirling,
Stirling, Scotland

1. INTRODUCTION

The majority of the studies on the metal ion-catalyzed decarboxylation of biological ligands have been restricted to studies of β-oxo acids (β-keto acids) and a few studies of α-amino acid decarboxylations.

A useful, if now somewhat dated, review of the general subject of decarboxylation is that of Brown [1] which covers much of the earlier literature on the topic. A more recent review of the subject is also available [2]. The present discussion will deal mainly with β-oxo acid decarboxylations as these have been the subject of many detailed kinetic investigations and some of these reactions provide simple models for the metal activated enzymic decarboxylations.

The enzymes catalyzing the decarboxylation of β-oxo acids are of two types: (a) the "active lysine" enzymes such as acetoacetate decarboxylase [3,4] which are not dependent on metal ions for activity, and (b) the metal-dependent enzymes such as oxaloacetate decarboxylase, which is activated by manganese(II) [5]. The "active lysine" enzymes operate via a Schiff's base mechanism (Section 9). A similar situation also occurs with the aldolases where enzymes of class II require metal ions for activity, whereas enzymes of class I react via Schiff's base intermediates [6].

The decarboxylation of all β-oxo acids is catalyzed by primary amines [7], aniline being a particularly effective catalyst. Secondary amines have some catalytic activity and tertiary amines are normally inactive. A Schiff's base (or possibly a carbinolamine) intermediate is involved in the amine catalysis [7], and the simple Schiff's base mechanism accounts for some 10^6 of the total 10^9 enzymic rate acceleration observed with acetoacetate decarboxylase

[3,8]. Cleavage of the carbon-carbon bond in acetoacetic acid is partially rate-determining in both the enzymic reaction and the reaction catalyzed by aminoacetonitrile [19].

The decarboxylation of a limited number of β-oxo acids is catalyzed by metal ions. Group 1 cations are normally inactive; however, lithium(I) has been reported [20] to catalyze the decarboxylation of oxaloacetic acid in ethanol. Some activity is generally observed with group II cations, but most investigations have been restricted to cations of the second half of the first transition series [$Mn^{(II)}$, $Fe^{(II)}$, $Co^{(II)}$, $Ni^{(II)}$, $Cu^{(II)}$, and $Zn^{(II)}$]. Some studies have been made with the lanthanides, lutetium(III) having a catalytic effect similar to that of copper(II) [9], the most active of the cations of the first transition series.

The β-oxo acids which are subject to metal ion catalysis are all dicarboxylic acids, e.g., oxaloacetic acid and its derivatives, 3-oxoglutaric acid (acetonedicarboxylic acid), dihydroxyfumaric acid, dihydroxytartaric acid, oxalosuccinic acid, and acetosuccinic acid. Simple monofunctional β-oxo acids such as acetoacetic acid are not subject to metal ion catalysis for the reasons discussed in Section 3.

2. DECARBOXYLATION MECHANISMS

The β-oxo acids decarboxylate spontaneously in aqueous solution as represented by equations (1) and (2). An enol (1) or an enolate ion (2) intermediate is involved in the process, the intermediate subsequently ketonizing [eq. (3)].

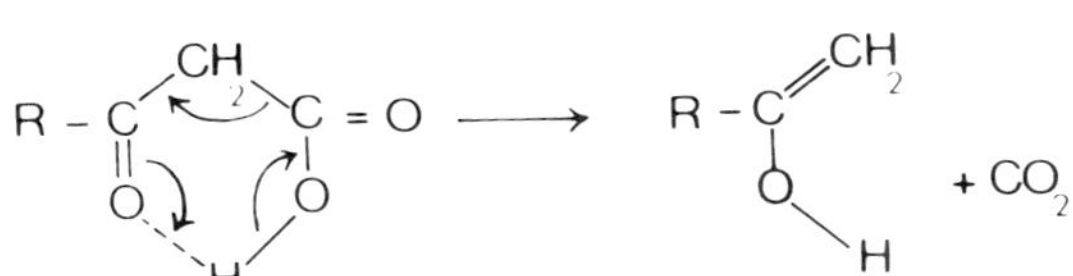

1 (1)

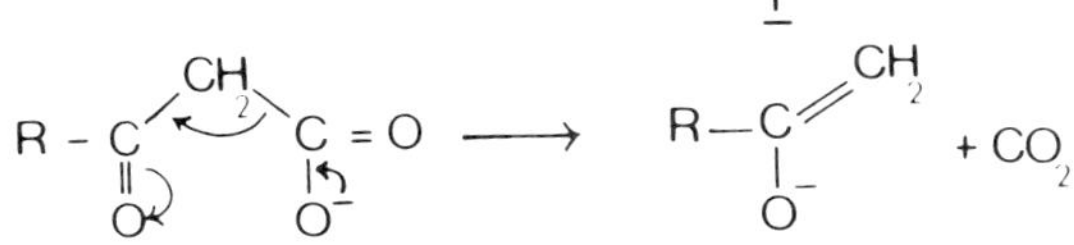

2 (2)

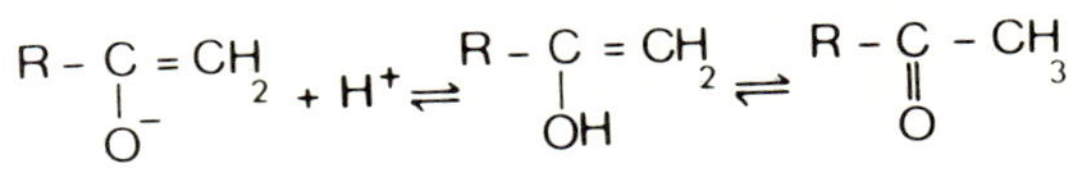
(3)

Pedersen [10] showed that α,α-dimethylacetoacetic acid which cannot enolize decarboxylates readily and hence concluded that the keto tautomers of β-oxo acids are kinetically unstable. The enol intermediate formed in the decarboxylation of β-oxo acids has been trapped by reaction with bromine [11] and has also been detected spectrophotometrically in the decarboxylation of α,α-dimethyloxaloacetic acid [12], oxaloacetic acid [13], and fluorooxaloacetic acid [14].

The instability of the β-oxo acids is a result of the electron-withdrawing carbonyl group which acts as an "electron sink" for the pair of electrons withdrawn from the carbon-carbon bond undergoing cleavage. Other electron-withdrawing groups, such as the nitro group and the trichloromethyl group, act in a similar manner [1].

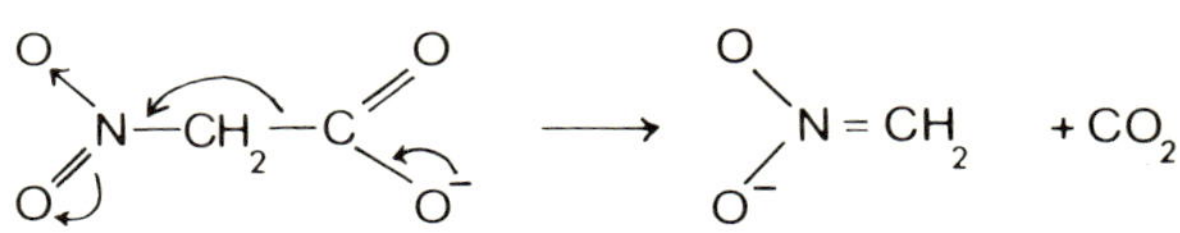
(4)

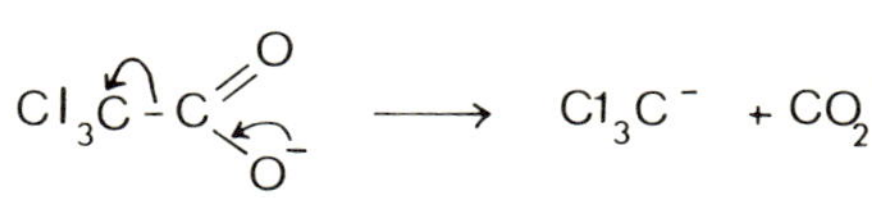
(5)

There has been considerable discussion on the question as to whether the unionized β-oxo acid decomposes via a zwitterion [eq. (6)] as initially suggested by Pedersen [10,11] or via the hydrogen-bonded structure [eq. (1)] suggested by Westheimer and Jones [15].

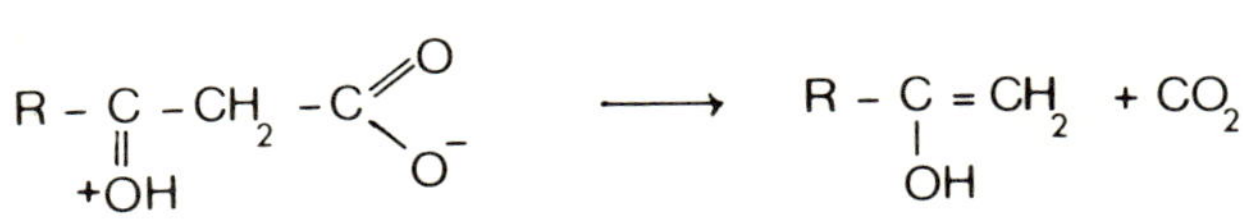
(6)

3

Various detailed discussions of the problem are available [16]. The possibility of a proton or a hydride ion transfer in β-oxo acid decarboxylations has been considered by Swain et al. [17]. These authors have concluded on the basis of isotope effects (k_{RCOOH}/k_{RCOOD}) that a cyclic proton transfer mechanism occurs.

The decarboxylation of a carboxylic acid involves the simultaneous decomposition of the free acid and the anion and the rate expression is given by

$$d(CO_2)/dt = k_1[HA] + k_2[A'] \qquad (7)$$

where k_1 and k_2 are the first-order rate constants for the decomposition of the acid and the anion, respectively.

Due to a close interplay of enthalpy and entropy factors quite marked differences can occur in the relative rates of decarboxylation of the acid and anion. Some thermodynamic parameters are given in Table 1. For oxaloacetic acid, $k_{acid}/k_{monoanion} = 0.02$ at 30°C, for malonic acid $k_{acid}/k_{monoanion} = 4.0$ at 30°C and for acetoacetic acid $k_{acid}/k_{anion} = 48$ at 25°C.

TABLE 1

Thermodynamic Parameters for the Decarboxylation of Some β-Oxo Acids[a]

	HA^- [b]		A^{2-} [b]	
Acid	$\Delta H^\dagger$	$\Delta S^\dagger$	$\Delta H^\dagger$	$\Delta S^\dagger$
Oxaloacetic	25.2	-1.3	22.5	-2.6
Malonic	30.2	+0.7	27.3	-9.7
Acetoacetic	23.3	-2.4	22.1	-14.0
Benzoylacetic	21.0	-6.0	25.1	+5.0
p-Nitrobenzoylacetic	20.7	-7.5	24.1	+9.3
p-Methoxybenzoylacetic	20.7	-6.5	24.6	+2.7

[a]Data from References 23, 59, and 60.
[b]$\Delta H^\dagger$ in kcal mol^{-1}; $\Delta S^\dagger$ in cal K^{-1} mol^{-1}. For the monobasic acids, the respective headings are HA and A^-.

3. METAL ION-CATALYZED DECARBOXYLATIONS

The decarboxylation of simple monofunctional β-oxo acids such as acetoacetic acid ($CH_3COCH_2CO_2H$) is not catalyzed by metal ions and a necessary prerequisite for metal ion catalysis is the presence of an additional functional group, in the α position to the labile carboxyl group, which is capable of interacting with metal ions. All the acids which are subject to metal ion catalysis have an α-carboxyl group so that formation of a five-membered chelate ring involving the carbonyl group of the acid can occur. In the case of oxaloacetic acid, the equilibria can be expressed as shown in reaction (8). The ketonic complex is the active species in the

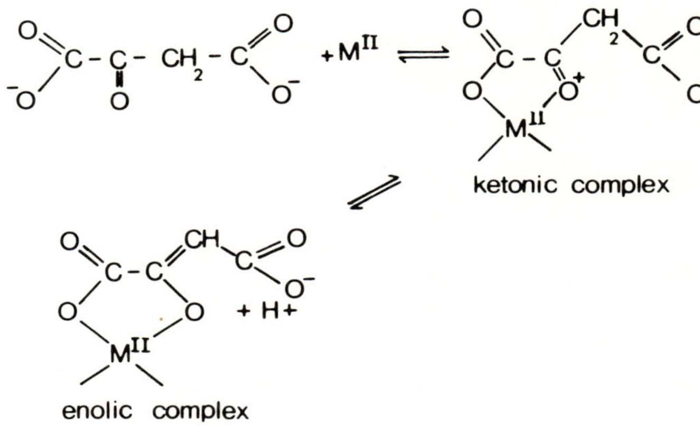

(8)

decarboxylation the enolic complex is kinetically stable [13]. The metal ion polarizes the carbonyl group and so assists the transfer of electrons from the carbon-carbon bond undergoing cleavage. The absence of catalysis by such highly charged cations as $[Co(NH_3)_6]^{3+}$ indicates that a purely electrostatic interaction involving an ion pair is not sufficient to promote decarboxylation [18]. Chelation involving a short-range interaction is a necessary requirement for catalysis to be observed. Metal ions appear to be much more effective catalysts than protons. Presumably at the acidities required to protonate the carbonyl group of the keto acid (pK_{BH^+} ∼-6) the carboxyl group is un-ionized. As the CO_2H^+ moiety is a very poor leaving group catalysis does not occur. It is noteworthy that in the protodecarboxylation of aromatic acids, decarboxylation always appears to proceed via the ampholyte species [21] [eq. (9)]. Con-

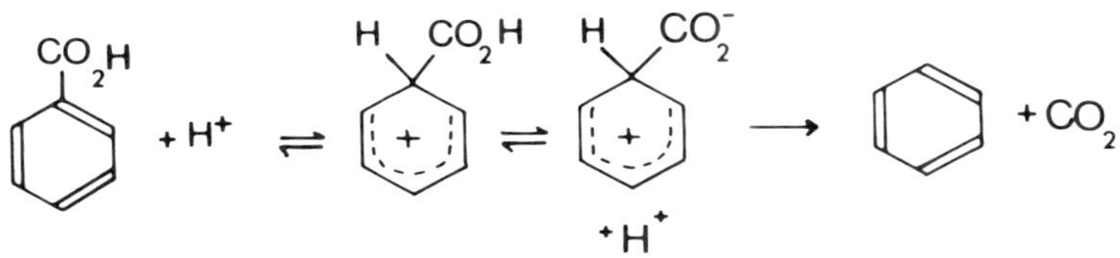

(9)

vincing evidence for this mechanism has been obtained by Longridge and Long for the protodecarboxylation of azulene-1-carboxylic acid [22].

3.1. Oxaloacetic Acid, α,α-Dimethyloxaloacetic Acid, and Fluorooxaloacetic Acid

The decarboxylation of oxaloacetic acid [eq. (10)] has been the subject of many investigations [23-26,43].

$$HO_2CCOCH_2CO_2H \rightarrow HO_2CCOCH_3 + CO_2 \qquad (10)$$

Ionization constants are available for oxaloacetic acid and pyruvic acid [27] and rate constants have been obtained for the H_2A, HA^-, and A^{2-} species over a range of temperature [23,24]. Oxaloacetic acid exists as the trans-enol tautomer in the solid state [44]. The tautomerism in solution is generally acid-base catalyzed and has been studied kinetically [45]. Tate et al. [46] have determined the ionization constants of the keto and enol tautomers of oxaloacetic acid in aqueous solution. The first and second ionization constants (at 25°C) of the keto ($pK_1 = 2.24$, $pK_2 = 3.90$) and enol ($pK_1 = 1.89$, $pK_2 = 3.72$) were obtained and the third ionization ($pK_3 = 13.06$) only occurred with the enol tautomer. The extinction coefficient of the enol is 3600 in water in the pH range 7.4 to 11.9 and the enol content of oxaloacetic acid is ca. 14% in the pH range 5.2 to 8.4 [46]. Similar data for the enol content have been obtained by Banks [44]. The decarboxylation of the monoethyl ester of oxaloacetic acid ($C_2H_5O_2CCOCH_2CO_2H$) has also been studied [47].

The decarboxylation of oxaloacetic acid is catalyzed by primary amines [7], polyvalent cations [9,12,13,14,25,28-38], and by enzymes [5,29,39]. Krampitz and Werkman [35] first observed that the decarboxylation was catalyzed by magnesium(II) and a further study of the magnesium catalysis has been carried out [25]. Formation constants for the magnesium complexes of oxaloacetic acid have also been determined [40]. Krebs [28] investigated the effect of a number of different cations and anions on the reaction, anions and monovalent cations were ineffective catalysts, but the decarboxylation was accelerated by various divalent and trivalent cations. Similar conclusions were reached by other workers.

Kornberg et al. [33] first observed the appearance and decline of a strong absorption band in the region 260 to 300 nm, on addition of metal ions to solutions of oxaloacetic acid under certain conditions of pH. They attributed this observation to the slow formation of the enolic complex of oxaloacetic acid which slowly disappeared on decarboxylation. Speck [29] studied the effect of various cations on the enzymic and the nonenzymic decarboxylation of oxaloacetic acid and noted that high metal ion concentrations could inhibit both the enzymic and nonenzymic reactions. The parsley root enzyme was activated by a number of divalent cations, but manganese(II) was most effective [29]. In a classical paper which resolved many of the mechanistic problems of these reactions Steinberger and Westheimer [12] showed that the decarboxylation of the monoethyl ester of α,α-dimethyloxaloacetic acid, $C_2H_5O_2CCOC(CH_3)_2CO_2H$ was not catalyzed by metal ions. This result supported the view that the catalytically active species formed by α,α-dimethyloxaloacetic acid (which was catalyzed) was the five-membered α-oxo carboxylate complex (4).

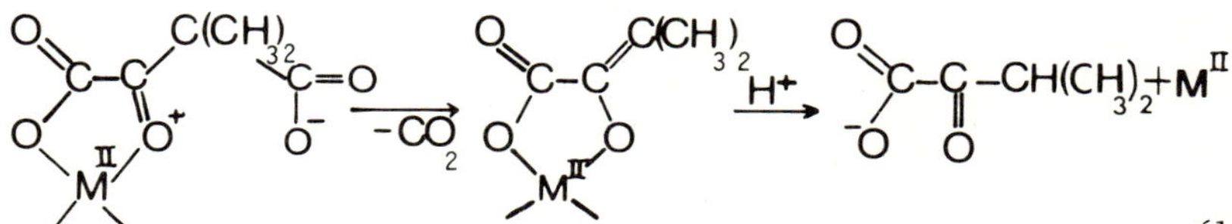

4 enolic intermediate (11)

In addition to the α-oxocarboxylate complex (5), the β-oxocarboxylate complex (6) and the dicarboxylate complex (7) could also form with, for example, oxaloacetic acid. NMR isotropic shifts of the

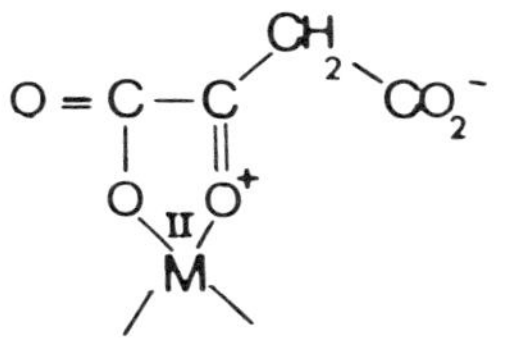

5

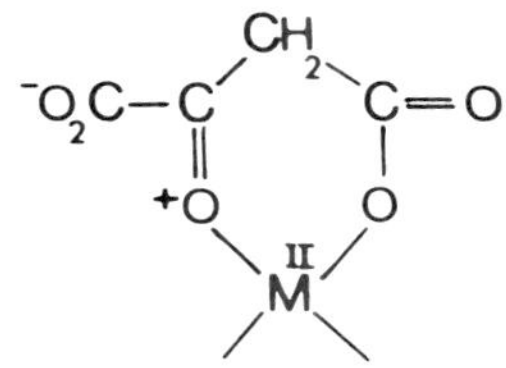

6

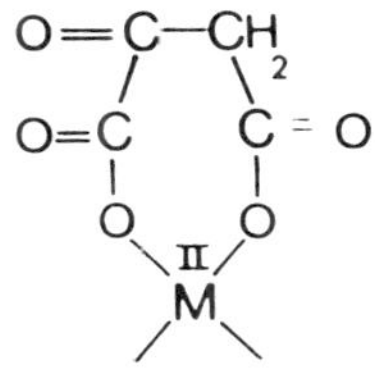

7

CH_2 group of the europium oxaloacetates provide spectral evidence for the formation of the α-oxocarboxylate complex between oxaloacetate and europium(III) in aqueous solution [41].

Steinberger and Westheimer [12] obtained chemical (bromine titration) and spectroscopic evidence for a mechanism involving an enolic intermediate. Visual observations of the $Fe^{(III)}$-catalyzed reaction provided confirmatory evidence for the reaction scheme:

$$\underset{\text{DMOAA-Fe}^{(III)}}{\text{Bright yellow}} \rightarrow \underbrace{\text{green} \rightarrow \text{blue} \rightarrow \text{deep blue}}_{\text{Fe}^{(III)}\text{-enolic intermediate}} \rightarrow \text{colorless products}$$

Dimethyloxaloacetic acid cannot form an enolic complex (9) and so by analogy a ketonic complex (8) seemed likely to be the catalytically active species in the decarboxylation of oxaloacetic acid.

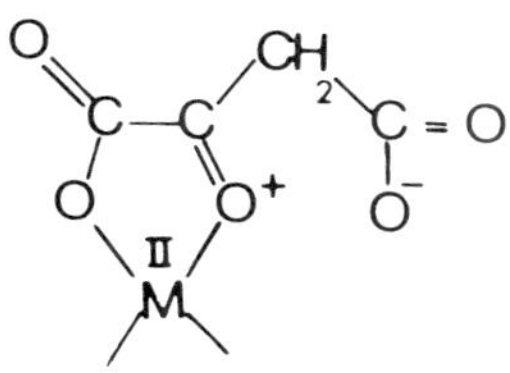

8

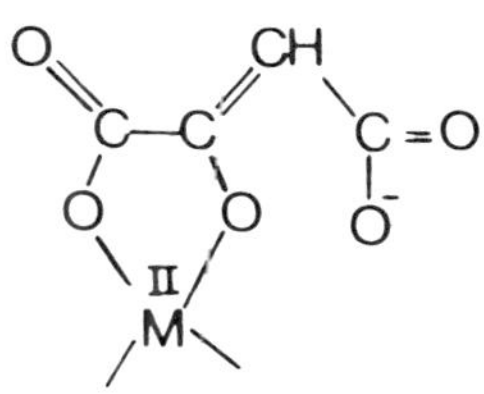

9

The work of Gelles and Hay [13] confirmed this view and established that the strongly absorbing complex first observed by Ochoa et al. [33] was in fact the enolic intermediate, not the enolic complex of oxaloacetic acid. The enolic complex is kinetically stable [13].

Leong and Leussing [42] have shown by stopped-flow measurements that the enolization of oxaloacetic acid (pH range 3.0 to 4.2) is complete within a few milliseconds in the presence of copper(II). A further increase in absorbance then occurs which reaches a maximum after 15 to 20 sec then slowly declines (enolic copper pyruvate intermediate).

The uncatalyzed and metal ion-catalyzed decarboxylation of monofluorooxaloacetic acid, $HO_2CCOCH(F)CO_2H$ has been studied [14] and rate constants obtained for the decarboxylation step and the ketonization step of the metal-fluoropyruvate intermediate. The electron withdrawing fluoro substituent appears to stabilize the enol intermediate so that even with copper(II), the most active catalyst of the bivalent metal ions, the formation and decay of the enolic copper-fluoropyruvate intermediate can be followed by conventional spectrophotometric techniques.

It is normally found that in simple metal ion-catalyzed "model systems," there is a close relationship between the catalytic activity of the metal ion and the formation constants of the metal complex, i.e., the catalytic activity follows the Irving-Williams natural stability order, $Cr^{(II)} < Mn^{(II)} < Fe^{(II)} < Co^{(II)} < Ni^{(II)} < Cu^{(II)} > Zn^{(II)}$ for metal complex formation. Gelles and Salama [32] have determined the rate constants k_{MA} for the 1:1 complex of a number of metal ions with the dianion (A^{2-}) of oxaloacetic acid and obtained the constants shown in Table 2. The copper(II) complex is some 10 times more reactive than the manganese(II) complex. The formation constants for the MA complexes, i.e.,

$$M^{2+} + A^{2-} \rightleftarrows MA \qquad K_{MA} = \frac{[MA]}{[M^{2+}][A^{2-}]} \tag{12}$$

have also been determined by Gelles and Salama [31] (Table 2). A linear free energy relationship might be expected to exist between

TABLE 2

Rate Constants and Formation Constants for Metal Oxaloacetates at 37°C and I = 0.1 M [a]

Species	10^3k (min^{-1})	log K_{MA}
H_2A	0.345	--
HA^-	15.4	--
A^{2-}	4.2	--
CaA	144	2.6
MnA	390	2.8
CoA	1440	3.1
ZnA	1860	3.2
NiA	1380	3.5
CuA	3960	4.9

[a]The rate constants for the uncatalyzed decarboxylation are from Reference 23, and for the catalyzed decarboxylation from Reference 32. Formation constants from Reference 31.

the logarithms of the appropriate rate constants and the formation constants, as the formation constants give a measure of the strength of the metal-ligand bonding. Such a relationship is not observed (Fig. 1). (It should be noted that this conclusion is rather dependent on the accuracy of the rate constant for CuA.) The rate constants do, however, parallel the stability constants of the corresponding oxalates. According to Hammond's postulate [48] the transition state of the decarboxylation (10) should more closely resemble the enolic intermediate (11) which is similar to the oxalate structure (12). A similar result has been noted in the transition metal ion catalyzed decarboxylation of 3-oxoglutaric acid (acetone dicarboxylic acid) where the catalytic activity of the metal ion follows the thermodynamic stability of the corresponding malonates [18].

The catalytic effect of rare earth ions on the decarboxylation

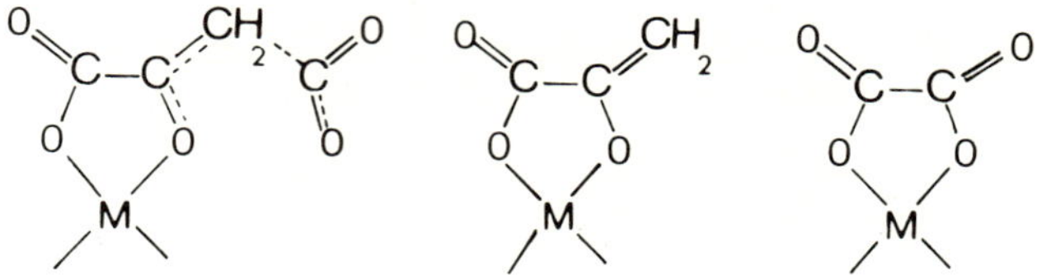

10 11 12

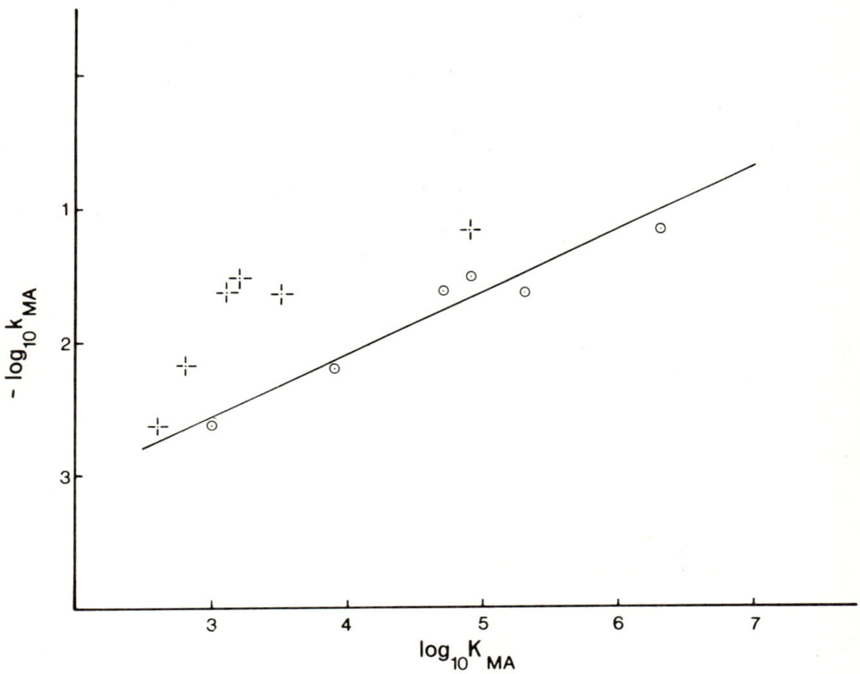

FIG. 1. Linear free energy relationship between log k_{MA} for the metal ion-catalyzed decarboxylation of oxaloacetic acid and log k_{MA} the formation constant of the corresponding metal oxaloacetate (-┆-) or metal oxalate (⊙). (From Ref. 32, with permission.)

of oxaloacetic acid has also been investigated [9]. In this case a linear free energy relationship does appear to exist between the logarithms of the appropriate rate constants and the formation constants. The paramagnetic ions $Gd^{(III)}$ and $Dy^{(III)}$ catalyze the decarboxylation more strongly than would be expected on the basis of this linear free energy relationship and a specific paramagnetic effect may be operating.

Rund and Plane [49] first observed that the catalytic activity of the enzymically important [5] manganese(II) ion in the decarboxylation of α,α-dimethyloxaloacetic acid could be significantly increased by incorporating a π-bonding ligand such as o-phenanthroline into the catalytic system. Further studies of this system have been made by Rund and Claus [50,51] and similar effects have been observed with oxaloacetic acid [36-38] and 3-oxoglutaric acid [52-54]. These reactions are discussed in detail in Section 5.

3.2. Acetonedicarboxylic Acid (3-Oxoglutaric Acid)

The decarboxylation of 3-oxoglutaric acid to give acetoacetic acid [eq. (13)] was first studied by Wiig [55], who determined rate

$$HO_2CCH_2COCH_2CO_2H \rightarrow CH_3COCH_2CO_2H + CO_2 \quad (13)$$

constants and activation parameters for the decarboxylation. However, many of his conclusions regarding the reaction are untenable as he did not consider the extent of ionization of the carboxylic acid. Krebs [28] first observed that metal ions catalyzed the decarboxylation and a number of studies of the metal ion catalysis have since been made [18,52-54,56,57]. The decarboxylation of acetoacetic acid is not catalyzed by metal ions, thus with 3-oxoglutaric acid in the presence of metal ions there is a rapid loss of 1 equivalent of carbon dioxide followed by a slower uncatalyzed decarboxylation.

The catalyzed decarboxylation can be represented by the reaction Scheme 1.

Yerhoff and Larsen [58] have isolated a number of metal complexes of 3-oxoglutaric acid with Na(I), Mg(II), Mn(II), Co(II), Ni(II), Cu(II), and Zn(II) and have studied their infrared spectra. No significant interaction between the metal ion and the carbonyl oxygen atom was observed in any of the complexes in the solid state.

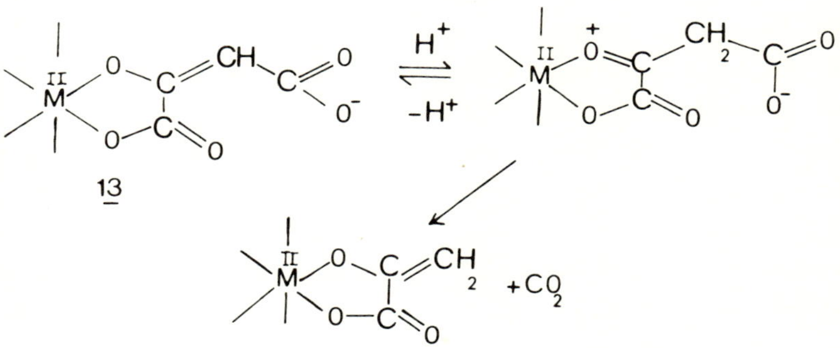

Scheme 1. Metal ion-catalyzed decarboxylation of 3-oxoglutaric acid.

Larsen and Lister [56] have studied the complexes formed in solution between a number of divalent metal ions and 3-oxoglutaric acid. It was established by the method of continuous variations that copper(II) forms a 1:1 complex. EMF measurements could not be rationalized on the basis of the single equilibrium

$$Cu^{2+} + A^{2-} \xrightleftharpoons{K_{CuA}} CuA \qquad (14)$$

and it was necessary to introduce the additional equilibrium

$$CuA \xrightleftharpoons{K_2} CuL^- + H^+ \qquad (15)$$

where A^{2-} is the dianion of 3-oxoglutaric acid. The values K_{CuA} = 112 M^{-1} and pK_2 = 3.81 were obtained at I = 0.6 M and 25°C. The additional proton ionization in the case of copper(II) may be due to the formation of the enolic complex (13) or to ionization of one of the coordinated water molecules, $[CuA(H_2O)_2] \rightleftarrows [CuA(H_2O)(OH)]^- + H^+$. The latter explanation seems least probable since the pK_a for such an ionization would be expected to lie in the range of 6 to 7.

Studies were also made of the interactions of zinc(II), cobalt(II), and nickel(II) with the dicarboxylic acid. A 1:1 complex, MA, predominates in each case, the values for K_{MA} for zinc(II), cobalt(II), and nickel(II) being 86.6, 45, and 60 M^{-1}, respectively, at 25°C. Interestingly no evidence was found for additional acid ionizations in these complexes. The formation of neutral MA type complexes seems to be a general feature of the β-oxodicarboxylic acids, although protonated MAH^+ species have been reported at low pH for oxaloacetic acid [32]. Mixed ligand complexes of the type MAB have been studied [52-54] and these are considered in detail in Section 5. Mixed ligand complexes involving a π-acid ligand such as 2,2'-bipyridyl which is capable of π-bonding with metal ions, enhances the catalytic activity of manganese(II) by a factor of 10 while the effect with copper(II) and nickel(II) is less marked (ca. 2 times) [52]. Acido ligands such as acetate or chloroacetate decrease the catalytic activity of the metal ion [18,52]. Rate constants and formation constants for the metal 3-oxoglutarates are summarized in Table 3.

TABLE 3

Rate Constants and Formation Constants for Metal 3-Oxoglutarates at 25°C [a]

Species	10^3k (min^{-1})	K_{MA} (M^{-1})	pK_a
H_2A	1.25	--	3.01
HA^-	2.90	--	3.82
A^{2-}	0.1	--	--
CoA	2.1	45	--
NiA	3.15	60	--
CuA	201	112	3.81
CuL^-	5.6	--	--
ZnA	1.5	87	--

[a]Data from Reference 56.

3.3. Dihydroxyfumaric Acid ("Dihydroxymaleic Acid")

The terms dihydroxyfumaric acid and dihydroxymaleic acid are often used interchangeably. It seems doubtful if the compound dihydroxymaleic acid exists and the problem has been discussed in detail by Hartree [61]. X-ray diffraction measurements [62] have confirmed that the acid occurs as the trans isomer in the solid state, and the relative values of the first and second ionization constants indicate that this is also so in solution [63]. Dihydroxyfumaric acid has a strong absorption band in the ultraviolet at 289 nm (ε = 9130) consistent with the view that the acid is almost completely enolized in solution [63].

Dihydroxyfumaric acid decomposes spontaneously in aqueous solution to give carbon dioxide and glycolaldehyde [reaction (16)].

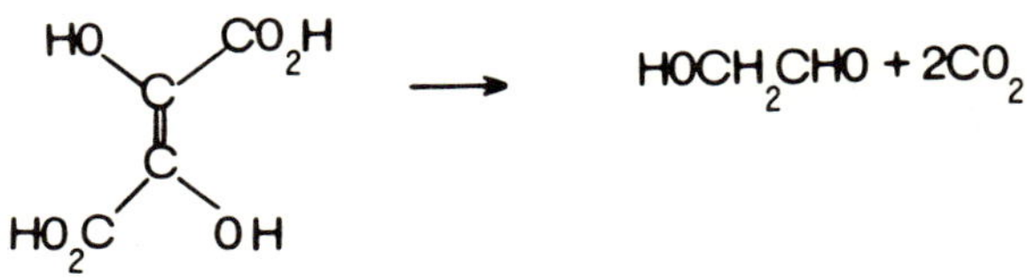

(16)

The kinetics of the reaction have been studied by Franke and Brathun [64]. These authors regarded the loss of the two moles of carbon dioxide as occurring simultaneously, but an analysis of their kinetic data indicates that the decarboxylation probably proceeds in two consecutive steps:

$$\mathrm{HO_2CCOCH(OH)CO_2H} \xrightarrow{k_1} \{\mathrm{HO_2CCOCH_2OH} + \mathrm{HO_2CCH(OH)CHO}\} \tag{17}$$

$$\mathrm{HO_2CCH(OH)CHO} \xrightarrow{k_2} \mathrm{HOCH_2CHO} + \mathrm{CO_2} \tag{18}$$

The first-formed enol intermediate can ketonize to give either hydroxypyruvic acid (14) or hydroxymalonic acid hemialdehyde (15). Hydroxypyruvic acid is an α-keto acid and is expected to be relatively stable in aqueous solution at 25°C, although it rapidly decomposes at 100°C [65]. The overall reaction releases almost the theoretical amount of carbon dioxide suggesting that ketonization of the enol intermediate gives predominantly hydroxymalonic acid hemialdehyde [63]. Tracer studies using ^{14}C have also indicated that not more than 10% of the glycolaldehyde could have been produced from hydroxypyruvic acid [66].

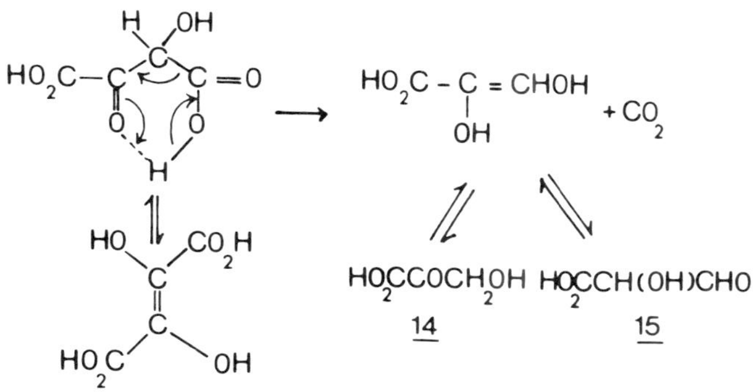

(19)

The reaction is catalyzed by nickel(II) [the copper(II)-catalyzed reaction cannot be studied as the acid is oxidized by copper(II)]. At a ligand concentration of ca. 2×10^{-2} M there is very pronounced catalysis at nickel(II) concentrations < 0.15 M, but at higher concentrations of nickel(II) the rate declines quite markedly [63].

Catalysis by nickel(II) is probably due to the formation of the chelate (16) [see reactions (20) and (21)]. At higher metal ion concentrations there is the possibility of the additional equilibrium shown in reaction (22). Formation of the complex 17 in which the labile carboxyl group is bound in the seven-membered chelate ring should decrease the decarboxylation rate. Similar explanations have been given for the inhibition observed [29] with oxaloacetic acid at high metal ion concentrations [31]. More detailed studies of the relevant equilibria are required before these views can be fully confirmed.

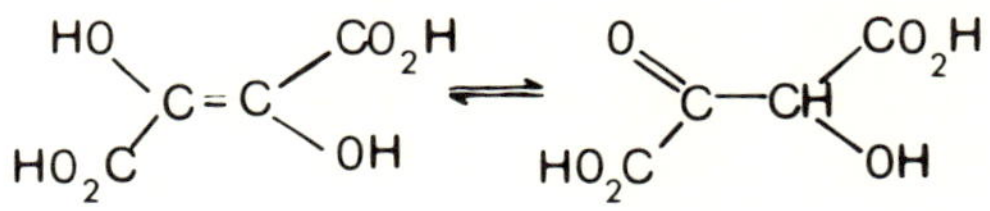

(20)

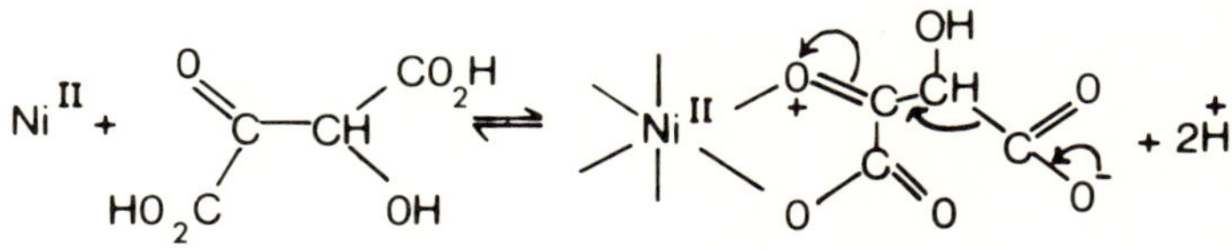

(21)

16

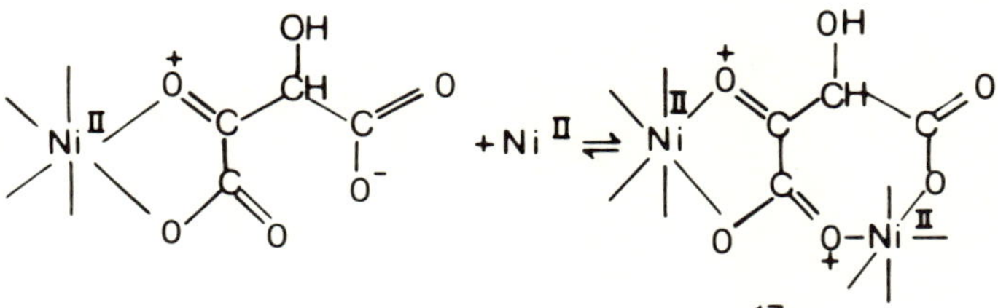

(22)

17

3.4. Dihydroxytartaric Acid

Dihydroxytartaric acid (18) is the dihydrate of diketosuccinic acid (19)

$$HO_2CCOCOCO_2H + 2H_2O \rightleftarrows HO_2CC(OH)_2CO_2H \quad (23)$$

19 18

The free acid $HO_2CCOCOCO_2H$ is unknown, although the diethyl ester is a well-established compound. The decarboxylation of dihydroxytartaric acid has been studied in detail by Pedersen [67] and rate constants obtained for the H_2A, HA^-, and A^{2-} species at 37°C. The product of the decarboxylation is tartronic acid (hydroxymalonic acid),

$$HO_2CC(OH)_2C(OH)_2CO_2H \rightarrow HO_2CCH(OH)CO_2H + CO_2 + H_2O \quad (24)$$

The decarboxylation is strongly catalyzed by copper(II) ions. Although detailed studies of the metal ion catalysis have not been carried out, it has been established that CuA is the active species in the decarboxylation.

It seems likely that dihydroxytartaric acid is in equilibrium in solution with the partially hydrated species,

$$HO_2CC(OH)_2C(OH)_2CO_2H \rightleftarrows HO_2CCOC(OH)_2CO_2H + H_2O \quad (25)$$

which can interact with metal ions in a similar manner to oxaloacetic acid (20).

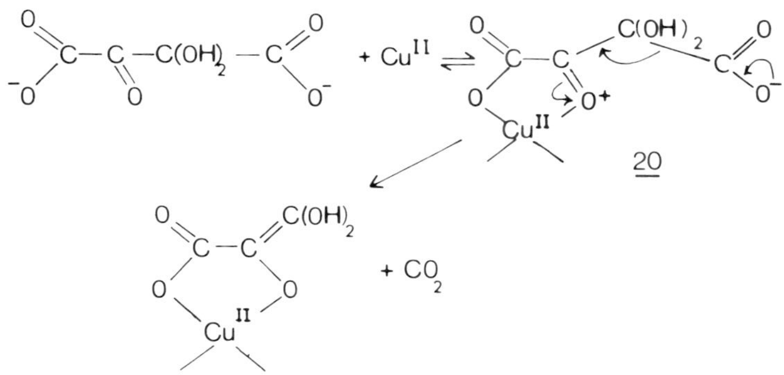

(26)

3.5. Acetosuccinic Acid and Oxalosuccinic Acid

Pedersen [68] has studied the uncatalyzed decarboxylation of acetosuccinic acid [eq. (27)] and has noted that copper(II) ions have a slight catalytic effect.

(27)

Two ketonic complexes, 21 and 22, can be considered. The labile carboxyl group could be bound in a six-membered chelate ring (21) which would tend to stabilize the ligand, or a seven-membered chelate ring of lower thermodynamic stability could occur (22). The latter complex could decarboxylate by the electron shifts shown. Both complexes are probably in equilibrium, thus accounting for the small catalytic effect observed.

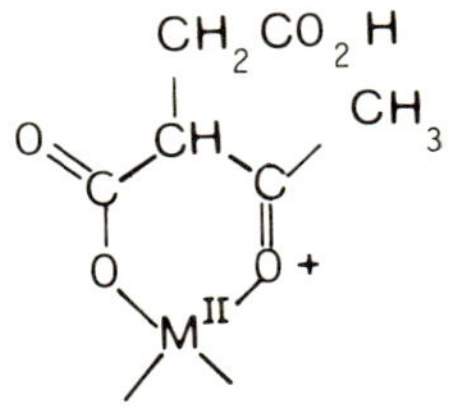

21

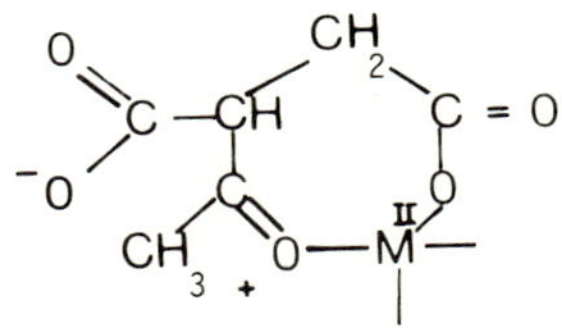

22

The decarboxylation of oxalosuccinic acid [23] to give 2-oxoglutaric acid (24) has also been shown to be catalyzed by metal ions [33]; however, the reaction has not been investigated in any detail. Formation of the α-oxocarboxylate complex (25) would lead to decarboxylation. Detailed kinetic and thermodynamic studies of reaction (29) are required.

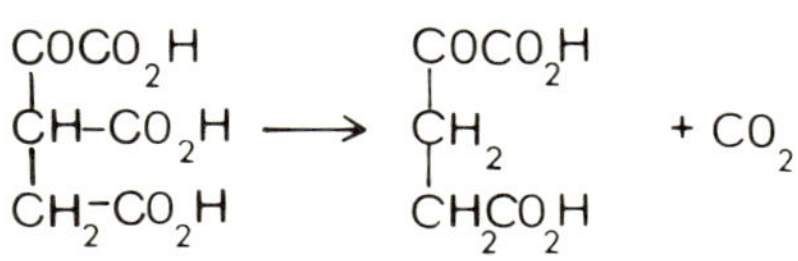

23 24 (28)

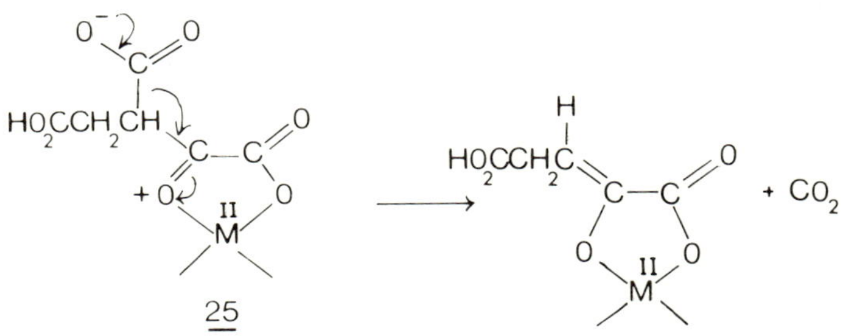

25 (29)

4. FORMATION CONSTANTS

The β-oxodecarboxylic acids tend to form neutral 1:1 metal complexes of the type MA with transition metal ions where M^{2+} is the divalent ion and A^{2-} is the dianion of the acid [31,56]. The interaction of Ca(II), Mn(II), Co(II), Ni(II), Cu(II), and Zn(II) with oxaloacetic acid has been studied potentiometrically at 25°C [31] and thermodynamic formation constants obtained for the equilibrium $M^{2+} + A^{2-} \rightleftarrows MA$. Evidence was also obtained for the equilibrium $M^{2+} + MA \rightleftarrows M_2A^{2+}$. Although kinetic evidence [32] has been obtained for the protonated complex MAH^+, this species was not detected potentiometrically at the pH's employed. Formation constants for the metal oxaloacetates are summarized in Table 2. The ionization constants of the keto and enol tautomers of oxaloacetic acid have been determined [46]. The enol content of the acid is less than 15% at pH values less than 8.4 [46]. The Mg(II) complexes of the keto and enol tautomers of oxaloacetic acid have been the subject of a detailed investigation [40]. The overall formation constant for the complex formed between Mg(II) and the various oxaloacetate dianions is log K_1 = 1.96 at 25°C and I = 0.1 M. This value may be compared with constants for similar complexes of other four-carbon dicarboxylic acids. Thus log K_1 = 1.2, 1.36, and 1.55 for the succinate, tartrate, and malate complexes, respectively, with Mg^{2+} at 25°C and I = 0.2 M [78]. A detailed investigation of the magnesium ion-catalyzed decarboxylation of oxaloacetic acid has been published [25].

The metal complexes of oxaloacetic acid are a mixture of ketonic (26) and enolic (27) species. The metal enolates (27) ab-

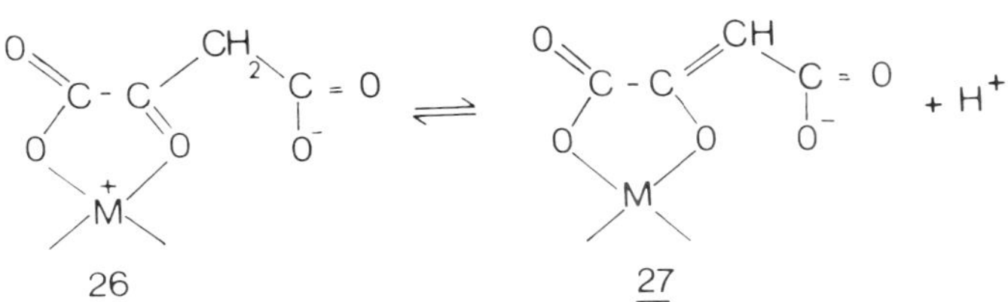

(30)

sorb strongly at 295 nm and thus the keto $\rightleftarrows$ enol equilibrium (30) can be investigated spectrophotometrically. In view of the facile decarboxylation of oxaloacetic acid in the presence of metal ions, use has been made of the half-ester $HO_2CCOCH_2CO_2C_2H_5$, which does not decarboxylate, as a suitable "model" compound for the equilibrium studies [13]. The proportion of enolic complex is pH-dependent and is greatest for copper(II) oxaloacetate (ca. 40%). The formation constant for the ketonic complex should therefore be about 40% less than the overall formation constant of 7.5×10^4 given in Table 2. A value of 4.5×10^4 is similar to that found for the formation constant of copper(II) α,α-dimethyloxaloacetate which cannot enolize [13], i.e., $K_{MA} = 4.5 \times 10^4$. As discussed in Section 3.2, Larsen and Lister [56] have determined the formation constants for the metal 3-oxoglutarates. Neutral MA-type species were found and enolization of CuA (pK 3.81) observed (Table 3).

5. MIXED LIGAND COMPLEXES

Rund and Plane [49] first observed that the decarboxylation of dimethyloxaloacetate in the presence of manganese(II) and nickel(II) could be enhanced by the addition of a π-acid ligand such as o-phenanthroline. The activity of manganese(II) was increased some 16-fold, while the effect with nickel(II) was less marked (2 times). Anionic oxygen base ligands such as 8-hydroxyquinoline-5-sulfonate lowered the catalytic activity of the metal ions. Further studies of this system have been made by Rund and Claus [50,51]. Similar effects have also been observed with oxaloacetic acid [36-38] and 3-oxoglutaric acid [52,53]. A mixed ligand complex of 5-nitro-o-phenanthroline and oxaloacetate has been isolated from solution [37]. Structure 28 is suggested for the compound on the basis of the ultraviolet, infrared, and analytical data. The mixed ligand complexes are of particular interest as enzyme models since oxaloacetate decarboxylase may be regarded as a special type of mixed ligand complex [5].

The amine ligand in isolation does not catalyze the reaction (only primary and secondary amines are active catalysts) and therefore the increased reactivity can be attributed to the formation of a 1:1:1 ternary complex. The relevant equilibria may be summarized in equations (31) and (32).

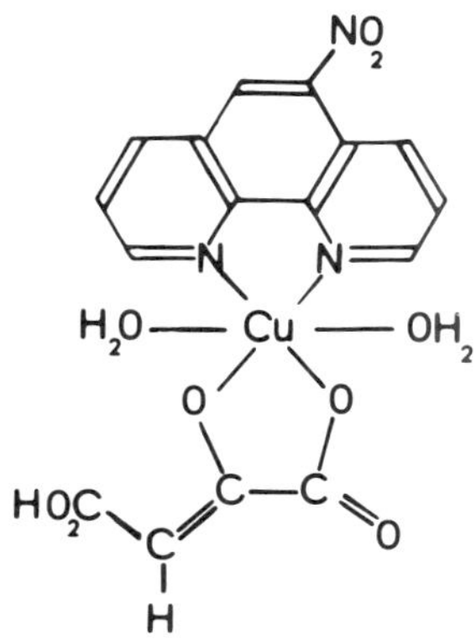

28

$$M^{2+} + A^{2-} \rightleftarrows MA \qquad K_{MA} = \frac{[MA]}{[M^{2+}][A^{2-}]} \tag{31}$$

$$MB^{2+} + A^{2-} \rightleftarrows MAB \qquad K_{MAB} = \frac{[MAB]}{[MB^{2+}][A^{2-}]} \tag{32}$$

The difference in stability between the binary and the ternary complexes may be expressed:

$$\Delta \log K = \log K_{MAB} - \log K_{MA} \tag{33}$$

Since more coordination positions are available for bonding of the first ligand to a given metal ion than for a second ligand, the stability constant for the formation of a 1:1 complex is usually greater than for a 2:1 complex [79], i.e., $\log K_{ML} > \log K_{ML_2}$, and one expects to find negative values for $\Delta \log K$ in equation (33). The difference ($\log K_{ML_2} - \log K_{ML}$) is ca. -0.5 log units for monodentate ligands and ca. -1 to -2 log units for bidentate ligands. In Table 4 are listed the $\Delta \log K$ values for several copper(II) systems.

A much more extensive compilation has been given by Sigel and McCormick [80]. Unexpectedly, it is found that positive $\Delta \log K$

TABLE 4

Formation Constants of Binary and Ternary Complexes and Their Respective Stability Differences, Δ log K[a]

Equilibrium[b]	log K	Δ log K
$Cu^{2+} + pyr^{2-} \rightleftarrows Cu(pyr)$	13.96	+0.4
$Cu(dipy)^{2+} + pyr^{2-} \rightleftarrows Cu(dipy)(pyr)$	14.39	+0.4
$Cu^{2+} + gly^{-} \rightleftarrows Cu(gly)^{-}$	8.27	-0.4
$Cu(dipy)^{2+} + gly^{-} \rightleftarrows Cu(dipy)(gly)^{-}$	7.88	-0.4
$Cu^{2+} + pyr^{2-} \rightleftarrows Cu(pyr)$	13.96	-0.8
$Cu(en)^{2+} + pyr^{2-} \rightleftarrows Cu(en)(pyr)$	13.20	-0.8
$Cu^{2+} + ssal^{3-} \rightleftarrows Cu(ssal)^{-}$	9.41	+0.5
$Cu(dipy)^{2+} + ssal^{3-} \rightleftarrows Cu(dipy)(ssal)^{-}$	9.86	+0.5

[a]Data from the compilation of H. Sigel and D. B. McCormick [80].
[b]pyr^{2-}, pyrocatechol dianion; en, 1,2-diaminoethane; dipy, 2,2'-dipyridyl; $ssal^{3-}$, trianion of 5-sulfosalicyclic acid.

values occur, indicating that the ternary complexes are more stable than the binary complexes in some situations. This appears to be the case for all systems which contain 2,2'-dipyridyl and a ligand having oxygen atoms as donors. The oxygen ligand is essential for a positive Δ log K value, thus in the 2,2'-dipyridyl-Cu^{2+}-1,2-diaminoethane system, Δ log K = -1.3. This discriminating behavior of the copper(II)-2,2'-dipyridyl 1:1 complex toward other ligands appears to be a quite general effect associated with the π system of the nitrogen ligand.

Lister and Leong [53,54] have studied the mixed ligand complexes formed between a number of metal ions with 3-oxoglutarate and aromatic amines at 25°C and I = 0.6 M (Table 5). Slightly negative Δ log K values are observed for the ternary complexes of cobalt(II), nickel(II), and zinc(II); however, positive Δ log K values are observed for those of manganese(II) and copper(II).

TABLE 5

Formation Constants of Binary and Ternary Complexes of Divalent Metal Ions with 3-Oxoglutarate and Aromatic Amines at 25°C[a]

Metal	Ligand A[b]	Ligand B[b]	log K_{MAB}	log K_{MA}	Δ log K
Mn^{2+}	OGA	Dipy	1.18	0.90	+0.28
Co^{2+}	OGA	Dipy	1.30	1.30	0
Co^{2+}	OGA	Phen	1.40	1.30	+0.10
Ni^{2+}	OGA	Dipy	1.26	1.40	-0.14
Ni^{2+}	OGA	Phen	1.30	1.40	-0.10
Cu^{2+}	OGA	Dipy	2.56	2.18	+0.38
Cu^{2+}	OGA	Phen	2.43	2.18	+0.25
Cu^{2+}	OGA	DMDP	2.05	2.18	-0.13
Zn^{2+}	OGA	Dipy	0.90	1.00	-0.10
Zn^{2+}	OGA	Phen	0.83	1.00	-0.17

[a]From Reference 54.
[b]Dipy, 2,2'-dipyridyl; DMDP, 4,4'-dimethyl-2,2'-dipyridyl; OGA, 3-oxoglutarate dianion; Phen, 1,10-phenanthroline.

The catalyzed decarboxylations may be represented by equations (34) and (35).

$$M^{2+} + A^{2-} \underset{}{\overset{K_{MA}}{\rightleftharpoons}} MA \xrightarrow{k_{MA}} CO_2 + \text{product} \quad (34)$$

$$MB^{2+} + A^{2-} \underset{}{\overset{K_{MAB}}{\rightleftharpoons}} MAB \xrightarrow{k_{MAB}} CO_2 + \text{product} \quad (35)$$

The enhanced catalytic effect caused by the amine in some cases, can arise from either changes in the equilibrium constants or in the rate constants or in both of these. Values of the rate constants k_{MA} and k_{MAB} have also been obtained [53,54] and the relevant data are summarized in Table 6. It is clear that in the 3-oxoglutaric acid system the small rate enhancement with copper(II) in the

TABLE 6

Formation Constants and Rate Constants for Binary (MA) and Ternary (MAB) Complexes of 3-Oxoglutaric Acid at 25°C and I = 0.6 M[a]

Catalyst[b]	K_{MA} or K_{MAB} (M^{-1})	10^3k_{MA} or 10^3k_{MAB} (min^{-1})	$10^2K_{MA}k_{MA}$ or $10^2K_{MAB}k_{MAB}$ (M^{-1} min^{-1})
Mn^{2+}	8	1.2	1
$Mn(dipy)^{2+}$	15	4.2	6
Co^{2+}	20	4.0	8
$Co(dipy)^{2+}$	20	9.3	19
$Co(phen)^{2+}$	25	12.0	31
Ni^{2+}	25	5.8	14
$Ni(dipy)^{2+}$	20	7.9	16
$Ni(phen)^{2+}$	20	11.0	23
Cu^{2+}	150	120	1800
$Cu(dipy)^{2+}$	360	160	5800
$Cu(DiMedipy)^{2+}$[b]	100	140	1400
$Cu(phen)^{2+}$	270	102	2700
Zn^{2+}	10	5.1	5
$Zn(dipy)^{2+}$	8	16.0	12
$Zn(phen)^{2+}$	6	31	18

[a]From Reference 54.

[b]See Table 5, footnote b for abbreviations; DiMedipy = 4,4'-dimethyl-2,2'-dipyridyl.

presence of 2,2'-dipyridyl is due primarily to a more favorable formation constant for the mixed ligand complex and the reaction has been discussed in detail [53]. In the case of manganese(II) catalysis, the formation constants for the ternary complex are higher than for the binary complex and the ternary complex also decarboxylates some 3 to 4 times more rapidly than the binary complex. The effects of 2,2'-dipyridyl are therefore due to both increased rate constants and formation constants.

The considerable stability of 2,2'-dipyridyl and o-phenanthroline complexes is believed to be due to the formation of $\pi(M \rightarrow L)$ bonds. The effect of such π-bonding would be to increase the effective charge on the central metal ion and so increase its activity as a Lewis acid catalyst. The importance of π-bonding is more pronounced with metal ions with d^4, d^5, and d^6 configurations, and possibly such effects occur in the d^5-manganese(II) system. It has also been suggested [5,49] that π-bonding may be of importance in the enzymic decarboxylation of oxaloacetic acid if the ligand-bonding sites in the apoenzyme are imidazole donors of histidine residues.

Rund and Plane [49] have shown that in the metal ion-catalyzed decarboxylation of α,α-dimethyloxaloacetic acid the decarboxylation rate can be increased by a factor of 10^3 by running the reaction in dioxane-water solutions. The lower dielectric constant of the mixed solvent increases the formation constant of the catalytically active metal complex.

6. CARBON ISOTOPE EFFECTS

A number of investigations of carbon isotope effects in metal ion-catalyzed decarboxylations have been reported [81-83]. In the case of oxaloacetic acid the relevant reactions are

$$HO_2CCO^{12}CH_2{}^{12}CO_2H \xrightarrow{k_{12}} HO_2CCOCH_3 + {}^{12}CO_2 \tag{36}$$

$$HO_2CCO^{12}CH_2{}^{13}CO_2H \xrightarrow{k_{13}} HO_2CCOCH_3 + {}^{13}CO_2 \tag{37}$$

$$HO_2CCO^{12}CH_2{}^{14}CO_2H \xrightarrow{k_{14}} HO_2CCOCH_3 + {}^{14}CO_2 \tag{38}$$

The kinetic isotope effect (KIE) is given by the ratios k_{12}/k_{13} and

k_{12}/k_{14}. The ratio is normally slightly greater than unity. The isotope effects can also be expressed as percentages, i.e., $100[(k_{12}/k_{13}) - 1]$ so that a 6% isotope effect indicates that $k_{12}/k_{13} = 1.06$. Most $^{12}C-^{13}C$ isotope effects fall within the range 3 to 5% and most $^{12}C-^{14}C$ isotope effects in the range 5 to 10% for decarboxylation reactions [84]. Typical KIE data for rare earth oxaloacetates are given in Table 7. In all cases cleavage of the carbon-carbon bond

TABLE 7

Isotope Effects in the Metal Ion-Catalyzed Decarboxylation of Oxaloacetic Acid at 25°C[a]

Species	k_{12}/k_{13}	k_{12}/k_{14}
Acid alone	1.0448	1.1337
YA^+ (diamagnetic)	1.0349	1.1075
DyA^+ (paramagnetic)	1.0336	1.1091
GdA^+ (paramagnetic)	1.0344	1.1093

[a]Data from References 81 and 82.

is rate-determining. Interest in catalysis by the lanthanides has been stimulated by the observation that paramagnetic ions possess an enhanced catalytic effect in the decarboxylation of phenylmalonic acid [85,86]. The catalytic effect of rare earth ions on the decarboxylation of oxaloacetic acid has also been investigated [9]. In this case there is an apparent linear free energy relationship between the logarithms of the appropriate rate constants and the thermodynamic formation constants. Paramagnetic ions Gd(III) and Dy(III) catalyzed the decarboxylation more strongly than would be expected on the basis of this linear free energy relationship and a specific paramagnetic effect may be operating. Some preliminary work [87] had indicated that the Y(III) chelate of oxaloacetic acid showed the "normal" $^{12}C/^{13}C$ isotope effect of 5 to 6%, but the isotope effect of the paramagnetic Dy(III) chelate was very considerably higher. The measurements of Wood [81,82] (Table 7) indicate that these initial measurements are in error. The whole concept of specific paramagnetic effects seems rather questionable and these reactions should be further investigated.

Carbon-13 isotope effects have also been determined for the decarboxylation of $HO_2CCOCH_2{}^{13}CO_2H$ catalyzed by manganese(II) and by the manganese(II) activated enzyme isolated from *Micrococcus lysodeikticus* [83]. These reactions are discussed in Section 9.

7. INHIBITION OF DECARBOXYLATION BY METAL IONS

The decarboxylation of a number of carboxylic acids is inhibited by metal ions. Inhibition of the decarboxylation of dihydroxyfumaric acid at high metal ion concentrations [63] has already been noted (Section 3.3). The decarboxylation of oxaloacetic acid is also inhibited at high metal ion concentrations [29,31], presumably due to the formation of the complex M_2A^{2+} containing both five- and seven-membered chelate rings (29). Inhibition is presumably due to binding of the labile carboxyl group in the second chelate ring.

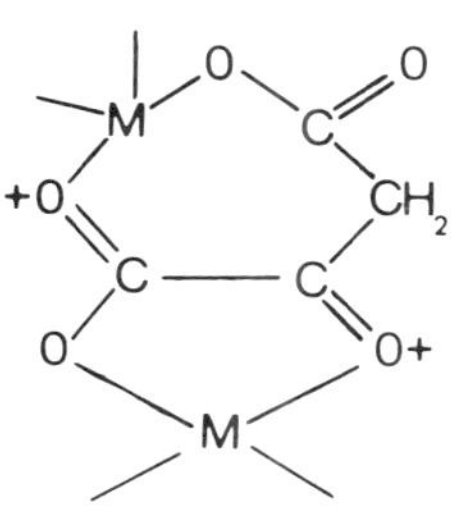

29

Pedersen [69] first noted that certain metal ions, for example, Al(III), Cu(II), and Mg(II), appreciably retarded the decarboxylation of nitroacetic acid. In the absence of metal ions the monoanion of nitroacetic acid (30) decomposes to give the anion of the aci-nitro tautomer of nitromethane (31). (The un-ionized acid and the dianion, $^-O_2N{=}CHCO_2{}^-$ do not decarboxylate.) Inhibition of decarboxylation in the presence of metal ions can be attributed to

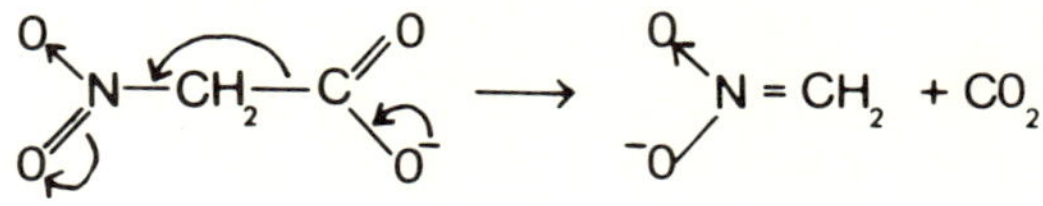

30 31

(39)

formation of complexes of the inactive aci-nitro tautomer (32).

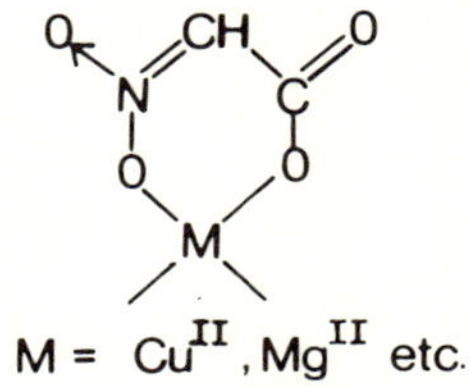

32

8. CARBOXYLATION

Stiles and Finkbeiner [70] found in 1959 that the Mg(II) chelate of nitroacetic acid could be prepared in methanol by adding an equivalent quantity of magnesium methoxide. The ultraviolet spectrum of the solution was consistent with the formation of the aci-nitro complex, and the solution was found to be stable for several hours.

Treatment of nitromethane with magnesium methoxide and carbon dioxide in methanol followed by careful hydrolysis gave a 63% yield of nitroacetic acid. (The actual reagent is methylmagnesium carbonate.) This observation has led to the development of methylmagnesium carbonate as a reagent for the carboxylation of primary nitroparaffins [70,71], methyl or methylene ketones [72], and 3-phenylhydantoins [73]. The latter reaction provides a general route for the synthesis of α-amino acids.

The carboxylation of a methylene ketone to give a β-oxo acid is shown in Scheme 2.

The general area of carboxylation mechanisms has been recently reviewed by Lynen [74]. Metal ions such as Mn(II) and Mg(II) are involved in a number of enzymic carboxylation systems and pyruvate carboxylase was the first Mn(II) metalloenzyme to be discovered [75]. The enzyme contains four Mn(II) ions and four biotin molecules per

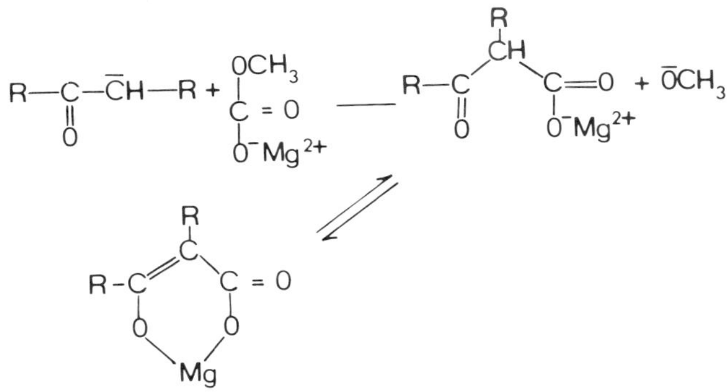

Scheme 2. Carboxylation of methyl and methylene ketones with methylmagnesium carbonate.

mole and is a tetramer [75]. In addition to other reactions the enzyme catalyzes the reaction (40).

$$\text{Enzyme-biotin-CO}_2 + \text{pyruvate} \rightleftarrows \text{Enzyme-biotin} + \text{oxaloacetate} \qquad (40)$$

Features of a possible active site complex due to Mildvan and coworkers [76,77] is shown in Scheme 3.

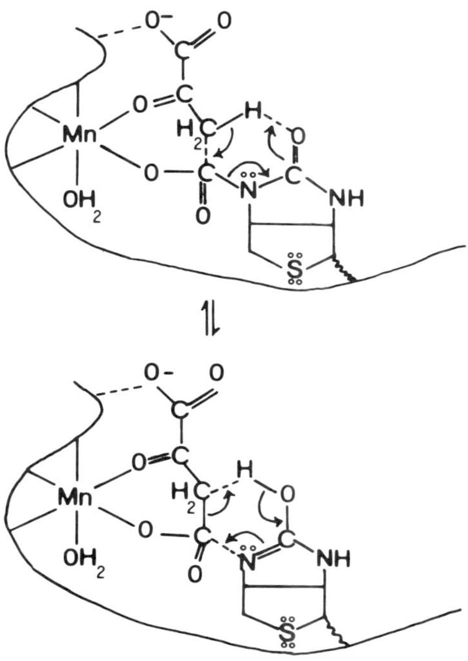

Scheme 3. Features of the active site of pyruvate carboxylase proposed by Mildvan et al. [76,77].

9. ENZYMIC DECARBOXYLATION

The enzymes catalyzing the decarboxylation of β-oxo acids are of two types: the "active lysine" enzymes such as acetoacetate decarboxylase which are not dependent on metal ions for activity, and the metal-dependent enzymes such as oxaloacetate decarboxylase.

9.1. Acetoacetate Decarboxylase

The enzyme from the microorganism *Clostridium acetobutylicum* has been studied in detail by Westheimer and his coworkers [3,4,88-90]. The enzyme is normally isolated as an aggregate containing subunits of molecular weight of approximately 29,000 [90], there are two such subunits per active site. A number of recent reviews of the enzyme are available [3]. The active intermediate in the reaction is a Schiff's base formed by condensation of the substrate with the ε-amino group of a lysine residue. The protonated Schiff's base acts as an electron sink thus assisting the transfer of electrons from the carbon-carbon bond undergoing cleavage [reaction (41)]. The initial product of the decarboxylation is the enamine (33) which must subsequently ketonize and hydrolyze to regenerate

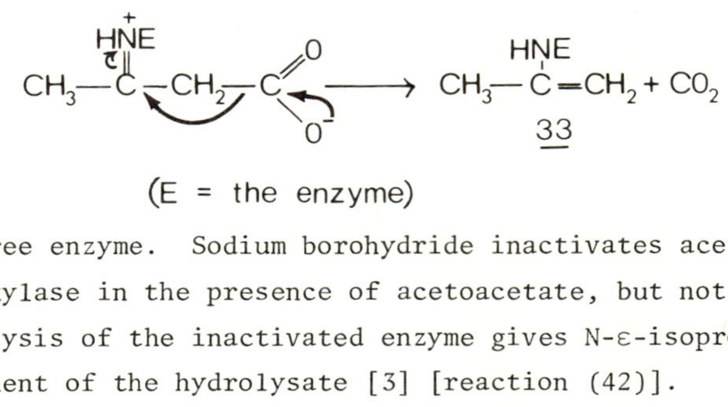

(41)

(E = the enzyme)

the free enzyme. Sodium borohydride inactivates acetoacetate decarboxylase in the presence of acetoacetate, but not in its absence. Hydrolysis of the inactivated enzyme gives N-ε-isopropyllysine as a component of the hydrolysate [3] [reaction (42)]. Other observations consistent with the Schiff base mechanism are (a) the oxygen of the keto group of acetoacetate is exchanged with that of the solvent (using an ^{18}O marker) more rapidly in the presence of the enzyme than in its absence, and (b) cyanide ions cause a progressive inactivation of the enzyme in the presence of substrate. Inhibition cannot be effected by incubating cyanide ion with either enzyme or substrate prior to the addition of substrate or enzyme,

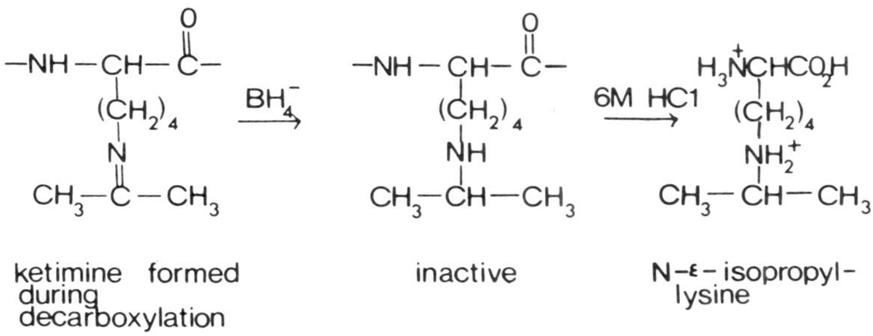

(42)

respectively. Cyanide ion must therefore interact with the enzyme substrate complex, presumably to give a cyanide adduct,

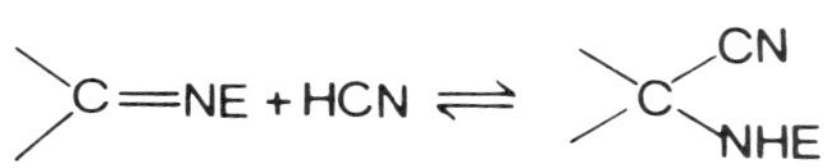

(43)

Acetoacetate decarboxylase is also inhibited by acetopyruvate which interacts with the active site but is not decarboxylated [90]. An absorption band at 325 nm in the enzyme-acetopyruvate complex has been interpreted as being due to an enamine resulting from tautomerism of the originally formed Schiff-base (34).

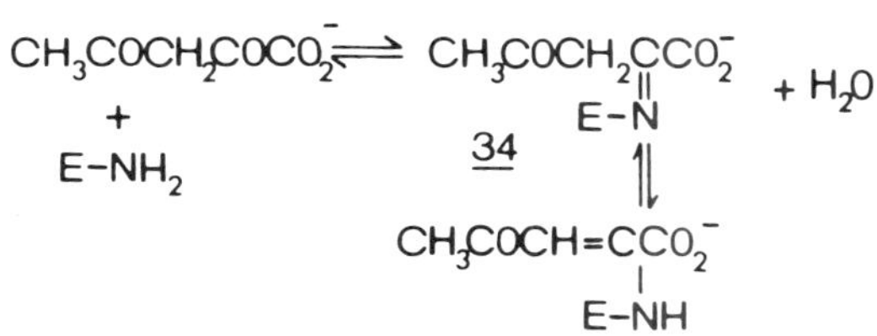

(44)

9.2 Oxaloacetate Decarboxylase

The oxaloacetate decarboxylases from *Micrococcus lysodeikticus* [39], cod [91], and pigeon liver [92] require metal ions for activation.

The oxaloacetate decarboxylase from cod has been crystallized by Kosicki [93] and has been shown to depend upon metal ions, preferably manganese(II), for activity. The mechanism of action of the crystalline oxaloacetate decarboxylase from cod has been studied by

Kosicki and Westheimer [5]. Sodium borohydride has no effect on the activity of the enzyme regardless as to the presence or absence of the substrate or of the product (pyruvate); however, the reduction of pyruvate by borohydride in the presence of manganese (II) and the enzyme leads to the formation of an excess of D-(-)-lactate [reaction (45)]. (The decarboxylation was conducted in the

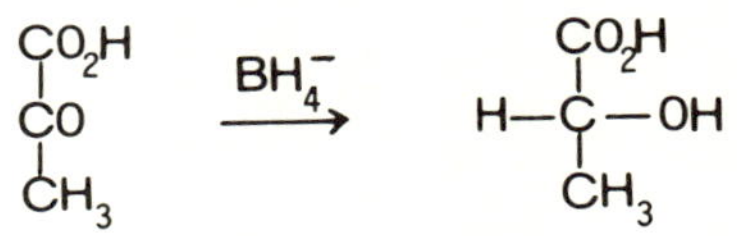

D-(-)- lactic acid

(45)

presence of borohydride in an attempt to trap a Schiff base as has been possible in the case of acetoacetate decarboxylase [89].) The enzymic catalysis of BH_4^- reduction requires the presence of metal ions and does not occur in the presence of EDTA.

The formation of an excess of D-(-)-lactic acid is consistent with the view that a mixed ligand complex of the enzyme and oxaloacetate with manganese(II) is involved in both the reduction and decarboxylation. As the enzyme is not inactivated by borohydride in the presence of substrate, a Schiff base mechanism as occurs with acetoacetate decarboxylase is excluded, and a possible mechanism for oxaloacetate decarboxylase is shown in Scheme 4.

Since a metal ion is required for the enzyme-catalyzed borohydride reduction, pyruvate is presumably bound for the reduction as it is as a product of the decarboxylation (Scheme 5). Formation of optically active lactic acid confirms that the reduction is enzymic. Seltzer et al. [83] have studied ^{12}C-^{13}C isotope effects in the decarboxylation of $HO_2CCOCH_2{}^{13}CO_2H$ catalyzed by manganese (II) and by the manganese(II)-activated enzyme isolated from *Micrococcus lysodeikticus*. The manganese(II)-catalyzed reaction shows a 6% carbon isotope effect ($k^{12}/k^{13} \sim 1.06$) indicating that cleavage of the carbon-carbon bound occurs in the rate-determining step (see also Section 6). The enzymic reaction does not show a carbon

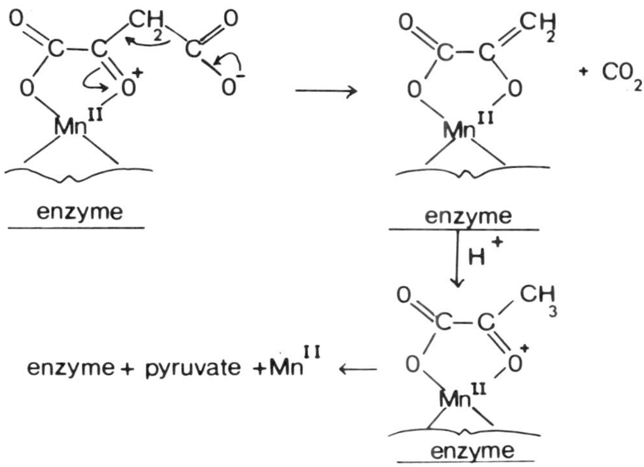

Scheme 4. The enzymic decarboxylation of oxaloacetic acid.

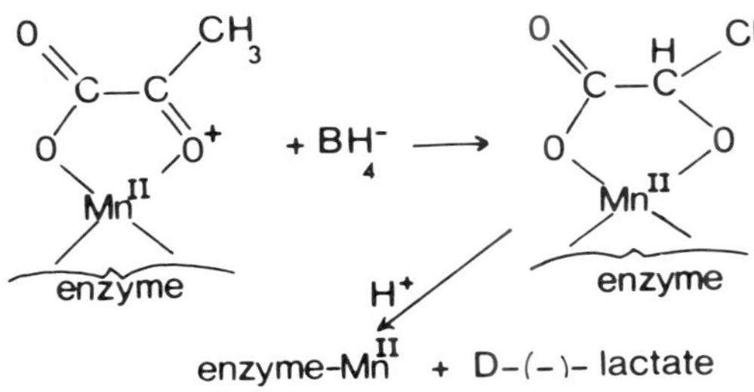

Scheme 5. Sodium borohydride reduction of pyruvate formed in the enzymic decarboxylation of oxaloacetate.

isotope effect, but the reaction is slower in D_2O than in H_2O, whereas the rate of the manganese-catalyzed reaction is unaffected by the solvent change.

It is clear that in the enzymic reaction, the rate-determining step is not cleavage of the carbon-carbon bond (which is considered to occur some 10^8 times faster in the enzymic reaction than for the metal ion-catalyzed reaction), but probably some subsequent step involving proton transfer, such as ketonization of the enol intermediate and release of pyruvic acid from the enzyme.

9.3. Catalytic effectiveness of Metal Ions

The rate constant for decarboxylation of the CuA species of oxaloacetic acid obtained by stopped-flow techniques is 0.36 sec^{-1} at 25°C [42], which compares reasonably well with the constant 1.52 sec^{-1} determined by Pedersen at 37°C [24]. The rate constant for the decarboxylation of the dianion A^{2-} of oxaloacetic acid is ca. 1.7×10^{-5} sec^{-1} at 25°C [23], and thus the metal complex decomposes some 2×10^4 times faster than the dianion. As enzymic reactions are usually at least some 10^9 times faster than the uncatalyzed reactions it is clear that additional effects must operate in the enzymic decarboxylation.

On mixing copper(II) chloride and oxaloacetate solutions at pH 3.0 to 4.2 in a stopped-flow spectrophotometer a rapid increase in absorbance is observed in the region of 315 nm which is complete within the deadtime of the instrument [42]. This increase in absorbance can be attributed to the very fast enolization of A^{2-} in the presence of copper(II). The absorbance continues to increase at a slower rate and reaches a maximum after 15 to 20 sec and then slowly decreases. In the presence of excess copper(II), the appearance and disappearance of this maximum arises from the complete decarboxylation of CuA followed by a much slower disappearance of the enolic copper pyruvate intermediate (Scheme 6).

The enolic copper(II) pyruvate intermediate is a very stable complex and in decarboxylation studies where an excess of oxaloacetate is employed, its decomposition and regeneration of the copper(II) catalyst may be rate-determining.

The metal ion-promoted dimerization of pyruvate (P^-) has been known for many years [94,95]. Recent work [96] has shown that the reaction proceeds via the rate-determining enolization of metal pyruvate (MP^+) [reaction (46)] followed by attack of a further pyruvate on the complexed metal pyruvate enolate [reaction (47)].

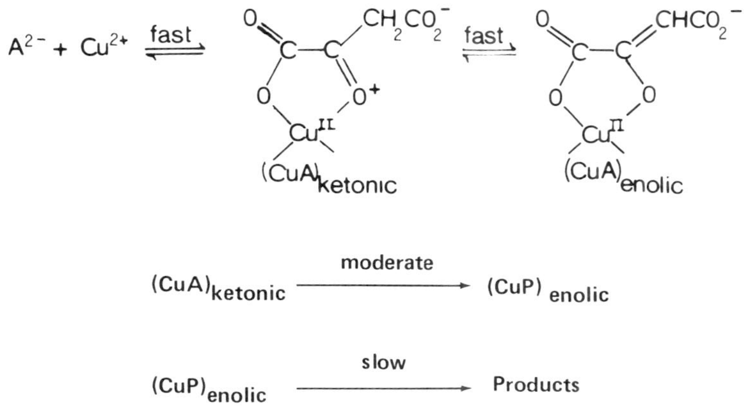

Scheme 6. Copper(II)-catalyzed decarboxylation of oxaloacetate.

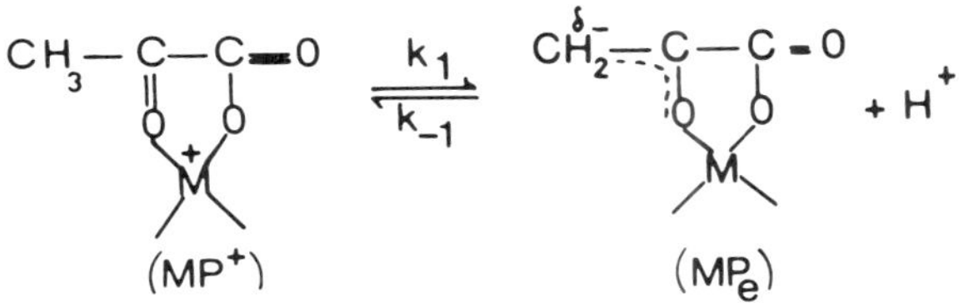

(46)

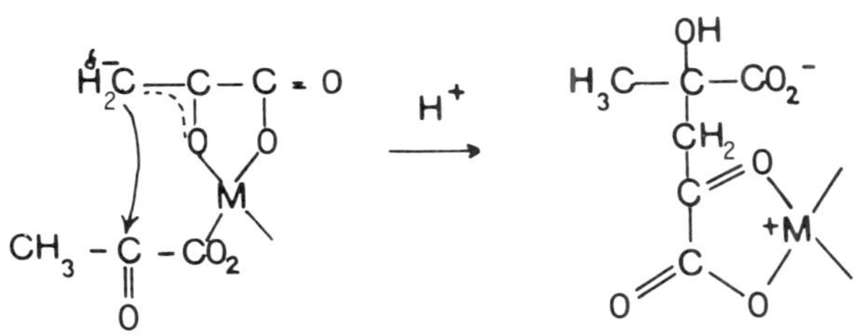

(47)

The inhibition of the decarboxylation of oxaloacetic acid has been studied at high copper(II) concentrations (Cu(II): ligand 100:1). Inhibition occurs near and above pH 3.5, but not at lower pH. It therefore appears that inhibition arises from the formation of the $Cu_2A^+_{enolic}$, rather than the $Cu_2A^+_{ketonic}$ complex. The rate constant for the decarboxylation of CuA is 0.8 sec^{-1} at 25°C and the formation constant of the complex is ca. 5000 M^{-1} [42]. The fate of CuP_e formed in the decarboxylation of oxaloacetate depends on the experimental conditions and both ketonization and condensation reactions have been observed [42]. The ketonization of CuP_e exhibits specific and general acid catalysis.

10. OXIDATIVE DECARBOXYLATIONS AND MISCELLANEOUS DECARBOXYLATIONS

If the metal ion used as a catalyst can also act as an oxidizing agent oxidative decarboxylation can occur [97]. For example, citrate in the presence of manganese(II) in basic solution gives 3-oxoglutarate [eq. (48)]. A large number of other α-hydroxypoly-

$$^{-}O\text{-}C(CH_2CO_2^-)_2\text{-}CO_2^- \xrightarrow[\text{Air, pH} \sim 11]{Mn^{(II)}} OC(CH_2CO_2^-)_2 + CO_2 + 2e \qquad (48)$$

carboxylic acids have been studied and the formation of oxo acids has been established as a fairly general reaction [eq. (49)]. A

$$R\text{-}C(CO_2^-)_2\text{-}O^- + \tfrac{1}{2}O_2 \xrightarrow[\text{Air, pH} \sim 10\text{-}12]{Mn^{(II)},\ Co^{(II)},\ \text{or } Cu^{(II)}} R\text{-}C(=O)\text{-}CO_2^- + CO_3^{2-} \qquad (49)$$

number of studies have been made on the metal ion-catalyzed decarboxylation of histidine [98-100], and the evidence for the involvement of a metal ion in the enzymic decarboxylation of histidine has been reviewed [101]. Andrews and Grundemeier [100] have studied the catalytic effect of Cu(II), Cr(III), Ni(II), Co(II), Al(III), Cd(II), Zn(II), and Pd(II) on the decarboxylation of histidine in ethyleneglycol as solvent at 137.2°C. For the uncatalyzed reaction, $k = 0.65 \times 10^{-6}\ sec^{-1}$, while for a 1:1 ratio of copper(II) to histidine, $k = 8.15 \times 10^{-6}\ sec^{-1}$. [See reaction (50).] The mechanistic details of this reaction are unclear, however, it appears that the -I effect (inductive withdrawal) exerted by the metal ion at the α-amino group is able to stabilize the carbanion developing

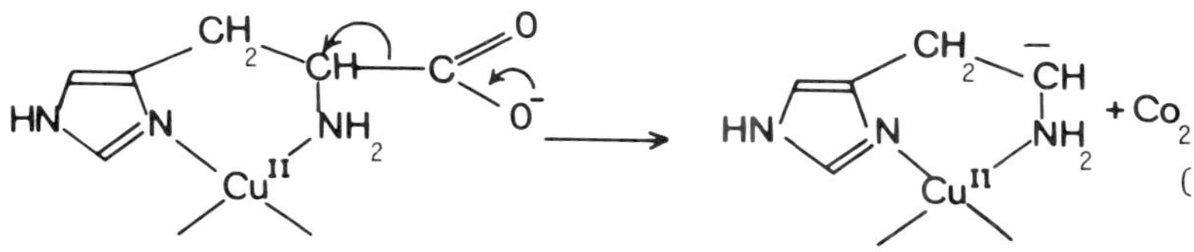

(50)

in the transition state of the decarboxylation. In addition to decarboxylation concomitant oxidation of the imidazole ring also occurs in the reaction.

An interesting metal chelate-catalyzed stereoselective decarboxylation has been reported by Asperger and Liu [102]. The L-cis-β isomer of the α,α'-dimethyltriethylenetetraminecobalt(III) ion reacts with DL-α-methyl-α-aminomalonate to give the mixed ligand diastereoisomeric chelate (35) [reaction (51)]. Decarboxylation

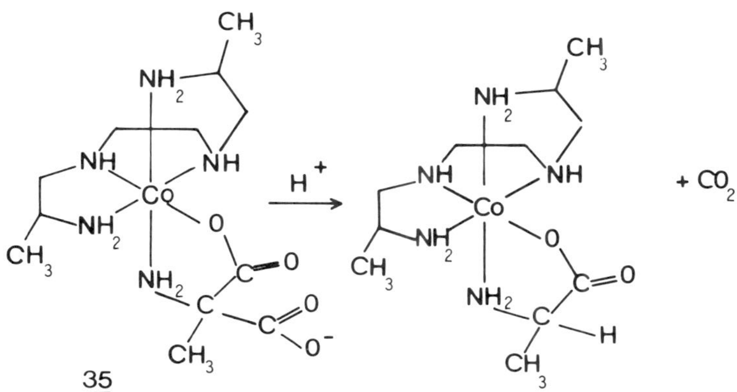

(51)

occurs to give a mixed ligand chelate of α-alanine in which the L-alanine enantiomer is enriched by ca. 25 to 30%. It is not known with certainty whether the stereoselectivity occurs in the formation of the mixed ligand complex or in the decarboxylation step, although the former seems more likely. This reaction is an interesting example of the type of stereoselectivity which occurs in any metal chelate-catalyzed reaction in which an optically active complex reacts with a racemic substrate.

Pyridoxal phosphate, vitamin B_6 (36) is involved in many

transformations of amino acids, for example, racemization, decarboxylation, transamination, β-substitution, and elimination [103]. Decarboxylation of an amino acid in the presence of pyridoxal and a metal ion is illustrated in Scheme 7.

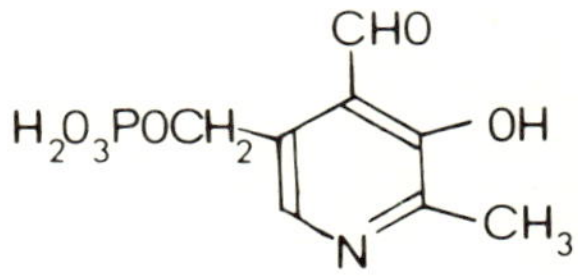

36

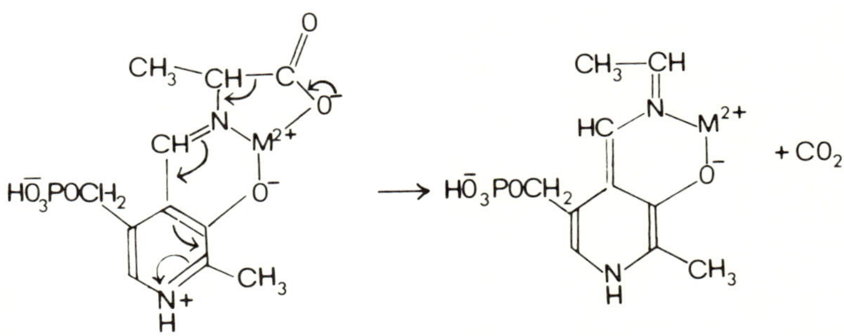

Scheme 7. Coenzyme reactions of pyridoxal-decarboxylation.

11. CONCLUSIONS

The model studies on the metal ion-catalyzed decarboxylation of β-oxo acids have been found to be relevant to the metal activated enzymic decarboxylation of oxaloacetic acid.

In spite of many kinetic studies the actual reactivities of the various metal complexes of β-oxo acid ligands are not known with precision, such data are required to account for the enzymic rate accelerations and the metal ion specificity of these reactions. Studies are also required of the ketonization rates of the enol intermediates in the metal ion-catalyzed reactions, since these reactions are quite slow, and are apparently rate-controlling in the enzymic systems.

The area of mixed ligand complexes in β-oxo acid decarboxylations is a promising subject for further detailed study.

ACKNOWLEDGMENT

I wish to thank Dr. K. N. Leong for his interest and assistance in the preparation of this review.

ADDENDUM

A number of detailed studies of the effect of metal ions (copper and zinc) on the rates of enolization and decarboxylation of oxaloacetate have appeared [104-106]. Covey and Leussing [104] have studied keto-enol tautomerism rates for oxaloacetate and zinc(II) oxaloacetate complexes in acetate buffer solutions at 25°C. Thus for the equilibrium $\text{oxac}^{2-}_{\text{keto}} \underset{k_b}{\overset{k_f}{\rightleftharpoons}} \text{oxac}^{2-}_{\text{enol}}$ the value of k_f for the reaction $\text{oxac}^{2-}_{\text{keto}} + \text{OAc}^-$ is $6.6 \times 10^{-3}\ M^{-1}\ \text{sec}^{-1}$, while for $\text{Zn(oxac)}_{\text{keto}} + \text{OAc}^-$, $k_f = 25\ M^{-1}\ \text{sec}^{-1}$. The rate acceleration at 25°C is ca. 4×10^3. Cox [107] has recently shown that a rate acceleration of 5×10^3 occurs in the zinc(II)-promoted acetate-catalyzed enolization of 2-acetylpyridine. Detailed studies of the zinc(II)-catalyzed decarboxylation of oxaloacetate have also been reported [105]. Two relaxations were observed spectrophotometrically. An initial absorbance increase being complete in about 30 sec and a subsequent absorbance decrease being complete after 15 to 30 min. The first relaxation arises from metal ion-promoted keto-enol tautomerization, and the second from catalyzed decarboxylation. The subsequent protonation of the zinc pyruvate enolate intermediate is rapid and unobservable. The rate constant for the decarboxylation of $\text{Zn(oxac)}_{\text{keto}}$ is $7.42 \times 10^{-3}\ \text{sec}^{-1}$ at 25°C and I = 0.1 M. The reaction scheme for the copper(II) system can be summarized [105] as follows:

$$\begin{array}{l} \text{Cu(oxac)}_{\text{keto}} \xrightarrow[k_{-CO_2}]{-CO_2} \text{Cu(pyruvate)}_{\text{enolate}} \xrightarrow[k_H]{H^+} \text{Cu(pyruvate)} \\ k_f \downarrow \uparrow k_b \\ \text{Cu(oxac)}_{\text{enol}} \end{array}$$

The release of CO_2 is biphasic as is also the rate of H^+ uptake in the step where pyruvate is formed. Only about 20% of the oxac^{2-}

in the system undergoes decarboxylation within the first 30 sec. The remaining 80% is converted to Cu(oxac)$_{enol}$ which does not decarboxylate. The slow rate of CO_2 evolution arises from the decarboxylation of a residual level of Cu(oxac)$_{keto}$ which is replenished by the reketonization of Cu(oxac)$_{enol}$. For the decarboxylation Cu(oxac)$_{keto}$ $\rightarrow$ CO_2 + Cu(pyruvate)$_{enolate}$, $k = 0.17\ sec^{-1}$ at 25°C and I = 0.1 M so that the copper(II) complex decarboxylates some 23 times faster than the zinc complex. The kinetics and mechanism of the copper(II)-promoted oxidative decarboxylation of aminomalonic acid in aqueous solution has been studied [108].

REFERENCES

1. B. R. Brown, *Quart. Rev., 5,* 131 (1951)
2. W. H. Richardson and H. E. O'Neal, in "Comprehensive Chemical Kinetics," Vol. 5 (C. H. Bamford and C. F. H. Tipper, eds.), pp. 456ff.
3. For review articles on acetoacetic decarboxylase see: I. Fridovich, in "The Enzymes," Vol. 6 (P. D. Boyer, ed.), Academic Press, New York and London 1972, pp. 255ff; F. H. Westheimer, IUPAC 23rd Intern. Congr. Pure Appl. Chem. (Boston), Butterworths, London, Vol. 5, 1971; F. H. Westheimer, *Proc. Chem. Soc.,* 253 (1963).
4. S. Warren, B. Zerner, and F. H. Westheimer, *Biochemistry, 5,* 817 (1966); R. A. Laursen and F. H. Westheimer, *J. Amer. Chem. Soc., 88,* 3426 (1966).
5. G. W. Kosicki and F. H. Westheimer, *Biochemistry, 7,* 4303 (1968).
6. E. Grazi, T. Cheng, and B. L. Horecker, *Biochem. Biophys. Res. Commun., 7,* 250 (1962).
7. R. W. Hay, *Aust. J. Chem., 18,* 337 (1965).
8. J. P. Guthrie and F. H. Westheimer, *Fed. Proc., 26,* 562 (1967).
9. E. Gelles and J. P. Clayton, *Trans. Faraday Soc., 52,* 353 (1956).
10. K. J. Pedersen, *J. Amer. Chem. Soc., 51,* 2098 (1929)
11. K. J. Pedersen, *J. Amer. Chem. Soc., 58,* 240 (1936).
12. R. Steinberger and F. H. Westheimer, *J. Amer. Chem. Soc., 73,* 439 (1951).
13. E. Gelles and R. W. Hay, *J. Chem. Soc.,* 3673 (1958).
14. R. J. Dummel, M. N. Berry, and E. Kun, *Mol. Pharmacol., 7,* 367 (1971).

15. F. H. Westheimer and W. A. Jones, *J. Amer. Chem. Soc., 63,* 3283 (1941).

16. See for example T. S. Straub, and M. L. Bender, *J. Amer. Chem. Soc., 94,* 8881 (1972).

17. C. G. Swain, R. F. W. Bader, R. M. Esteve, and R. N. Griffin, *J. Amer. Chem. Soc., 83,* 1951 (1961).

18. J. E. Prue, *J. Chem. Soc.,* 2331 (1952).

19. M. H. O'Leary and R. L. Baughn, *J. Amer. Chem. Soc., 94,* 626 (1972).

20. G. W. Kosicki, S. N. Lipovac, and R. G. Arnett, *Can. J. Chem., 42,* 2806 (1964).

21. J. M. Los, R. F. Rekker, and C. H. T. Tonsbeek, *Rec. Trav. Chim., 86,* 622 (1967).

22. J. L. Longridge and F. A. Long, *J. Amer. Chem. Soc., 90,* 3092 (1968).

23. E. Gelles, *J. Chem. Soc.,* 4736 (1956).

24. K. J. Pedersen, *Acta Chem. Scand., 6,* 285 (1952).

25. G. W. Kosicki and S. N. Lipovac, *Can. J. Chem., 42,* 403 (1964).

26. E. Bamann and V. S. Sethi, *Arch. Pharm., 301,* 12 (1968).

27. K. J. Pedersen, *Acta Chem. Scand., 6,* 243 (1962).

28. H. A. Krebs, *Biochem. J., 36,* 303 (1942).

29. J. F. Speck, *J. Biol. Chem., 178,* 315 (1949).

30. P. M. Nossal, *Australian J. Exptl. Biol. Med. Sci., 27,* 143 (1949); *27,* 312 (1949).

31. E. Gelles and A. Salama, *J. Chem. Soc.,* 3683 (1958).

32. E. Gelles and A. Salama, *J. Chem. Soc.,* 3689 (1958).

33. A. Kornberg, S. Ochoa, and A. Mehler, *J. Biol. Chem., 174,* 159 (1948).

34. E. Bamann and V. S. Sethi, *Arch. Pharm., 301,* 78 (1968).

35. L. O. Krampitz and C. H. Werkman, *Biochem. J., 35,* 595 (1941).

36. P. R. Bontchev and V. Michaylova, *J. Inorg. Nucl. Chem., 29,* 2945 (1967).

37. Y. Yamane, M. Miyazaki, N. Yamaji, and T. Ogashiwa, *Chem. Pharm. Bull. (Tokyo), 19,* 2342 (1971).

38. M. Munakata, M. Matsui, M. Tabuschi, and T. Shigematsu, *Bull. Chem. Soc. (Japan), 43,* 114 (1970).

39. D. Herbert, in Methods in Enzymology, Vol. 1 (S. P. Colowick and N. O. Kaplan, eds.), Academic, New York, 1955, p. 753.

40. S. S. Tate, A. K. Grzybowski, and S. P. Datta, *J. Chem. Soc.,* 1381 (1964).

41. C. Reyes-Zamora and C. S. Tsai, *Chem. Commun.*, 1047 (1971).

42. K. N. Leong and D. L. Leussing, *Proc. 14th Intern. Conf. Coordination Chemistry, Toronto, 1972,* p. 608.

43. C. S. Tsai, *Can. J. Chem., 45,* 873 (1967).

44. B. E. Banks, *J. Chem. Soc.,* 5043 (1961).

45. B. E. Banks, *J. Chem. Soc.,* 63 (1962).

46. S. S. Tate, A. K. Grzybowski, and S. P. Datta, *J. Chem. Soc.,* 1372 (1964).

47. C. S. Tsai, Y. T. Lim, and E. E. Sharkawi, *J. Org. Chem., 37,* 85 (1972).

48. G. S. Hammond, *J. Amer. Chem. Soc., 77,* 334 (1955).

49. J. V. Rund and R. A. Plane, *J. Amer. Chem. Soc., 86,* 367 (1964).

50. J. V. Rund and K. G. Claus, *Inorg. Chem., 7,* 860 (1968).

51. K. G. Claus and J. V. Rund, *Inorg. Chem., 8,* 59 (1969).

52. R. W. Hay and K. N. Leong, *J. Chem. Soc.* (A), 3639 (1971).

53. K. N. Leong and M. W. Lister, *Can. J. Chem., 50,* 1461 (1972).

54. K. N. Leong, Ph.D. Thesis, University of Toronto, 1971.

55. E. O. Wiig, *J. Phys. Chem., 32,* 961 (1928); *34,* 596 (1930).

56. D. W. Larsen and M. W. Lister, *Can. J. Chem., 46,* 823 (1968).

57. O. Toyama, Y. Kubokawa, and Y. Yoshida, *J. Chem. Soc. Japan,* (Pure Chem. Section), *74,* 955 (1953); *75,* 879 (1954).

58. F. W. Yerhoff and D. W. Larsen, *Can. J. Chem., 50,* 826 (1972).

59. R. W. Hay and M. A. Bond, *Aust. J. Chem., 20,* 1823 (1967).

60. R. W. Hay and K. R. Tate, *Aust. J. Chem., 23,* 1407 (1970).

61. E. F. Hartree, *J. Amer. Chem. Soc., 75,* 6244 (1953).

62. M. P. Gupta, *Can. J. Chem., 33,* 450 (1955).

63. R. W. Hay and S. J. Harvie, *Aust. J. Chem., 18,* 1197 (1965).

64. M. Franke and G. Brathun, *Liebigs. Ann., 487,* 1 (1931).

65. F. Dickens and D. H. Williamson, *Biochem. J., 68,* 74 (1958).

66. C. T. Chow and B. Vennesland, *J. Biol. Chem., 233,* 997 (1958).

67. K. J. Pedersen, *Acta Chem. Scand., 9,* 1640 (1955).

68. K. J. Pedersen, *Acta Chem. Scand., 12,* 919 (1958).

69. K. J. Pedersen, *Acta Chem. Scand., 3,* 676 (1949), and references cited therein.

70. M. Stiles and H. L. Finkbeiner, *J. Amer. Chem. Soc., 81,* 505 (1959).

71. H. L. Finkbeiner and M. Stiles, *J. Amer. Chem. Soc.*, *85*, 616 (1963).

72. M. Stiles, *J. Amer. Chem. Soc.*, *81*, 2598 (1959); *Ann. N.Y. Acad. Sci.*, *88*, Art. 2, 332 (1960).

73. H. L. Finkbeiner, *J. Amer. Chem. Soc.*, *86*, 961 (1964).

74. F. Lynen, *Biochem. J.*, *102*, 381 (1967).

75. J. E. Coleman, in "Progress in Bioorganic Chemistry," Vol. 1, (E. T. Kaiser and F. J. Kezdy, eds.), Wiley-Interscience, New York, 1971, pp. 246ff.

76. A. S. Mildvan, M. C. Scrutton, and M. F. Utter, *J. Biol. Chem.*, *241*, 3488 (1966).

77. A. S. Mildvan and M. C. Scrutton, *Biochemistry*, *6*, 2978 (1967).

78. R. K. Cannan and A. Kibrick, *J. Amer. Chem. Soc.*, *60*, 2314 (1938).

79. L. G. Sillen and A. E. Martell, "Stability Constants of Metal Ion Complexes," Special Publication, No. 17, The Chemical Society, London, 1964.

80. H. Sigel and D. B. McCormick, *Accounts Chem. Res.*, *3*, 201 (1970).

81. A. Wood, *Trans. Faraday Soc.*, *60*, 1263 (1964).

82. A. Wood, *Trans. Faraday Soc.*, *62*, 1231 (1966).

83. S. Seltzer, G. A. Hamilton, and F. H. Westheimer, *J. Amer. Chem. Soc.*, *81*, 4018 (1959).

84. J. Bigeleisen and M. Wolfberg, *Advan. Chem. Phys.*, *1*, 15 (1958).

85. K. S. Pitzer and E. Gelles, *J. Amer. Chem. Soc.*, *75*, 5132 (1953).

86. E. Gelles and K. S. Pitzer, *J. Amer. Chem. Soc.*, *77*, 1974 (1955).

87. E. Gelles and R. I. Reed, *Nature*, *176*, 1262 (1955).

88. G. A. Hamilton and F. H. Westheimer, *J. Amer. Chem. Soc.*, *81*, 6332 (1959).

89. I. Fridovich and F. H. Westheimer, *J. Amer. Chem. Soc.*, *84*, 3208 (1962).

90. W. Tagaki and F. H. Westheimer, *Biochemistry*, *7*, 891 (1968).

91. A. Schmitt, I. Bottke, and G. Siebert, *Z. Physiol. Chem.*, *347*, 18 (1966).

92. S. Ochoa and E. Weisz-Tabori, *J. Biol. Chem.*, *174*, 128 (1948); J. B. Veiga Salles, and S. Ochoa, *J. Biol. Chem.*, *187*, 849 (1950).

93. G. W. Kosicki, *Biochemistry*, *7*, 4299 (1968).

94. L. Wolff, *Ann. Chem., 305,* 154 (1899).

95. A. W. K. De Jong, *Rec. Trav. Chim., 20,* 81 (1901).

96. D. E. Tallman and D. L. Leussing, *J. Amer. Chem. Soc., 91,* 6253, 6256 (1969).

97. S. P. Datta, A. K. Grzybowski, and S. S. Tate, *Nature, 207,* 1047 (1965).

98. A. Goudat, *C. R. Acad. Sci. Paris, 240,* 975 (1955).

99. J. Hearon, *Fed. Proc., 6,* 260 (1957).

100. A. C. Andrews and E. W. Grundemeir, *J. Inorg. Nucl. Chem., 28,* 455 (1966).

101. P. Boyer, H. Lardy, and K. Myrback, "The Enzymes," Vol. 2, (2nd ed.), Academic Press, New York, 1960, p. 115.

102. R. G. Asperger and C. F. Liu, *Inorg. Chem., 6,* 796 (1967).

103. E. E. Snell, A. E. Braunstein, E. S. Severin, and Yu. M. Torchinsky (eds.), "Pyridoxal Catalysis: Enzymes and Model Systems," Interscience, New York, 1968.

104. W. D. Covey and D. L. Leussing, *J. Chem. Soc. (Chem. Commun.), 31,* 1974.

105. W. D. Covey and D. L. Leussing, *J. Amer. Chem. Soc., 96,* 3860 (1974).

106. N. W. Raghavan and D. L. Leussing, *J. Amer. Chem. Soc., 96,* 7147 (1974).

107. B. G. Cox, *J. Amer. Chem. Soc., 96,* 6823 (1974).

108. J. H. Fitzpatrick and D. Hopgood, *Inorg. Chem., 13,* 568 (1974).

Chapter 4

METAL ION-PROMOTED HYDROLYSIS OF AMINO ACID ESTERS AND PEPTIDES

R. W. Hay and P. J. Morris
Department of Chemistry, University of Stirling,
Stirling, Scotland
and
Department of Chemistry, University of Waikato,
Hamilton, New Zealand

1. INTRODUCTION

The metal ion-catalyzed (or, more correctly, promoted) hydrolysis of amino acid esters and peptides has been a subject of continuing interest over the past two decades [1]. In the early 1930s the hydrolysis of peptides was found to be subject to metal ion catalysis, but the discovery by Kroll [2] in 1952 that the hydrolysis of α-amino acid esters was catalyzed by metal ions stimulated considerable interest in the area. Many of these reactions provide simple models for the much more complex metalloenzymes such as carboxypeptidase A, leucine aminopeptidase, and glycylglycine dipeptidase [3].

A major problem in studying these reactions has been the determination of the actual binding sites of the ligand to the metal ion, and the identification of the catalytically active species in solution. Many of the early kinetic studies were carried out with monoamino esters where the formation constants are quite low. Solutions containing, for example, copper(II) and the ester ligand (E), will contain a large variety of complexes such as CuE^{2+}, CuE_2^{2+}, CuE_3^{2+}, and CuE_4^{2+}. The most abundant species will depend on the copper (II): ester ratio, and inevitably there is more than one complex present. In addition there will be mixed species $CuEA^+$, CuE_2A^+, etc. formed as the hydrolysis of E to A^- (the anion of the amino acid) occurs. It is therefore extremely difficult to unequivocally establish the nature of the hydrolytically most important species. A number of these problems can be overcome by using polydentate amino acid ester derivatives containing additional functional groups capable of interacting with a metal ion, and considerable success has been obtained in interpreting the kinetic data in such systems (Sections 3.3 and 3.4).

Many of the initial studies were carried out using buffer systems which led to additional problems due to metal ion-buffer interactions. The advent of pH-stats has solved this problem, although the lack of a suitable technique for monitoring peptide hydrolysis (other than the use of automatic amino acid analyzers) has hindered developments in the peptide field.

The most important recent developments in the area have come from the use of inert cobalt(III) systems and it has been possible to resolve many of the mechanistic problems which have arisen from studies of the kinetically labile systems (Section 4). Amino acid esters can be monodentate or bidentate, and examples of both types of complex have been isolated and their solid state infrared spectra studied [4-6].

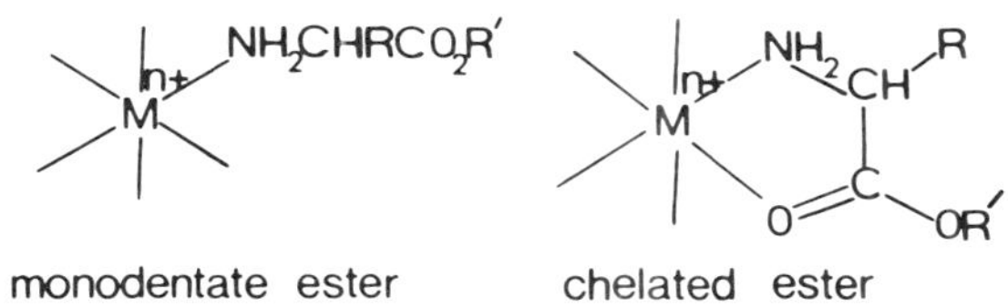

Formation of the chelated ester species leads to a considerable reduction in the carbonyl stretching frequency, e.g., from 1740 to 1600 cm^{-1} in the case of copper(II), indicating significant polarization of the carbonyl bond by the metal ion. Kroll [2] initially suggested that such chelated species were involved in the metal ion-catalyzed hydrolysis of glycine methyl ester (Scheme 1). A scheme of this type would be similar to the classical $A_{AC}2$ mechanism [7] for the acid-catalyzed hydrolysis of a carboxylic ester (Scheme 2). Such reactions involve the formation of a tetrahedral intermediate. The tetrahedral intermediate has not been observed directly but has been inferred from the results of $^{18}_{6}O$ experiments. The ester labeled with $^{18}_{6}O$ in the carbonyl group undergoes isotopic replacement by $H_2{}^{16}O$ in the presence of H^+ [8]. The base hydrolysis ($B_{AC}2$) of carboxylic esters involves a similar type of tetrahedral intermediate rather than an S_N2 displacement process [reaction (1)].

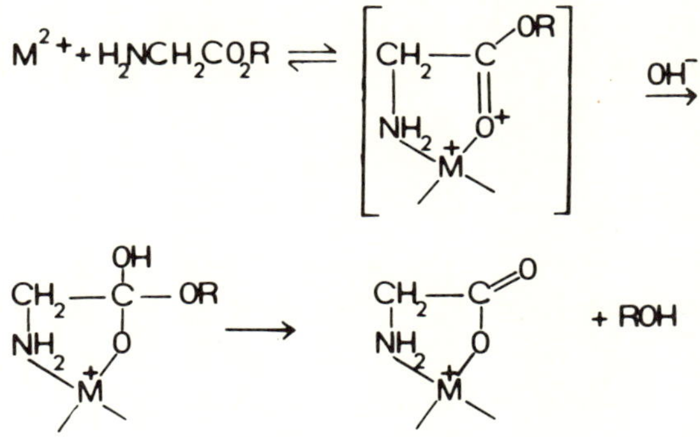

Scheme 1. A possible mechanism for the metal ion-catalyzed hydrolysis of glycine esters.

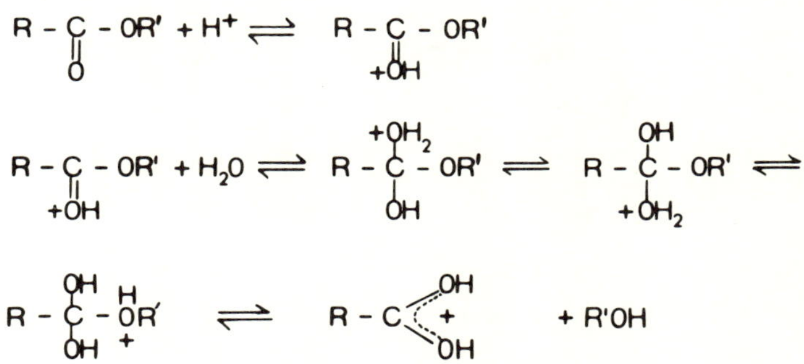

Scheme 2. The $A_{AC}2$ mechanism for the acid-catalyzed hydrolyses of a carboxylic ester.

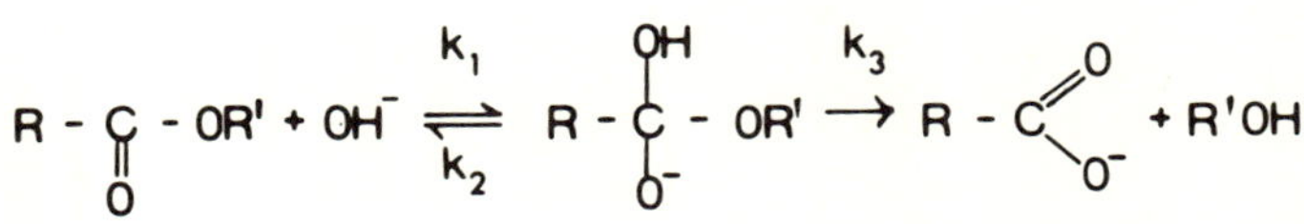

(1)

Bender and Turnquest [9] established that carbonyl oxygen exchange occurred during the copper(II)-catalyzed hydrolysis of DL-phenylalanine ethyl ester-carbonyl-^{18}O at pH 7.3 in glycine buffers. On the basis of the kinetic and oxygen-exchange evidence a mechanism was suggested involving the transient formation of a chelated ester species which facilitated nucleophilic attack at the carbonyl carbon atom by a water molecule (Scheme 3). An alternative mechanism may involve intramolecular attack by OH^- bound to copper(II)

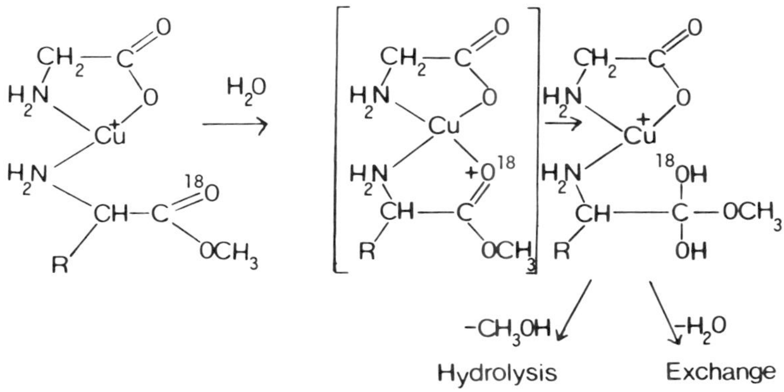

Scheme 3. Metal ion-catalyzed hydrolysis and carbonyl oxygen exchange.

[reaction (2)]. Such a possibility was strengthened by the observation of Gurd et. al. [10] that hydroxy complexes of copper(II)-glycylglycine catalyzed the hydrolysis of a p-nitrophenyl acetate.

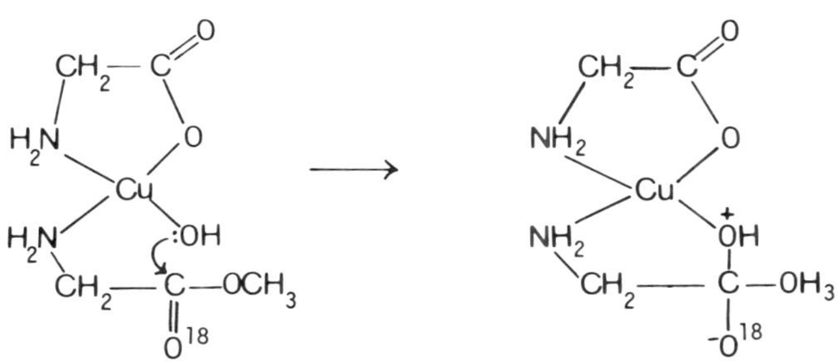

(2)

In addition, the nature of the nucleophile (H_2O or OH^-) was not clearly established in many of the early studies.

In the peptide field the early studies of Datta and Rabin [11] did much to clarify the nature of the species formed in solution between simple dipeptides and metal ions. The deprotonation of peptide nitrogen atoms has been found to occur in the copper(II) complexes of peptides such as glycylglycine, sarcosylglycine, L-prolylglycine, glycyl-L-valine, and L-valylglycine [10]. At high pH similar observations have been made for the nickel complexes of glycinamide, triglycine, and tetraglycine [12].

The structure of the 1:1 complex of copper(II) and glycylglycine may be represented (1) where the initial binding sites are the terminal amino group, the carboxyl group, and the carbonyl oxygen

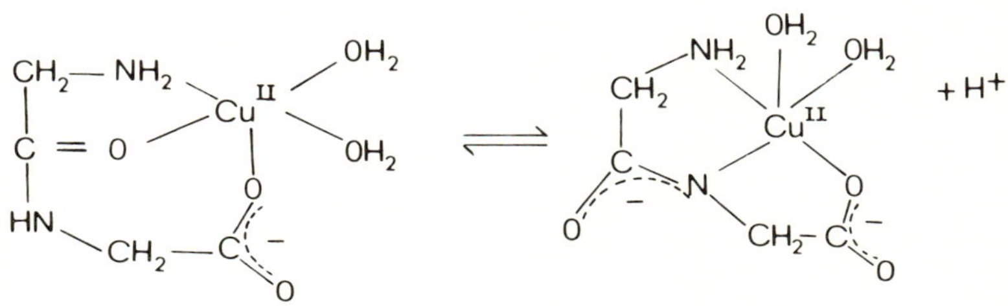

(3)

1 2

atom of the peptide bond. As the pH of the solution is raised (>5) the single peptide nitrogen is deprotonated to give 2 in which the terminal amino group, peptide nitrogen atom, and carboxyl group act as ligand donors. The carbonyl-bonded species (1) would be expected to be susceptible to nucleophilic attack by hydroxide ion, while the deprotonated species (2) should be inert to hydroxide ion. These conclusions have been fully verified in both nonlabile and labile systems.

2. BASE HYDROLYSIS OF AMINO ACID ESTERS

In order to provide information on the magnitude of the catalytic effect of metal ions on amino acid ester hydrolysis it is necessary to have data on the base hydrolysis of amino acid esters.

The reaction scheme for the base hydrolysis of an amino acid ester can be summarized in equations (4)-(7).

$$\underset{(EH^+)}{H_3\overset{+}{N}CHRCO_2R'} + H_2O \underset{}{\overset{K_E}{\rightleftharpoons}} \underset{(E)}{H_2NCHRCO_2R'} + H_3O^+ \quad (4)$$

$$H_2NCHRCO_2R' + OH^- \xrightarrow{k_E} H_2NCHRCO_2^- + R'OH \tag{5}$$

$$H_3\overset{+}{N}CHRCO_2R' + OH^- \xrightarrow{k_{EH^+}} H_3NCHRCO_2^- + R'OH \tag{6}$$

$$H_3NCHRCO_2^- + H_2O \underset{}{\overset{k_A}{\rightleftharpoons}} H_2NCHRCO_2^- + H_3O^+ \tag{7}$$

2.1. Proton Ionization Constants

For most α-amino acid methyl and ethyl esters [13-30] the values of pK_E fall within the range 7 to 7.6 (Table 1) so that at physiological pH values significant amounts of both the protonated and unprotonated ester are present. The pK_E values are usually some 2 pK units lower than the corresponding pK_A values for the amino acids themselves. A result which reflects the larger -I effect of the $-CO_2CH_3$ (or $-CO_2C_2H_5$) group compared with $-CO_2^-$. Stepwise removal of the amino group from the α to the β and γ positions leads to a steady rise in the value of pK_E, while the introduction of a second -I group (NH_2, OH, SH, etc.) lowers the pK_E value.

The temperature dependence of the pK_E values has been studied and the resulting thermodynamic parameters discussed [13,14,19] (Table 2).

2.2. Hydrolysis Kinetics

The base hydrolysis of an amino acid ester follows the rate expression:

$$-d[\text{total ester}]/dt = k_E[E][OH^-] + k_{EH^+}[EH^+][OH^-] \tag{8}$$

where E and EH^+ represent the protonated and unprotonated forms of the ester. The hydrolysis is best monitored at constant pH using a pH-stat. Quite precise values of k_E are available for a large number of amino acid esters [13-15] (Table 3). Rate constants have also been reported for a number of N-acyl amino acid esters [31].

TABLE 1

Thermodynamic Proton Ionization Constants, $(pK^{H})^{T}$, of Amino Groups of Amino Acids and Esters

(In water, at 25°C and 0.1 M (KCl) ionic strength) [13-30]

Monoamino	Acid	Ester	Ref.[a]
Glycine	9.60	7.66 (methyl)	13 (15)
		7.69 (ethyl)	13 (15,16,17)
α-Alanine	9.916	7.743 (methyl)	14 (13)
α-Amino-n-butyric	9.852	7.640 (methyl)	14
Norvaline	9.869	7.662 (methyl)	14
Norleucine	9.879	7.682 (methyl)	14
Valine	9.736	7.483 (methyl)	14 (13)
Leucine	9.60	7.62 (methyl)	14 (13,15)
		7.66 (ethyl)	13 (15)
Isoleucine	9.785	7.542 (methyl)	14 (13)
β-Alanine	10.333	9.170 (methyl)	14
4-Amino-n-butyric	10.565	9.839 (methyl)	14
Serine	9.260	7.030 (methyl)	14 (13,15)
β-Phenylalanine	9.13	7.05 (methyl)	14 (13)
		7.12 (ethyl)	14 (15)
Methionine	--	7.09 (methyl)	14
Trytophan	--	7.29 (methyl)	14
S-Methylcysteine	--	6.70 (methyl)	14
Tyrosine	--	7.09 (methyl)	14
Glutamic	9.67	7.03 (dimethyl)	14 (15)

TABLE 1 (continued)

Monoamino	Acid		Methyl ester		Ref.[a]
	α-$\overset{+}{N}H_3$	ω-$\overset{+}{N}H_3$	α-$\overset{+}{N}H_3$	ω-$\overset{+}{N}H_3$	
2,3-Diamino-propionic	6.674	9.623	4.412	8.250	18 (19,22-25)
2,4-Diamino-butyric	8.169	10.424	5.868	8.655	19 (24)
Ornithine	8.739	10.770	6.501	8.323	19
Lysine	9.116	10.902	6.964	10.251	19
Histidine	9.313	6.024 (ImH^+)	7.230	5.007 (ImH^+)	20 (15,26-30)
Cysteine	8.5[b]	10.5[b]	6.52[b]	9.15[b]	21 (15)

[a]The value quoted is from the first reference given, additional references are given in parentheses.
[b]Ionization of SH and $\overset{+}{N}H_3$ groups occurs simultaneously.

Methyl esters undergo base hydrolysis at roughly twice the rate of ethyl esters and methyl glycinate undergoes the most rapid base hydrolysis. The rate constants k_E can be correlated with Taft's polar substituent constant σ^*. The methyl esters may be considered to be derivatives of methyl α-alaninate, $RCH(NH_2)CO_2CH_3$ (R = CH_3 for methyl α-alaninate.) If the relative reactivities are mainly controlled by polar effects, a plot of $\log(k_E/k_E^o)$ vs σ^*, where k_E^o is the second-order rate constant for the hydrolysis of the α-alaninate ester, would be expected to be linear. A reasonable relationship of this type is observed [14] with the slope $\rho^* = +0.65$ (Fig. 1). The small substituent constant reflects the insensitivity of the reaction rate to the nature of R, the dominant polar effect being exerted by the α-amino group. Specific steric effects occur with the methyl esters of valine and isoleucine where chain branching at the β position occurs. The rate constants k_{EH^+} cannot

TABLE 2

Temperature Dependence of Amino Group $(pK^H)^T$ Values for Amino Acid Esters

(In water at 0.1 M (KCl) ionic strength; ΔH° in kcal $mole^{-1}$ and ΔS° in cal deg^{-1} $mole^{-1}$)

	25°C	37°C	50°C	ΔH°	ΔS°_{298}	Ref.
Monoamino ester						
Methyl glycinate	7.66	7.33	7.00	11.50	3.5	13
Ethyl glycinate	7.69	7.40	7.09	10.60	0.4	13
Methyl α-alaninate	7.743	7.421	7.091	11.50	3.2	14
Methyl α-amino-n-butyrate	7.640	7.327	6.989	11.45	3.6	14
Methyl valinate	7.483	7.16	6.85	11.29	3.6	13
Ethyl leucinate	7.66	7.35	7.06	10.46	0.0	13
Methyl β-alaninate	9.170	8.804	8.454	12.62	0.4	14
Methyl 4-amino-n-butyrate	9.839	9.461	9.091	13.19	-0.8	14
Methyl serinate	7.030	6.724	6.412	10.90	4.4	14
Methyl β-phenyl alaninate	7.05	6.73	6.41	11.11	5.0	13
Methyl S-methyl cysteinate	6.70	6.39	6.07	10.99	6.2	13

Diamino ester						
Methyl 2,3-diaminopropionate (α-$\overset{+}{N}H_3$)	4.412	4.176	3.897	9.08	10.3	18 (19)[b]
(β-$\overset{+}{N}H_3$)	8.250	7.959	7.632	10.90	-1.2	18 (19)[b]
Methyl lysinate (α-$\overset{+}{N}H_3$)	6.964	6.658	6.304	11.64	7.2	19
(ε-$\overset{+}{N}H_3$)	10.251	9.898	9.520	12.89	-3.7	19
Methyl histidinate (ImH^+)	5.01	4.78	4.54	8.2	4.6	20
(α-$\overset{+}{N}H_3$)	7.23	6.96	6.67	9.9	0.3	20
Methyl cysteinate[a] pK_1	6.52	6.265	6.00	9.15	0.8	21
pK_2	9.15	8.995	8.85	5.28	-24	21

[a]Ionization of SH and $\overset{+}{N}H_3$ groups occurs simultaneously, so the pK's are composite values.
[b]Additional reference in parentheses.

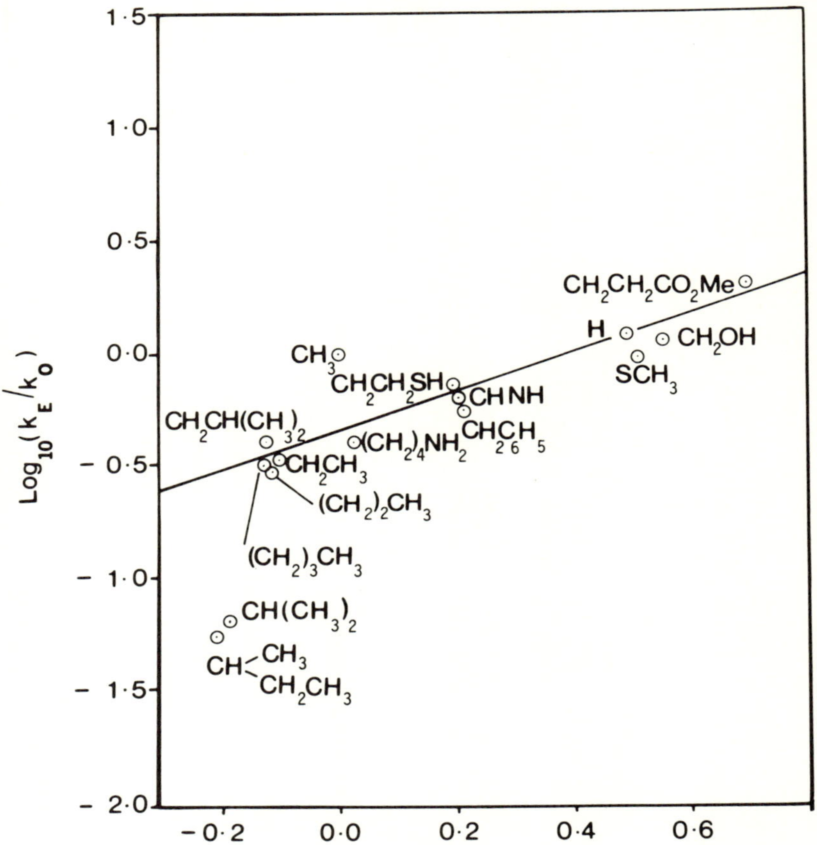

FIG. 1. Taft plot for the base hydrolysis of α-amino acid esters $RCH(NH_2)COOCH_3$ in water at 25°C and I = 0.1 M (KCl). The value of R is given for each point.

be determined with high precision by pH-stat techniques; however, a number of rate constants have been reported (Table 3). The effect of charge on the rate of base hydrolysis of carboxylic esters has been extensively studied [32,34,42-47]. Generally, the hydrolysis rate of the 1+ charged form of an ester is ca. 100 to 1000 times that of the neutral form. Values of k_{EH^+}/k_E are ca. 100 for α-amino acid esters. Withdrawal of the amino group to the β position

TABLE 3

Typical Rate Constants for the Base Hydrolysis of Amino Acid Esters in Aqueous Solution at 25°C and I = 0.1 M (KCl)[a]

Amino ester	k_E (M^{-1} min^{-1})	k_{EH^+} (M^{-1} min^{-1})	Ref.
Methyl glycinate	77	1700	13 (15)
Ethyl glycinate	38.1	1376	15 (16,17,32-37)
Methyl α-alaninate	66.4	4820	14 (13)
Methyl α-amino-n-butyrate	23.4	2670	14
Methyl norvalinate	24.1	2410	14
Methyl norleucinate	22.1	2450	14
Methyl valinate	4.57	--	14 (13)
Methyl leucinate	27.3	--	15 (13)
Ethyl leucinate	11.2	--	15
Methyl isoleucinate	4	--	13
Methyl β-alaninate	8.14	412	14
Methyl serinate	59.3	--	14 (13,15)
Methyl β-phenyl alaninate	33	--	13 (14)
Ethyl β-phenyl alaninate	14.1	--	15
Methyl methioninate	46	--	13
Methyl tryptophanate	17.5	--	13

TABLE 3 (Continued)

Amino ester	k_E (M^{-1} min^{-1})	k_{EH^+} (M^{-1} min^{-1})		Ref.
Methyl S-methyl cysteinate	64	--		13
Methyl tyrosinate	16	--		13
Me				
		k_{EH^+}	$k_{EH_2^{2+}}$	
Methyl 2,3-diaminopropionate	43.8	3440	--	19
Methyl lysinate	27.6	75.4	4410	19
Methyl histidinate	37.1	2790	--	38 (26-30)
Methyl cysteinate[b]	4.4	64 (EH)	230 (EH^+)	39
		(I = 0.1 M)		
Ethyl cysteinate[b]	2.5	36 (EH)	400 (EH^+)	29
		(I = 0.16 M)		

[a]Rate constants are not listed for amino esters such as methyl 4-amino-n-butyrate where lactamization also occurs; see, for example, Reference 19.

[b]In the case of the cysteine esters $E^- = \bar{S}CH_2CH(NH_2)COOCH_3$; $EH = HSCH_2CH(NH_2)COOCH_3$; $EH^{\pm} = \bar{S}CH_2CH(\overset{+}{N}H_3)COOCH_3$.

(methyl β-alaninate, methyl 2,3-diaminopropionate) lowers k_{EH^+}/k_E to about 50 and further withdrawal to the ε-position (methyl lysinate) reduces the ratio to 2.7. It can be shown that k_{EH^+}/k_E should approximately equal the ratio of the proton ionization constant of the ester to that of the parent amino acid [49], a relationship which fits the experimental data quite well [14]. It is not clear whether electrostatic or inductive facilitation of the reaction is involved when the amino group is protonated. The recent work of Holmquist and Bruice [41] suggests that inductive facilitation of the reaction is of considerable importance.

2.3. Thermodynamic Parameters

Thermodynamic parameters are available for the base hydrolysis of a number of amino acid esters. Some typical data is listed in Table 4. Values of $\Delta H^{\ddagger}$ are similar to those generally obtained for the base hydrolysis of carboxylic acid esters (∿10 kcal mol^{-1}). Little difference is generally observed between $\Delta H^{\ddagger}$ for the unprotonated (E) and protonated (EH^+) species, a result which is expected theoretically [16,17,44]. Values of $\Delta S^{\ddagger}_{298}$ are all large and negative, presumably a result of the ordering effect of forming a transition state (one particle) from two particles. In most cases $\Delta S^{\ddagger}_{298}$ for EH^+ is less negative than $\Delta S^{\ddagger}_{298}$ for E. The reaction between EH^+ and OH^- gives a transition state which is overall charge neutral, thus producing more desolvation than for the reaction of a neutral molecule with an ion (E and OH^-).

2.4. Intramolecular Catalysis

The hydrolysis of amino acid esters can be subject to several types of intramolecular catalysis. Such reactions have been the subject of recent reviews [50]. (See structures below.) Intramolecular nucleophilic catalysis (Mechanism A), general base catalysis (Mechanism B), and specific base-general acid catalysis (Mechanism C) can occur. The hydrolysis of amino acid esters of

TABLE 4

Thermodynamic Data for Base Hydrolysis of Amino Acid Esters

(In water at 0.1 M (KCl) ionic strength; all rate constants are in liter $mole^{-1}$ min^{-1})

Ester		25°C	37°C	50°C	$\Delta H^{\ddagger}$ (kcal $mole^{-1}$)	$\Delta S^{\ddagger}_{298}$ (cal deg^{-1} $mole^{-1}$)	Ref.
Monoamino ester							
Methyl glycinate	k_E	77	145	285	9.5	-28	14
Ethyl glycinate	k_E	38	70.5	130	8.8	-32	13
	k_{EH^+}	1200	--	7500	11.8	-5.9	13
Methyl α-alaninate	k_E	66.4	132.9	249.1	9.5	-26.3	14
	k_{EH^+}	4815	9216	19471	10.1	-15.9	14
Methyl α-amino-n-butyrate	k_E	23.3	46.3	88.2	9.6	-28.2	14
	k_{EH^+}	2667	5610	11667	10.7	-15.1	14
Methyl valinate	k_E	4.5	7.8	13.5	7.8	-39.5	13
Ethyl leucinate	k_E	11	19	31.5	7.5	-39	13
Methyl β-alaninate	k_E	8.13	22.3	56.9	14.3	-14.5	14
	k_{EH^+}	412.1	933.3	2167	12.1	-14.1	14

Methyl 4-amino-n-butyrate	k_E[a]	1758	2613	4069	5.8	-32.2	19
	k_{EH^+}	125.0	216.0	395.8	8.2	-29.4	19
Methyl serinate	k_E	59.3	124.8	271.1	11.0	-21.6	14
Methyl β-phenyl alaninate	k_E	33	63	118	9.2	-31	13
Methyl S-methyl cysteinate	k_E	64	114	194	7.4	-34	13
Diamino esters							
Methyl 2,3-diaminopropionate	k_E	43.8	93.7	214.3	11.6	-20.3	19
	k_{EH^+}	3440	7050	15440	10.9	-13.9	19
Methyl lysinate	k_E	27.6	60.1	125.9	11.0	-23.1	19
	k_{EH^+}	75.4	137.5	251.8	8.7	-29.0	19
	$k_{EH_2^{2+}}$	4410	9800	24000	12.4	-8.4	19
Methyl histidinate	k_E	37.1	72.3	142.7	9.7	-26.9	38
	k_{EH^+}	2790	5439	10690	9.7	-18.4	38

[a]This constant involves intramolecular cyclization.

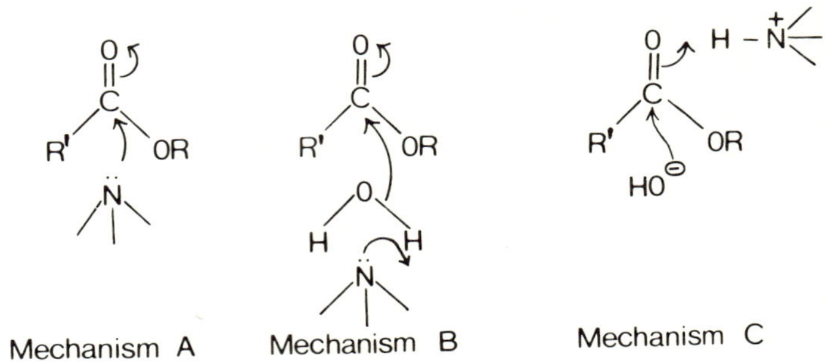

the type $H_2N(CH_2)_nCOOCH_3$ is subject to intramolecular nucleophilic catalysis when n = 3 or 4 (i.e., where a 5- or 6-membered lactam can be formed [19,51]). Similar effects are observed with the α,ω-diamino esters $NH_2(CH_2)_nCH(NH_2)COOCH_3$. The E forms of the methyl esters of 2,3-diaminopropionic acid (n = 1) and lysine (n = 4) hydrolyze with the expected rates while the corresponding derivatives of 2,4-diaminobutyric acid (n = 2) and ornithine (n = 3) hydrolyze very rapidly [19]. The product lactams have been characterized by chromatography and UV and IR spectroscopy [19,51]. Similar cyclizations have been observed for the ethyl esters of glutamic acid, glutamine [51], asparagine, and N-(2-aminoethyl)-glycine [52]. Such cyclizations are often prevented by the presence of a metal ion, since the amino nucleophile is bound to the metal ion [52].

3. METAL ION PROMOTION: LABILE SYSTEMS

Studies of the metal ion catalyzed or promoted hydrolysis of amino acid esters have been carried out with a variety of substrates. A number of investigations have been made of the hydrolysis of glycine esters and of the esters of cysteine, histidine, and 2,3-diaminopropionic acid. These latter three esters are capable of chelate ring formation without any involvement of the carbonyl group of the ester (see Sections 3.3.1 and 3.3.2).

In addition Angelici and his coworkers have studied the hydrolysis of esters such as ethyl glycinate-N,N-diacetic acid [68] and the corresponding derivatives of ethyl leucinate [69] and ethyl valinate [69] (see Section 3.4.1).

3.1. Formation Constants

A necessary prerequisite to any study of metal ion-catalyzed ester hydrolysis in which labile complexes are used, is a knowledge of the requisite formation constants for the system. In the case of an ester such as ethyl glycinate, the constants may be defined as follows:

$$Me^{2+} + E \rightleftarrows MeE^{2+} \qquad K_{MeE}^{Me} = \frac{[MeE^{2+}]}{[Me^{2+}][E]} \tag{9}$$

$$MeE^{2+} + E \rightleftarrows MeE_2^{2+} \qquad K_{MeE_2}^{MeE} = \frac{[MeE_2^{2+}]}{[MeE^{2+}][E]} \tag{10}$$

The symbol E represents the unprotonated form of the ester, $NH_2CH_2CO_2C_2H_5$. Some typical constants for amino acid esters are listed in Table 5. In the case of the glycine esters it is difficult to obtain precise values for the formation constants due to quite rapid hydrolysis of the ligand at relatively low pH values. Hix and Jones [37] have reported constants for ethyl glycinate with nickel(II), cobalt(II), and zinc(II) at 30° and I = 1.0 M. The constants are much lower than the corresponding values for glycine [53].

More precise data (Table 5) are available for the bidentate esters, e.g., histidine methyl ester [20], cysteine methyl ester [21], methyl 2,3-diaminopropionate [18], and for such esters as ethyl glycinate-N,N-diacetic acid (see Section 3.4.1).

Values of the thermodynamic parameters $\Delta H°$ and $\Delta S°_{298}$ have been determined for the CuE^{2+} and CuE_2^{2+} species of methyl DL-2,3-diaminopropionate [18] and methyl L-histidinate [20].

Stereoselective effects associated with the asymmetric center of the ligand have not as yet been reported for amino acid ester

TABLE 5

Stability Constants for Some Metal Ion-Amino Acid Ester Complexes[a]

Ester	Metal ion	$\log K^{Me}_{MeE}$	$\log K^{MeE}_{MeE_2}$	$\log \beta^{Me}_{MeEA}$	Ionic strength	Temp. (C°)	Ref.
Ethyl glycinate	Cu(II)	4.04	3.89[b]	--	0.1	25	36
		3.83	--	4.51	--	25	59
	Ni(II)	2.49	2.09	--	0.15	25	33
		2.30	1.92	--	1.0	30	37
Methyl D,L-2,3-diaminopropionate	Cu(II)	8.986	7.763	18.77	0.1	25	18
	Hg(II)	6.377	5.104	--	0.1	25	18
Methyl L-histidinate	Cu(II)	8.523	5.976	16.93	0.1	25	20
		8.55	6.02	--	0.1	25	27
		8.32	5.82	--	0.25	25	28
		8.48	5.90	--	0.16	25	29
		9.10	6.70	--	0.20	25	33
	Ni(II)	6.1	4.8	--	0.10	25	20
		6.21	4.94	--	0.10	25	27
		6.02	4.30[c]	--	0.25	25	28
		6.73	5.11	--	0.20	25	33
	Hg(II)	5.33	4.14	--	0.10	25	20
Methyl L-cysteinate	Ni(II)	7.61	9.63	--	0.10	25	21
	Zn(II)		15.91[d]	--	0.10	25	21
	Pb(II)	9.13	6.16	--	0.10	25	21
	Cd(II)	8.89	7.35	--	0.10	25	21

[a] Reference 53 contains an extensive summary of stability constants of metal ion-amino acid complexes.

[b] $\log K^{MeE_2}_{MeE_3} = 4.80$.

[c] $\log K^{MeE_2}_{MeE_3} = 2.50$.

[d] Me

complexes [54-61]. The 1:1:1 complex of nickel(II) with histidine and methyl histidinate has the same formation constant for all combinations of D and L stereoisomers [54]. Kinetic stereoselectivity has, however, been observed in this system [54] (see Section 3.5).

Thermodynamic stereoselectivity appears to be restricted to tridentate ligands containing bulky side chains which are coordinated to a small metal ion [58]. The complex M(D-ligand)(L-ligand) has a higher formation constant than the complex $M(D\text{-ligand})_2$ or $M(L\text{-ligand})_2$ since in the D,L complex the repulsion of the bulky side chains is minimized. Morris and Martin [58] have observed stereoselectivity in the coordination of histidine to cobalt(II), nickel(II), and zinc(II), but not for copper(II) (6-coordination unfavored) and cadmium(II) (too large an ion). Barnes and Pettit [62] have, however, reported stereoselectivity in the copper(II) system.

In addition to the simple mono and bis complexes ME and ME_2 other species have been reported. Polymeric species arise where there is a potential bridging atom such as sulfur in the ligand. Formation constants for M_2E_3 and higher polymers have been reported for M = Ni(II), Cd(II), and Hg(II) in the case of methyl L-cysteineate [21]. Similar polymeric species occur with other amino thiols such as penicillamine and cysteine [60,61].

Hydroxy complexes tend to form from the $ME(H_2O)_x$ complexes on raising the pH. Equilibria of the type

$$MeE(H_2O)_x^{n+} \rightleftarrows MEOH(H_2O)_{x-1}^{(n-1)^+} + H^+ \tag{11}$$

have been studied for a variety of amino esters and metal ions. Reported pK values at 25°C range from 7 to 9 [18,20,21]. Such equilibria are of importance in kinetic studies of 1:1 complexes.

3.2. Monoamino Esters

Following the work of Kroll [2], studies of amino acid ester hydrolysis were centered on simple mono amino esters [2,9,33]. However, the results obtained were rather conflicting and subject to such criticisms as: interaction between the buffer system employed and the metal ion-ester complexes [2]; failure to study the pH-dependence of the reaction [9]; failure to control pH [33]. Some subsequent work [36] has been reinterpreted [37]. More recent studies (Table 6) [35,37,59] have provided rate constants which show only order of magnitude agreement; however, uncertainty from the earlier studies [2,9,33] regarding the nature of the nucleophile has been resolved. Hydroxide ion is the predominant nucleophile in spite of the reaction being studied at pH values of ∿5 (higher pH's lead to precipitation of metal hydroxides, due to the rather low formation constants of the metal complexes). Evidence for nucleophilic attack by water has been obtained [35,37,59]. The precise structure of the catalytically active species is not as yet fully resolved. However, it is clear that the rate enhancements obtained in the copper(II)-glycine ester systems is very large, ca. 10^5, which is considerably greater than observed for the bidentate esters (see Section 3.3). These large rate accelerations are

TABLE 6

Rate Constants for the Metal Ion-Promoted Hydrolysis of Methyl Glycinate

Process[a]	Rate constant (Ionic strength; temp.)	Ref.
$CuE^{2+}+OH^-$	7.6×10^4 liter mole^{-1} sec^{-1} (0.16 M, 25°C)	35
$CuE^{2+}+H_2O$	4.3×10^{-5} sec^{-1} (0.16 M, 25°C)	35
$NiE^{2+}+OH^-$	3.98×10^3 liter mole^{-1} sec^{-1} (1.0 M, 30°C)	37[b]
$NiE^{2+}+H_2O$	1.9×10^{-4} sec^{-1} (1.0 M, 30°C)	37[b]
$CuE^{2+}+OH^-$	1.4×10^5 liter mole^{-1} sec^{-1} (?, 25°C)	59[b]
$CuEA^{+}+OH^-$	8.4×10^3 liter mole^{-1} sec^{-1} (?, 25°C)	59[b]
$CuE^{2+}+H_2O$	2.5×10^{-5} sec^{-1} (?, 25°C)	59[b]

[a]E is $H_2NCH_2COOC_2H_5$; A^- is $H_2NCH_2COO^-$.

[b]The values refer to glycine ethyl ester.

similar to the rate accelerations of ca. 10^6 observed in the inert cobalt(III) systems where direct metal-ester carbonyl bonding occurs. It is thus very probable that the glycine ester systems involve such species as 3 in which there is a direct interaction between the metal ion and the carbonyl group of the ester. The

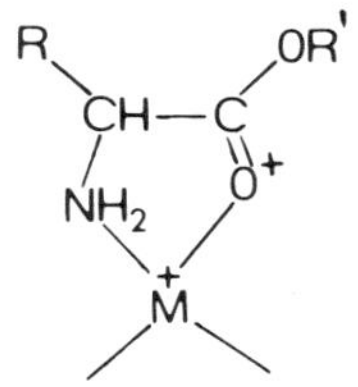

3

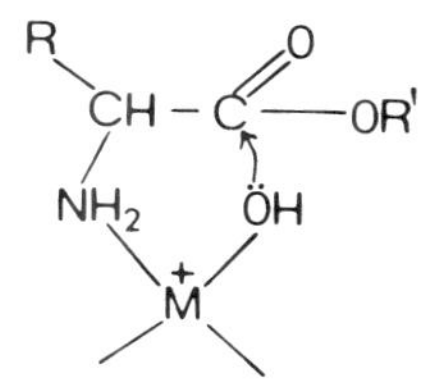

4

importance of coordinated nucleophiles as illustrated in 4 is still unresolved in the labile complexes. In the cobalt(III) systems (Section 4) rate accelerations of the order of 10^9 to 10^{11} occur with internal nucleophiles, so that on this basis it must be concluded that the carbonyl-bonded structures provide an adequate explanation for the observed rate accelerations in the copper-glycine ester systems. A discussion of the relative effectiveness of proton catalysis and metal ion catalysis has been given by Martin [48].

Metal-ester-carbonyl bonding is unlikely with the bidentate esters (Section 3.3) thus accounting for their lower rates of hydrolysis. A number of metal complexes of amino acid esters have been isolated and their solid state infrared spectra studied [4-6]. Examples of both monodentate (amino-coordinated) and bidentate (amino- and carbonyl-coordinated) species have been obtained. Chelation leads to a characteristic decrease in the carbonyl stretching frequency indicating significant polarization of the carbonyl group by the metal ion.

Pilbrant has studied the copper(II) promoted hydrolysis of the monoethyl esters of aspartic and iminodiacetic acids [63]. The esters are chelated to copper(II) via the free carboxyl group and the amino nitrogen. Inductive effects operating via the coordinated amino group account for the rate accelerations observed in the 1:1 complexes of copper(II) with the two aspartic esters; however, more pronounced rate accelerations were observed with the monoethyl ester of iminodiacetic acid suggesting that metal-ester carbonyl bonding is involved.

The ring-closure reactions and the copper(II) promoted hydrolysis of asparagine ethyl ester and N-(2-aminoethyl)glycine ethyl ester have also been studied [52]. In basic solution N-(2-aminoethyl)glycine ethyl ester (5) can undergo ring closure to 2-ketopiperazine (6) or hydrolysis to N-(2-aminoethyl)glycine (7). The

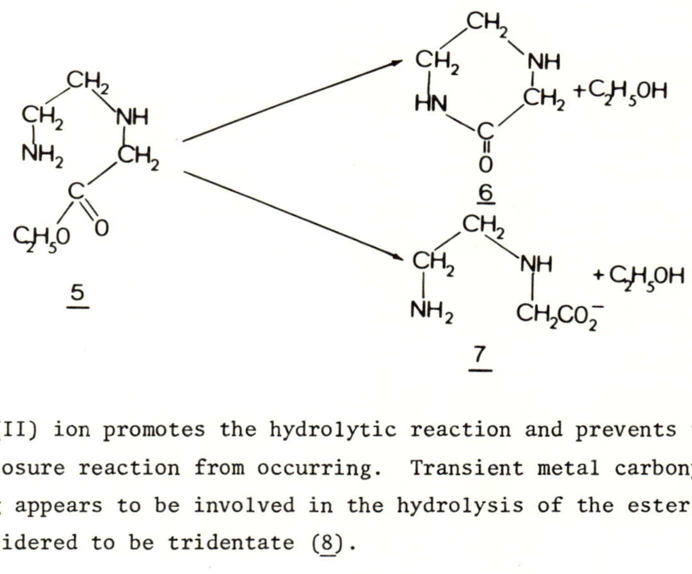

copper(II) ion promotes the hydrolytic reaction and prevents the ring-closure reaction from occurring. Transient metal carbonyl bonding appears to be involved in the hydrolysis of the ester which is considered to be tridentate (8).

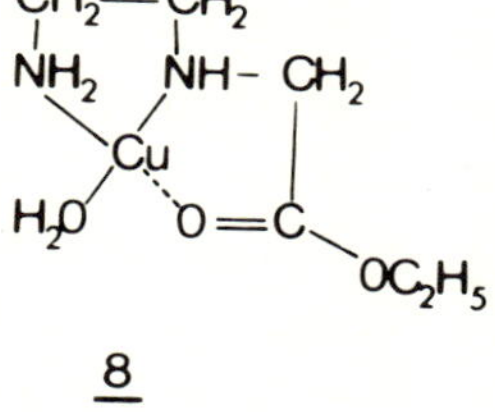

8

3.3. Bidentate Esters

3.3.1. Esters with 2N Donors

Both methyl 2,3-diaminopropionate (9) and methyl histidinate (10) form metal complexes where the formation constants are large [18,20,26,27,29].

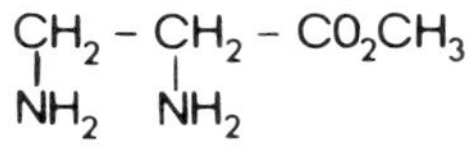

9

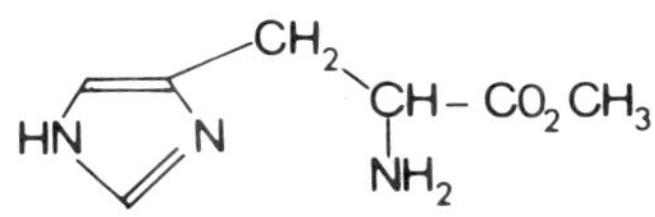

10

Metal complexes of methyl 2,3-diaminopropionate contain a five-membered chelate ring and they are the most stable of the complexes of α,ω-diamino acid esters. The histidine complexes contain six-membered chelate rings, but the complexes are more stable than the analogous five-membered ring complexes of 4-imidazolylglycine [24].

Metal Ion:Ester = 1:2. A number of kinetic studies at this metal to ligand ratio have been reported for copper(II) and nickel(II) with L-methyl histidinate [26,27,29,38] and for copper(II) and mercury(II) with methyl DL-2,3-diaminopropionate [64]. In the case of the copper(II) complexes, solutions of pH > 6.5 contain mainly the bis complexes CuE_2^{2+} (>99%). An excess of the ester ligand (E) can be used in the methyl 2,3-diaminopropionate system to ensure essentially complete formation of HgE_2^{2+}. However, dissociation of NiE_2^{2+} to NiE^{2+} cannot be repressed by adding excess ligand as NiE_3^{2+} is formed [38]. The kinetic data for the nickel catalysis are therefore approximate.

The kinetic data, obtained by pH-stat on the hydrolysis of the bis complexes is consistent with two consecutive second-order (pseudo-first order at constant pH) reactions [38,64]:

$$MeE_2^{2+} + OH^- \xrightarrow{k_{MeE_2}} MeEA^+ + CH_3OH \qquad (12)$$

$$MeEA^+ + OH^- \xrightarrow{k_{MeEA}} MeA_2 + CH_3OH \qquad (13)$$

Initial more rapid hydrolysis of MeE_2^{2+} is followed by a slower hydrolysis of $MeEA^+$. The various rate constants obtained are summarized in Table 7. Some of the early results are in error [26] since at the pH values employed (∿11) the half-life of the catalyzed hydrolysis of methyl histidinate is ∿1 min and only the last stages of the reaction could be observed, so that low rate constants were obtained. Use of the initial rate method does not uncover the two-step nature of the reaction [29]. The rate data reported in a number of investigations [27,38,64] were obtained by the Swain time-ratio method [65]. The time-ratio treatment gives a good approximation of the overall reaction but ignores disproportionation of $MeEA^+$:

$$2MeEA^+ \rightleftharpoons MeE_2^{2+} + MeA_2 \qquad (14)$$

<u>Metal Ion:Ester = 1.1</u>. Kinetic measurements have also been carried out at ligand:metal ratios of 1:1 with methyl histidinate and methyl 2,3-diaminopropionate. The formation of hydroxy complexes was observed in some cases, e.g., reaction (15) and the kinetic data for the CuE^{2+} and $CuEOH^+$ species is available (Table 7).

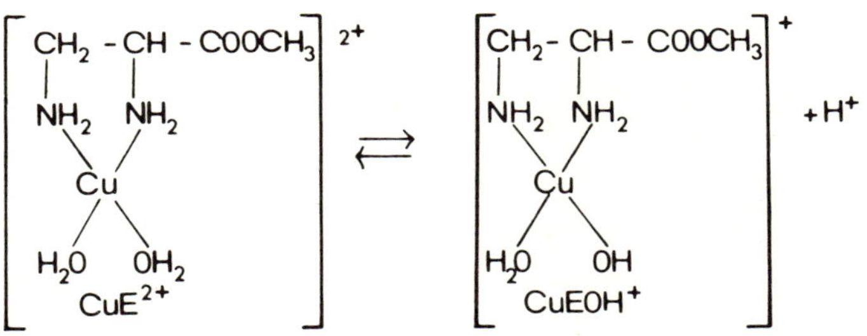

(15)

<u>Conclusions</u>. The order of decreasing reactivity of the various L-(+)-histidine methyl ester species [38] toward nucleophilic attack by hydroxide ion at 25°C and I = 0.1 M is $CuE_2^{2+} > NiE_2^{2+} > CuE^{2+} > NiE^{2+} > EH^+ > CuEA^+ > CuEOH^+ > NiEA^+ > E$. A very similar situation occurs with methyl 2,3-diaminopropionate [64] where the reactivity order is $CuE^{2+} > CuE_2^{2+} > CuEOH^+ > CuE(en)^{2+}$ (en = 1,2-diaminoethane) $> HgE_2^{2+} > CuEA^+ > HgE^{2+} > EH^+ > HgEOH^+ > HgEA^+ > E$.

TABLE 7

Rate Constants for the Metal Ion-Promoted Hydrolysis of Amino Acid Esters[a]

System	Complex				Ref.
	MeE_2^{2+}	$MeEA^+$	MeE^{2+}	$MeEOH^+$	
Cu(II)-L-methyl histidinate	1.704×10^2	--	--	--	26[b]
	1.94×10^2	2.32×10^3	--	--	27
	1.193×10^4	--	--	--	29[c]
	1.969×10^4	2.560×10^3	1.05×10^4	2.02×10^3	38
Ni(II)-L-methyl histidinate	9.42×10^1	--	--	--	26[b]
	7.09×10^3	9.92×10^2	4.7×10^3	--	27
	2.250×10^3	--	1.935×10^4	--	29[c]
	1.13×10^4	1.7×10^3	5.26×10^3	--	38
Cu(II)-D,L-methyl 2,3-diaminopropionate	1.83×10^4	5.32×10^3	3.71×10^4	8.11×10^3	64
Hg(II)-D,L-methyl 2,3-diaminopropionate	6.90×10^3	1.46×10^3	4.92×10^3	3.00×10^3	64

Table 7 (Continued)

System	Complex		Ref.
	MeE_2	$MeEA^-$	
Ni(II)-L-methyl cysteinate	6.50×10^2	2.15×10^2	39
Pb(II)-L-methyl cysteinate	3.40×10^2	5.8×10^1	39
Zn(II)-L-methyl cysteinate	2.80×10^2	8.3×10^1	39
Cd(II)-L-methyl cysteinate	2.10×10^2	4.2×10^1	39
Hg(II)-L-methyl cysteinate	1.40×10^2	4.6×10^1	39
Ni(II)-L-ethyl cysteinate	1.53×10^2		29[c]
Zn(II)-L-ethyl cysteinate	1.13×10^2		29[c]
Cd(II)-L-ethyl cysteinate	6.3×10^1		29[c]

[a]All constants refer to: complex + OH^- → products, and are in liter $mole^{-1}$ min^{-1} at 25°C and I = 0.1 M unless otherwise stated.

[b]I is uncertain.

[c]I = 0.16 M.

The CuE_2^{2+} species of methyl histidinate hydrolyzes 265 times more rapidly than the unprotonated species when statistical corrections are made, and EH^+ hydrolyzes 75 times more rapidly than E. The rate accelerations are very much less than observed in the glycine systems, and metal-ester-carbonyl bonding is apparently unimportant. The main factor determining the rate of hydrolysis in these systems is the charge carried by the substrate. A species carrying a double positive charge is more reactive than a species carrying a single positive charge which in turn is more reactive than the neutral ester molecule E.

Thermodynamic parameters $\Delta H^{\ddagger}$ and $\Delta S^{\ddagger}_{298}$ have been reported for the hydrolysis of CuE_2^{2+} and $CuEA^+$ (E = L-methyl histidinate [38] and methyl DL-2,3-diaminopropionate [64]). The metal ion reactions have $\Delta H^{\ddagger}$ values 1 to 2 kcal mol^{-1} higher than for the uncomplexed ester, and the rate increaes are due to more positive values of $\Delta S^{\ddagger}_{298}$. These effects have been discussed in detail [38].

3.3.2. L-Methyl Cysteinate

A number of investigations of the interaction of metal ions with methyl and ethyl esters of cysteine have been carried out [2, 21,29,66,67]. Various complexes have been isolated and their solid state infrared spectra studied [5,6,21]. Infrared spectral measurements provide no evidence for bonding between the metal ion and the ester function of methyl cysteinate in both the solid state [5,6, 21] and in aqueous solution [6]. Methyl cysteinate (11) forms charge neutral complexes (12) with bivalent metal ions (protonated and polynuclear species can also occur [21]). The results obtained

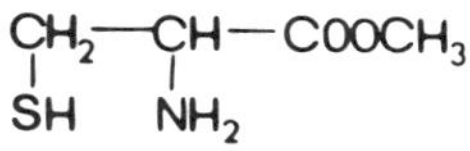

11

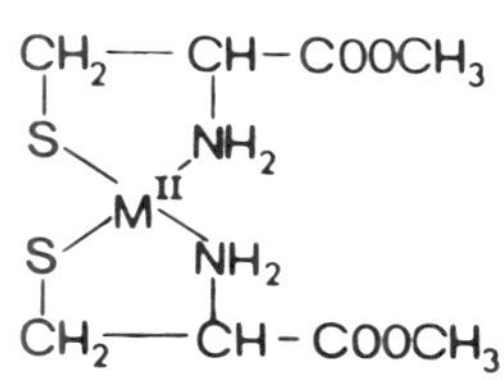

12

in early studies of the metal ion-promoted ester hydrolysis [2,66,67] are open to criticism since the pH dependence of the reaction was not studied and the kinetically active species were not identified.

A detailed study of the various complexes formed in solution by methyl L-cysteinate and nickel(II), zinc(II), lead(II), and mercury(II) has been published [21]. Two recent studies of the metal ion-catalyzed hydrolysis of the ethyl ester [29] and the methyl ester [39] have been made. The data for the ethyl ester were obtained by the initial rate method which limits their usefulness.

The hydrolysis kinetics of the methyl ester at a ligand:metal ratio of 2:1 [Me = Ni(II), An(II), Cd(II), and Pb(II)] in the pH range 9 to 11 are consistent with the kinetic scheme [39]:

$$MeE_2 + OH^- \xrightarrow{k_1} MeEA^- + CH_3OH \tag{16}$$

$$MeEA^- + OH \xrightarrow{k_2} MeA_2^{\,2-} + CH_3OH \tag{17}$$

where E^- is the methyl cysteinate anion and A^{2-} is the dianion of cysteine. Rate constants obtained in various investigations are summarized in Table 7. The complexes are uncharged and undergo slower base hydrolysis than the analogous complexes of methyl histidinate and methyl 2,3-diaminopropionate. The catalytic activity of the various metal ions (based on k_1) is Ni > Pb > Zn > Cd > Hg which correlates quite well with the order of metal nitrogen bond strengths obtained from infrared spectral measurements on the complexes, viz., Ni > Zn > Pb > Cd > Hg [21,29]. This rough correlation of hydrolytic reactivity with metal-nitrogen bond strength probably implies that the metals catalytic activity in these neutral complexes is due to the inductive effect of the metal ion acting through the amino nitrogen. Withdrawal of electrons from the acyl carbon atom facilitates nucleophilic attack by hydroxide ion.

Rate constants have also been reported [39] for the base hydrolysis of some mixed ligand complexes of nickel(II), L-methyl cysteinate and a second ligand L. For L = 2-aminoethanethiol and cysteine, kinetic evidence for mixed ligand complexes was obtained. Other ligands (1,2-diaminoethane, histamine, o-phenanthroline) did not give mixed ligand complexes.

3.4. Polydentate Amino Ester Systems

The difficulties associated with the low formation constants for monoamino ester complexes (see Section 3.2) can be overcome in two ways [68-75]. The ester moiety can be incorporated as part of a larger polydentate ligand, e.g., by using ligands such as ethyl glycinate-N,N-diacetic acid (13) which hydrolyzes to give nitrilotriacetic acid (14). Reactions of this type are described in

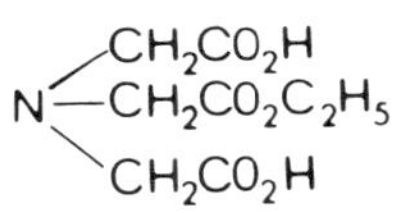

13

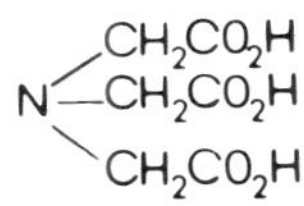

14

Section 3.4.1. An alternative approach is to use mixed ligand complexes of the monoamino ester and a polydentate ligand (Sections 3.4.2 and 3.4.3). Of course, there is the possibility of direct interaction between the metal ion and ester carbonyl group.

3.4.1. Metal Ion-Amino Ester N,N-Diacetic Acids

Reactions of ligands of this type have been studied in detail by Angelici and his coworkers [68-70]. Kinetic measurements have been reported [68-70] for ligands involving ethyl glycinate (15a), ethyl alaninate (15b), ethyl valinate (15c), ethyl leucinate (15d), butyl glycinate, ethyl β-alaninate (16a), and ethyl 4-aminobutyrate (16b). The catalytic effects of the metal ions Cd(II), Ni(II), Mn(II), Fe(II), Co(II), Cu(II), Zn(II), Pb(II), La(III), Nd(III), Gd(III), Dy(III), Er(III), Ye(III), Lu(III), and Sm(III) have been studied for a few of the ligands. The 1:1 complexes have large

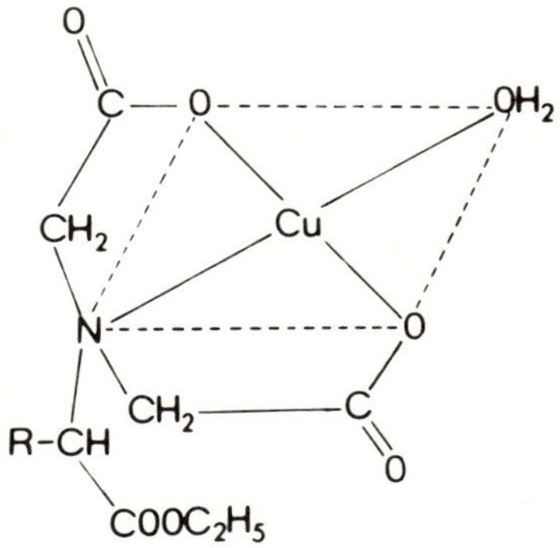

15

a, R = H
b, R = CH_3
c, R = $CH(CH_3)_2$
d, R = $CH_2CH(CH_3)_2$

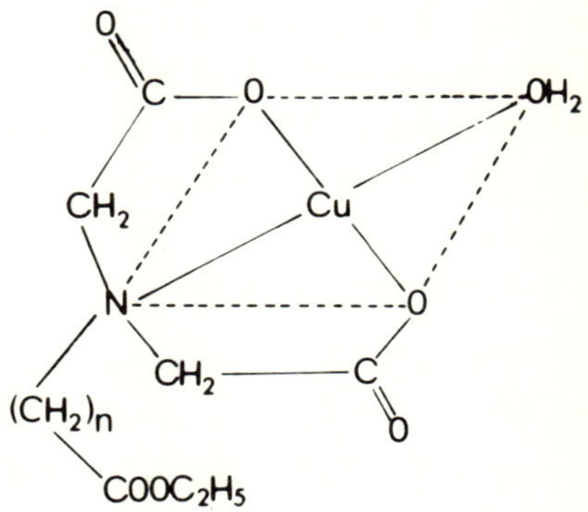

16

a, n = 2
b, n = 3

formation constants and show little tendency to add a second ligand. Hydroxo complexes may, however, be formed by deprotonation of coordinated water molecules. In the case of the ethyl glycinate derivative, which may be taken as a specific example of these reactions, pH-stat measurements have been carried out in the pH range 5.0 to 7.0 for the reaction

$$\mathrm{Cu(EGDA)} + \mathrm{OH}^- \xrightarrow{k} \mathrm{Cu(NTA)}^- + \mathrm{C_2H_5OH} \tag{18}$$

where EGDA = ethyl glycinate-N,N-diacetate and NTA = nitriloacetate. The rate expression has the form, rate = $k[\mathrm{Cu(EGDA)}][\mathrm{OH}^-]$ with $k = 2.18 \times 10^4\ M^{-1}\ \mathrm{sec}^{-1}$ at 25°C [68]. The kinetic measurements are equally consistent with bimolecular hydroxide ion attack on the 1:1 aquo chelate or to intramolecular attack by coordinated hydroxide ion (ca. 0.01 to 20% of $[\mathrm{Cu(EGDA)(OH)}]^-$ exists in solution in the pH range 5.0 to 7.0). The hydrolytic reactions are extremely rapid with reaction rates of a similar order to those observed for the monoamino ester-metal ion systems (see Section 3.2). A result which suggests that direct interaction between the metal ion and the carbonyl group of the ester occurs (Scheme 4). Formation constant measurements [69] suggest that any such interaction must involve only a very small proportion of chelate molecules. Nucleophilic catalysis (by acetate, HPO_4^{2-}, pyridine, 4-picoline, and nitrite) of some of these reactions has been studied [69] using the esters derived from ethyl valinate and ethyl leucinate which are sterically hindered to base hydrolysis. It has been concluded that ester hydrolysis occurs via OH^- attack on the ester carbonyl group transiently coordinated to the metal ion rather than to intramolecular attack by bound hydroxide ion. The complexes undergoing hydrolysis are uncharged so that the large rate accelerations do not involve charge effects.

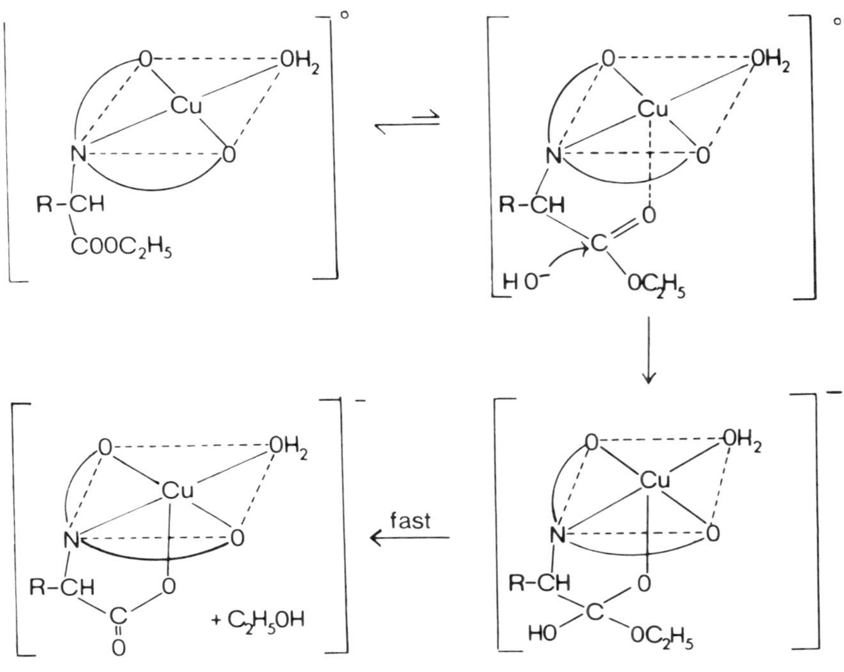

Scheme 4. Metal ion-catalyzed hydrolysis of amino ester N,N-diacetic acids.

3.4.2. Metal Ion-Nitrilotriacetate-Amino Ester Complexes

Nitrilotriacetate (14) NTA^{3-} forms very stable complexes $M(NTA)^{-}$ with many divalent metal ions [71]. Monoamino acids coordinate to $M(NTA)^{-}$ in a bidentate fashion, but their esters are essentially monodentate (esters such as methyl histidinate are bidentate). At a Cu(II):NTA^{3-}:monoamino ester = 1:1:1 mole ratio, the major species present in solution is the mixed ligand complex [Cu(NTA)(monoamino ester)]$^{-}$.

Hydrolysis of the uncomplexed ester is insignificant at the low pH values (generally <9) where hydrolysis of the mixed ligand complex occurs. The hydrolysis can be monitored by pH-stat, and as with the N,N-diacetate systems (Section 3.4.1) rapid hydrolysis of the ester function is observed [72]. It is suggested that catalysis is due to activation of the carbonyl group to intermolecular nucleophilic attack by OH^{-} as a result of transient coordination of the carbonyl oxygen atom to copper(II). The hydrolysis of methyl histidinate is much slower, as expected if there is no direct metal-carbonyl group interaction with the bidentate ester. For base hydrolysis of the complex $[Cu(NTA)(NH_2CH_2CO_2C_5H_5)]^{-}$ the activation parameters are $\Delta H^{\ddagger}$ = 4.9 kcal mol^{-1} and $\Delta S^{\ddagger}$ = -33 eu. Catalysis is primarily due to a substantial lowering of $\Delta H^{\ddagger}$ (by 5 kcal mol^{-1}). The reverse result is observed with the bidentate esters (Section 3.3.1).

3.4.3. Copper(II)-Amino Diacetate-Amino Ester Complexes

Base hydrolysis of the ester function in complexes of the type $Cu[RN(CHR'CO_2)(CH_2CO_2)]$ (monoaminoester) (i.e., mixed ligand complexes of the substituted imino diacetates with copper and the ester ligand) have also been studied [73]. Similar rate enhancements occur to those discussed in Sections 3.4.1 and 3.4.2. However, nucleophilic attack by both OH^{-} and H_2O is observed. The ester carbonyl group is considered to be coordinated to the metal ion when hydrolysis occurs.

Some similar studies have also been reported on mixed ligand complexes of the type $[Cu(dien)(NH_2CH_2CO_2CH_3)]^{2+}$ [75]. Hydrolysis in the dien complexes is considerably slower than observed in the other mixed ligand systems and the reasons for this behavior have been discussed [75].

3.5. Kinetic Stereoselectivity

The general topic of stereoselectivity in metal complexes of amino acids and their derivatives has been reviewed [76]. The rates of metal ion promoted amino ester hydrolysis are usually independent of the configuration of the ligand (D or L); however, some stereoselective effects have been observed. The hydrolysis of L-ethyl β-phenylalaninate occurs more rapidly with trans-$[Co(D(-)pn)_2Cl_2]Cl$ than with trans-$[Co(L(+)pn)_2Cl_2]Cl$, (pn = 1,2-diaminopropane) [77]. The 1:1 nickel(II) complex of histidine promotes the hydrolysis of methyl histidinate more effectively if the two ligands are of opposite configurations [54]. $[Ni(D\text{-}HistOMe)(L\text{-}Hist)]^+$, for example, hydrolyzes some 40% faster than $[Ni(D\text{-}HistOMe)(D\text{-}Hist)]^+$. Stereoselectivity is not observed for the analogous copper(II) complexes [38,78], but the nickel(II) result has been confirmed [38,78]. The $NiEA^+$ complex probably involves tridentate A^- and primarily bidentate E (A^- = the amino acid anion, E = the amino acid ester). The octahedral coordination of nickel(II) in these systems results in the possibility of a direct metal ion-ester carbonyl interaction (<u>17</u>). Such an interaction can only occur when the two ligands are of opposite configurations since a trans arrangement of the bulky imidazole donors is preferred [58,79]. The interaction must be very weak in view of the small rate differences between the diastereoisomeric complexes. Such transient coordination is unlikely for copper(II) which favors tetragonal coordination geometry. In $CuEA^+$, both E and A^- are probably bidentate with possibly weak apical A^- carboxylate coordination. Hix and Jones [54] have established that this is a kinetic stereoselective effect since the formation constants of the diastereoisomeric complexes are identical. Leach and

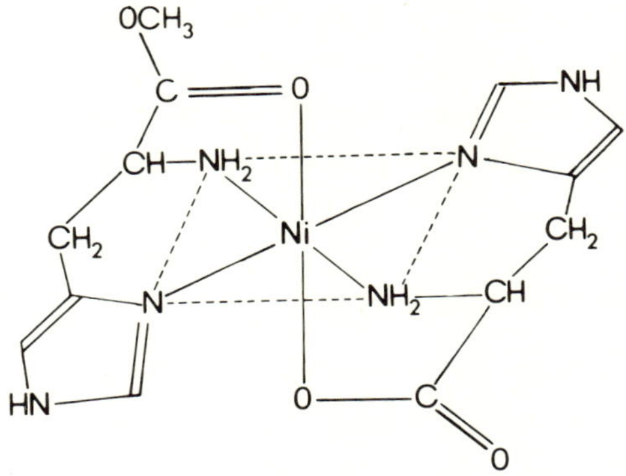

17

Angelici [74] have measured the equilibrium constants for the coordination of optically active amino acidates (A^-) and their esters (E) to (L-valine-N-monoacetato)copper(II), i.e., reactions of the type

$$\text{Cu(L-ValMA)} + \text{L- or D-A}^- \overset{K}{\rightleftarrows} \text{Cu(L-ValMA)(A}^-\text{)} \qquad (19)$$

In the case of ethyl leucinate, the L enantiomer has an equilibrium constant K some 3.4 times greater than for the D enantiomer. The rate of hydrolysis of methyl leucinate and methyl phenylalaninate in the complexes Cu(L-ValMA)(E) is greater for the D-amino acid ester than for the L enantiomer. The asymmetrically selective hydrolysis of phenylalanine esters catalyzed by poly-L-lysine copper-(II) complexes has recently been reported [80].

3.6. Ester Exchange Reactions

A number of metal ion-catalyzed ester exchange reactions have been studied in recent years. The chelate bis(pentane-2,4-dionato)-μμ'-dimethoxydicopper(II) (18) is a very effective and selective catalyst for ester exchange of ethyl picolinate and ethyl 2-pyridylacetate. It seems likely that the chelate dissociates into the monomeric species which reacts reversibly with the ester to form the chelate 19 which undergoes rapid ester exchange [81].

Sigman and Jorgenson [82] have found that zinc(II) catalyzes the transesterification reaction between 4-(β-hydroxyethyl)ethylenediamine and p-nitrophenyl picolinate. The reaction is considered to involve a reactive tertiary complex in which the zinc(II) ion perturbs the pK_a of the hydroxy ethyl group of N-(β-hydroxyethyl)-ethylenediamine to provide a high effective concentration of the nucleophile. Intramolecular nucleophilic attack then occurs at the carbonyl group of p-nitrophenyl picolinate (20).

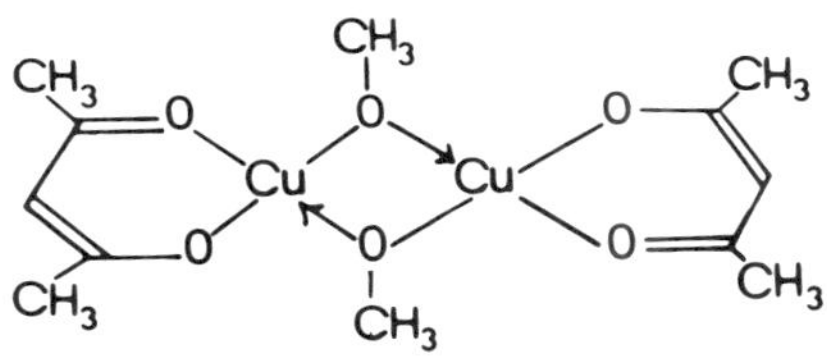

18

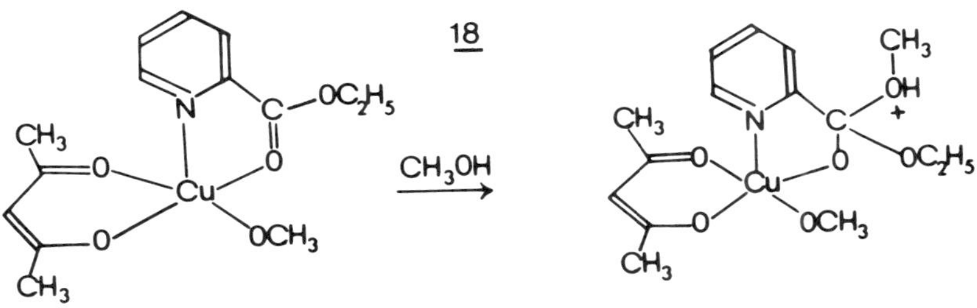

19

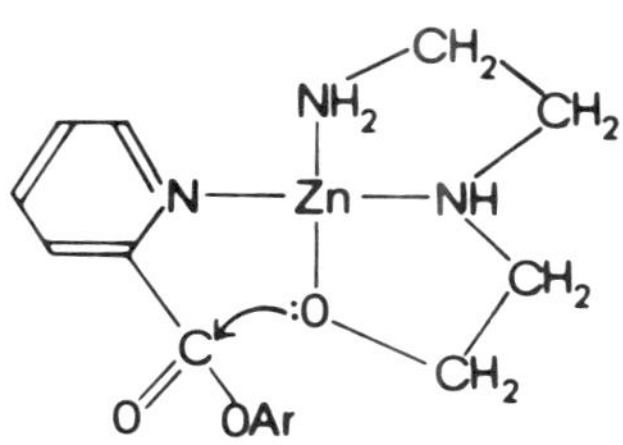

20

3.7. Peptides and Amides

Very limited data exist on the base hydrolysis of amino acid amides and simple peptides. A number of papers by Long [83] and his coworkers do, however, provide data on the acid-catalyzed hydrolysis of peptides determined using an automatic amino acid analyzer.

3.7.1. Metal Peptide Complexes

A number of discussions are available on the general topic of metal peptide complexes [76,84-87] and the subject is currently reviewed [88].

The complex diaquoglycylglycinatocopper(II) hydrate has been crystallized from solution and its crystal structure determined [87,90]. The peptide molecule behaves as a tridentate chelate via its terminal amino group, peptide nitrogen atom, and carboxyl groups, with a water molecule completing the square planar structure. A much more loosely bound water molecule occurs perpendicular to the plane and the third water molecule links the structure by means of hydrogen bonds. Formation of the deprotonated complex is understandable on several counts: (a) no loss of resonance energy is entailed, as would be the case if the peptide nitrogen were not deprotonated (the resonance energy of the peptide bond is ca. 40 kJ mol^{-1}); (b) only five-membered chelate rings are involved.

Nickel(II) also induces the ionization of the amide hydrogen atoms of peptides to give complexes with peptide nitrogen donors. In the presence of peptides containing three or more amino acid residues, nickel(II) undergoes a stereochemical change from octahedral to planar upon peptide hydrogen ionization [12,91]. Thus these complexes are similar in structure to the corresponding tetragonal copper(II) tripeptide complexes.

Palladium(II) induces the ionization of peptide hydrogens near pH 3.5 [92] to give tetragonal complexes with amide nitrogen donor atoms similar to the copper(II) and nickel(II) peptide complexes. Promotion of peptide hydrogen ionization increases in the series

Co(II) < Ni(II) < Cu(II) > Pd(II) [92]. A number of bisdipeptide-cobalt(III) complexes containing deprotonated amide groups have been investigated and their spectral properties studied [85,93-95]. Protonation in acidic media occurs on oxygen rather than nitrogen to give the iminol tautomer of the dipeptide ligand [95,96].

In spite of some references to the contrary it is now well established by NMR and x-ray diffraction studies that the oxygen atom is the site of both protonation and metal ion coordination to the neutral amide function, while the nitrogen atom becomes the metal ion bonding site if ionization of an amide hydrogen occurs.

3.7.2. Hydrolytic Reactions

An extensive series of studies of the metal ion-catalyzed hydrolysis of peptides and related compounds, including both homogeneous and heterogeneous reactions have been carried out by Bamann and his collaborators [97].

It was found that highly charged ions such as thorium(IV) catalyzed the hydrolysis of leucylglycylglycine at pH values as low as 5. The thorium(IV) species is very extensively hydrolyzed at this pH and the reaction is presumably heterogeneous. The hydrolysis of phosphate esters catalyzed by metal hydroxide gels was also first reported by the same group [98]. Gel hydrolysis is effected at relatively low temperatures (37°C), while observable effects are only obtained with such ions as copper(II) at temperatures of ca. 70°C.

The hydrolysis of di- and tripeptides is also catalyzed by rare earth ions at pH 8.6 [99]. Cerium(IV) and cerium(III) were particularly effective catalysts even at a temperature of 37°C. The same reactions with La(III) as a catalyst are much slower and only occur at an appreciable rate at 70°C. A comprehensive review of the work of Bamann and his coworkers [100] is available, which covers their work up to 1959.

In view of the heterogeneous nature of these reactions it is difficult to come to firm mechanistic conclusions; however, it seems likely that metal-bound hydroxide nucleophiles are probably involved. Copper(II), nickel(II), and cobalt(II) promote the hydrolysis of glycine amide in the pH range 9.35 to 10.25 and temperatures of 65 to 75°C [101]. In the absence of metal ions, phenylalanylglycine amide undergoes ring closure to 3-benzyl-2,5-diketopiperazine, but in the presence of copper(II) at pH 5 hydrolysis at both the amide and peptide bonds is competitive with ring closure.

The copper(II)-catalyzed hydrolysis of glycylglycine has been studied in some detail. Copper(II) ions catalyze the hydrolysis of glycylglycine in the pH range 3.5 to 6 at 85°C [102]. The pH rate profile has a maximum at pH 4.2, consistent with the view that the catalytically active species in the reaction is the carbonyl-bonded complex. The decrease in rate at higher pH values is associated with the formation of a catalytically inactive complex produced by ionization of the peptide hydrogen atom [180]. The reaction has also been studied [103] in basic solution (pH > 11) and copper(II) has been found to inhibit the base hydrolysis of glycylglycine. Conley and Martin [35] have found that, at pH values in excess of 11, copper(II) inhibits the hydrolysis of glycine amide due to amide hydrogen ionization. Similar results were obtained with picolinamide and a bispicolinamide complex of nickel(II) containing deprotonated amide groups was isolated [35].

It is clear from these studies that metal ions catalyze the hydrolysis of amides and peptides at pH's where the carbonyl-bonded species is present. At higher pH values where deprotonated complexes are formed the hydrolysis is inhibited [see reaction (20)]. These conclusions have been amply confirmed in subsequent studies involving inert cobalt(III) complexes (Section 4.3). Polarization

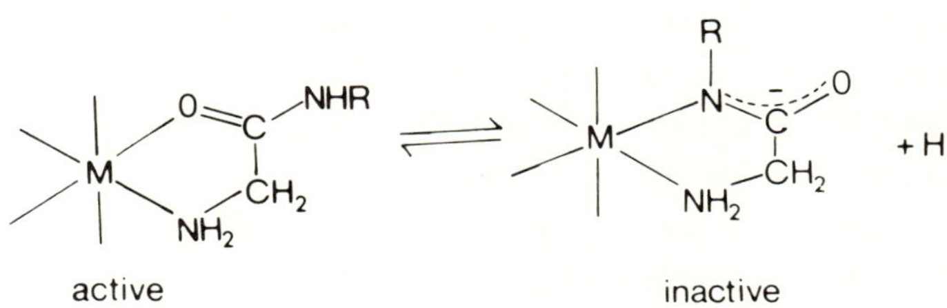

(20)

of the carbonyl group by the metal ion in the carbonyl-bonded complex promotes nucleophilic attack at the carbonyl carbon by hydroxide ion (and possibly water).

Two interesting hydrolytic reactions have been noted as a result of preparative work for x-ray crystallographic studies. Attempts to prepare crystals of the hexaglycine complex of copper(II) in molar base solution, gave the pentaglycine complex of copper(II) and free glycine [104]. Similar reactions were also observed with nickel(II). Over a longer period, further, less specific, hydrolysis occurs. Solution studies suggested that peptide is coordinated to the copper(II) ion via nitrogen donors, so that hydrolytic cleavage probably occurred at the C-terminal residue.

The compound prepared by treating aqueous diketopiperazine and potassium hydroxide (molar ratio 1:2) with an excess of "copper(II) hydroxide" and originally considered to be a complex of diketopiperazine, was found, as a result of elemental and x-ray analyses to contain two glycylglycinate molecules per copper(II), the glycylglycine being produced by hydrolysis of the diketopiperazine [105].

An additional metal-promoted hydrolytic reaction has been observed during an investigation of Schiff base complexes of copper (II) with glycine, diglycine, and triglycine [107]. Diglycine reacted with bis(salicylaldehydato)copper(II) and bis(benzoylacetaldehydato)copper(II) to give the glycylglycinato-Schiff base copper(II) complexes. The similar reaction with triglycine at pH 4.5 gave however, the copper(II) complex of the diglycine Schiff base (21)

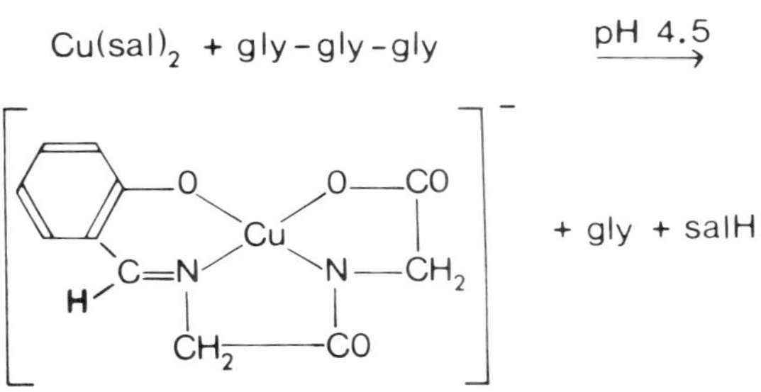

(21)

21

and free glycine [reaction (21)]. Subsequent work [108] has shown that hydrolytic cleavage occurred at the carboxyl end of the tripeptide, although not with 100% specificity.

The reactions of $Cu(Sal)_2$ (Sal = salicylaldehyde monoanion) with glycinamide, glycine esters and the amide and esters of glycylglycine have been studied [109] and the crystal structure of N-salicylideneglycinatoaquocopper(II) tetrahydrate has recently been reported [110]. Reactions of this type merit detailed study.

3.7.3. Peptide Bond Formation

Peptide bond formation using nonlabile cobalt(III) complexes has been studied in detail (Section 4.4). The formation of glycine peptide esters in the presence of copper(II) has recently been studied [111]. Treatment of glycine esters with copper(II) (other metal ions can also be used) in a nonaqueous solvent at room temperature gave di-, tri-, and tetraglycine peptide esters. After carbobenzyloxylation, the peptide esters were separated by column chromatography, and no evidence was obtained for diketopiperazine formation. Presumably carbonyl-metal-bonded intermediates are involved as in the cobalt(III) systems, and the lability of the copper(II) complexes allows the formation of tri- and tetrapeptides.

4. METAL ION-PROMOTED HYDROLYSIS: NONLABILE COBALT(III) SYSTEMS

4.1. Introduction

The use of kinetically inert cobalt(III) complexes rather than for example the kinetically labile nickel(II) and copper(II) systems has led to significant developments in recent years. The elegant work of the Buckingham and Sargeson group in Canberra has greatly clarified the mechanistic pathways available in reactions of this type. It is convenient to discuss these processes under the headings: ester hydrolysis, peptide (amide) hydrolysis, and peptide bond formation; however, it should be borne in mind that the intermediates are similar in each case and the discussion overlaps the three sections.

4.2. Ester Hydrolysis

Alexander and Busch first described the preparation of cobalt (III) complexes of glycine [112] and glycine esters [113]. Complexes of the type cis-$[Co(en)_2(NH_2CH_2CO_2R)Cl]Cl_2$ (23) were prepared by reacting the appropriate amino acid ester hydrochloride with trans-dichlorobis(ethylenediamine)cobalt(III) chloride (22) in aqueous solution. The free amino acid ester was generated in situ by the presence of a weakly coordinating base such as diethylamine.

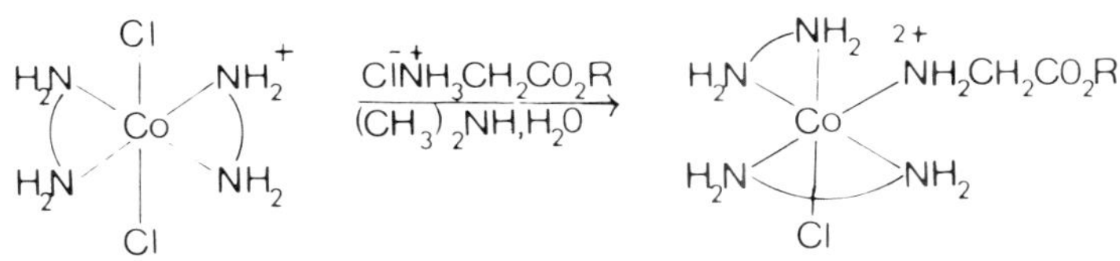

22 23 (22)

The N-coordinated amino acid derivatives (24) could be prepared by acid hydrolysis of the ester function [reaction (23)]. Alexander

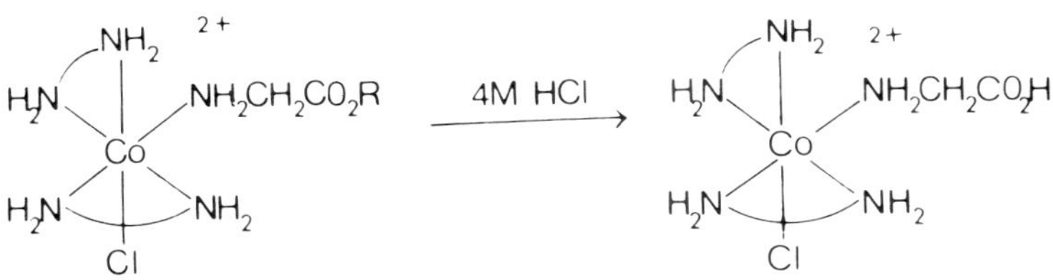

(23)

24

and Busch [114] studied the Hg^{2+}-promoted hydrolysis of the cis-$[Co(en)_2X(glyOR)]^{2+}$ ion in acid solution:

$$[Co(en)_2X(glyOR)]^{2+} + Hg^{2+} + H_2O \rightarrow [Co(en)_2gly]^{2+} + ROH + H^+ + HgX^+ \quad (24)$$

$(X = Cl, Br; R = CH_3, C_2H_5, i\text{-}C_3H_7)$

and proposed that a chelated ester species $[Co(en)_2(glyOR)]^{3+}$ (25) was the reactive intermediate in the Hg^{2+}-promoted reaction.

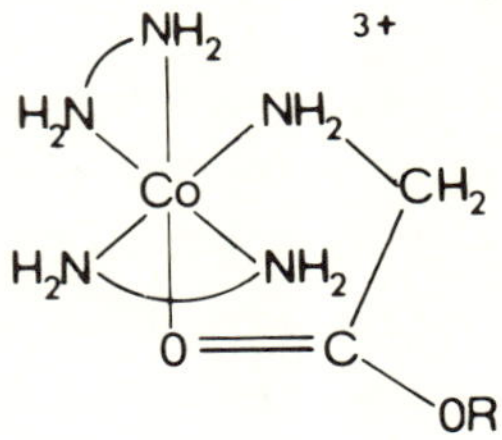
(25)

25

Evidence for such a species was obtained primarily from the change in the C=O stretching frequency as the monodentate ester (1735 cm^{-1}) was first chelated (1610 cm^{-1}) and then hydrolyzed to $[Co(en)_2gly]^{2+}$ (1640 cm^{-1}). The mercury(II) ion assists removal of the coordinated halide ion to give a 5-coordinate species, and the ester carbonyl oxygen competes so effectively with solvent water for the vacant coordination site, that the chelated ester complex is formed exclusively [reaction (26)]. Wu and Busch [115] have

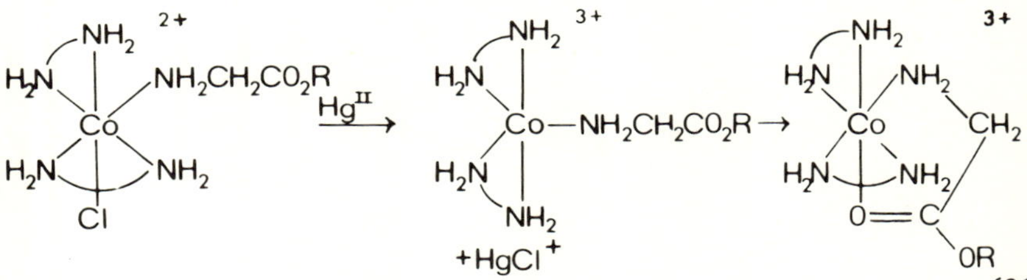
(26)

studied the analogous reactions of t-butyl glycinate in the complex $[Co(en)_2(NH_2CH_2CO_2C(CH_3)_3)Br]^{2+}$. The intermediate chelated ester complex was detected. The rate acceleration is lower by a factor of $\sim10^4$ for the t-butyl ester compared with the ethyl ester and hydrolysis of the t-butyl ester occurs with alkyl-oxy rather than acyl-oxy fission.

The perchlorate salts of the chelated ester species $[Co(en)_2(glyOR)](ClO_4)_3$ were subsequently isolated by Buckingham et al. [116] and used for the synthesis of peptide esters. Treatment of $[Co(en)_2(glyOCH_3)](ClO_4)_3$ with amino acid or peptide esters in anhydrous sulfolane, dimethylsulfoxide, or acetone solutions give the $[Co(en)_2(peptide\text{-}OR)]^{3+}$ ion [reaction (27)]. The isolation of

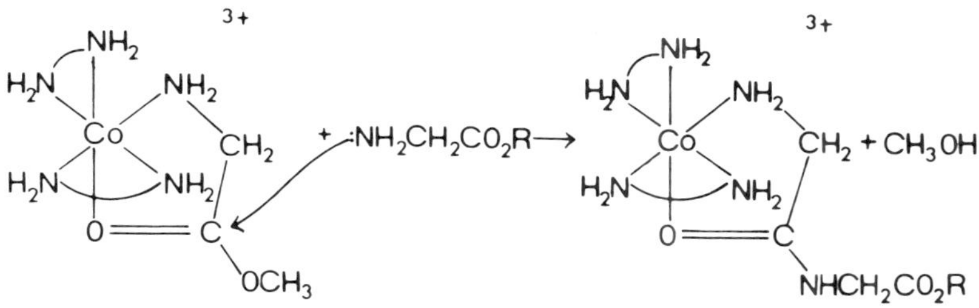
(27)

the chelated ester intermediate led to kinetic studies on such species. Kinetic data for the hydrolysis of $[Co(en)_2(glyOC_3H_7)](ClO_4)_3$ have been reported [117] and the results confirmed the proposal of Alexander and Busch [114] that a chelated ester intermediate is formed following the Hg^{2+} induced removal of halide ion from the chloropentammine complex. A similar intermediate [117] is generated following HOCl oxidation of the coordinated Br^- in the $[Co(en)_2Br(glyOR)]^{2+}$ ion (R = CH_3, C_2H_5, C_3H_7). The chelated ester is formed directly in the Hg^{2+}-assisted reaction, but some of the aquo complex $[Co(en)_2(H_2O)(glyOC_3H_7)]^{3+}$ occurs in the HOCl reaction. Isotopic studies using ^{18}O confirm that hydrolysis of the chelated ester occurs without rupture of the chelate ring.

In view of the possible biochemical significance of these reactions, it is important to study the reactions of the $Co^{(III)}$-ester systems at pH values more closely approaching physiological conditions than employed in the induced acid hydrolyses. The base hydrolysis of halopentammines of the type $[CoN_5X]^{2+}$ (X = Cl, Br; N_5 = a series of nitrogen donors) is considered to occur via a 5-coordinate intermediate which is different from those formed in the Hg^{2+}-induced reactions [118]. In the case of the ester complexes the ester carbonyl group and solvent water can compete for the vacant coordination site. The latter pathway leads to the formation of hydroxy ester complexes and the possibility of subsequent internal nucleophilic displacement of the ester group by the bound hydroxide ion. In a related study [119] of the base hydrolysis of $[Co(NH_3)_5NH_2CH_2CO_2R]^{3+}$ the importance of the role of internal nucleophiles was established. Treatment of the complex with hydroxide ion in the pH range 9 to 14 resulted in the formation of both $[Co(NH_3)_5(NH_2CH_2CO_2^-)]^{2+}$ containing the monodentate glycinate anion and $[Co(NH_3)_4(NH_2CH_2CONH)]^{2+}$, containing glycine amide chelated through both nitrogen atoms. The formation of chelated glycine amide can be rationalized as a result of the formation of the amido complex (26) [reaction (28)]. N-Coordinated aziridines also occur

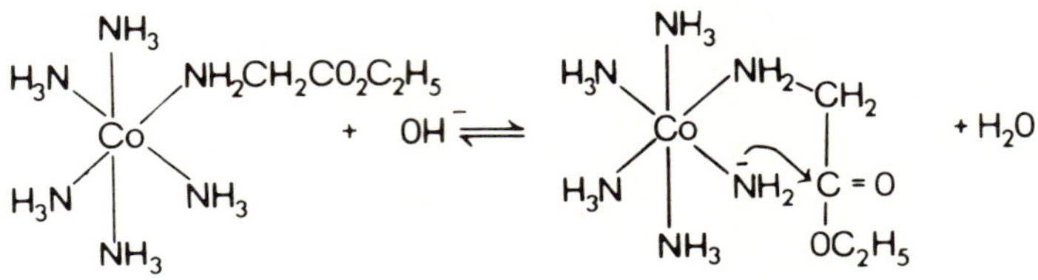

26

(28)

in the base hydrolysis of $[Co(en)_2Br(NH_2CH_2Br)]^{2+}$ by a similar intramolecular pathway [120]. The base hydrolysis of the $[Co(en)_2X(glyOR)]^{2+}$ ions have been studied in detail by Buckingham et al. [121], where (X = Cl, Br; R = CH_3, C_2H_5, $(CH_3)_2CH$, $CH_3(CH_2)_3$, $(CH_3)_3C$, and $C_6H_5CH_2$). The ion $[Co(en)_2(gly)]^{2+}$ is formed primarily, along with numerous side products, some of which have not yet been characterized. Tracer experiments with ^{18}O have established that the two main pathways to the chelated glycine complex are: (a) internal nucleophilic attack by the bound hydroxide ion, and (b) incorporation of the ester carbonyl group into the vacant coordination site of the 5-coordinate intermediate (Scheme 5). The $Hg^{(II)}$-induced acid hydrolysis and the base hydrolysis of the β_2-$[Co(trien)Cl(glyOEt)]^{2+}$ ion have also been studied [122]. Base hydrolysis of the Cl^- occurs some 10^3 times more rapidly than in the analogous bisethylenediamine system, and only chelated glycine products are formed:

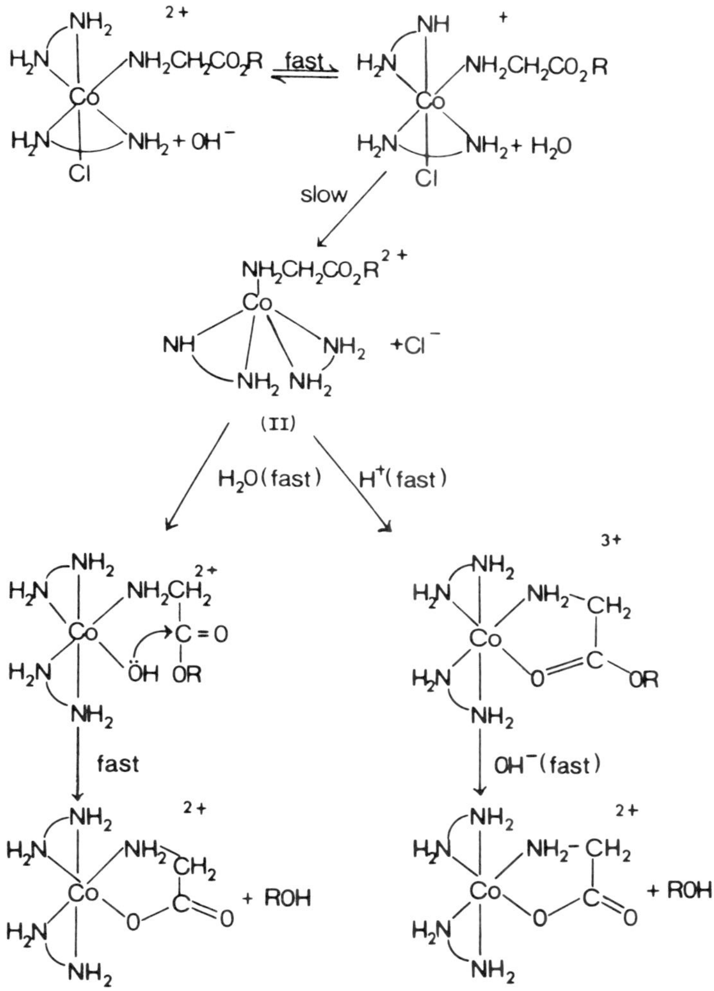

Scheme 5. Pathways to $[Co(en)_2gly]^{2+}$.

$$\beta_2\text{-}[Co(trien)Cl(glyOC_2H_5)]^{2+} + Hg^{2+} + H_2O \rightarrow$$
$$\beta_2\text{-}[Co(trien)gly]^{2+} + C_2H_5OH + HgCl^+ + H^+ \qquad (29)$$

$$\beta_2\text{-}[Co(trien)Cl(glyOC_2H_5)]^{2+} + OH^- \rightarrow$$
$$\beta_2\text{-}[Co(trien)gly]^{2+} + Cl^- + C_2H_5OH \qquad (30)$$

Base hydrolysis in the ^{18}O-labeled solvent establishes that some 84% of the β_2-$[Co(trien)gly]^{2+}$ product arises via the chelated ester and the remainder by intramolecular nucleophilic attack by the coordinated hydroxide ion.

Studies have also been made of the base hydrolysis of coordinated glycine amide derivatives cis-$[Co(en)_2Br(glyNR_1R_2)]^{2+}$ [123]. These reactions are discussed in Section 4.3.

Base hydrolysis of cis-$[Co(en)_2Cl(NH_2(CH_2)_5CO_2Me]^{2+}$ has recently been investigated [124]. Two consecutive reactions occur in base. Following loss of chloride ion, a slower base hydrolysis of the ester function occurs. The final product of the reaction is the hydroxypentammine and the reaction is free from the intramolecular effects observed with the glycine esters.

Detailed studies have now been made of the hydrolysis and aminolysis of the cobalt(III) carbonyl-bonded species (27) [125].

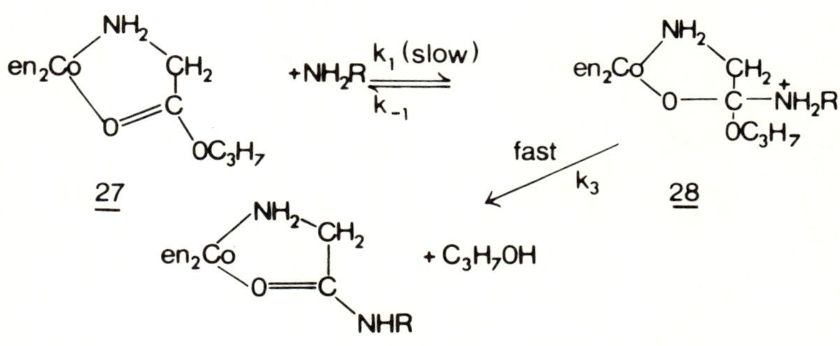

27 28 (31)

It has been found that all oxygen and nitrogen bases act as specific nucleophiles rather than as general bases toward acyl-activated esters or amides of the general structural type <u>27</u>, irrespective of their basicity toward a proton [125].

The tetrahedral intermediate involved in the aminolysis of organic esters has not been directly observed; however, the tetrahedral intermediate <u>28</u> has now been detected spectrophotometrically [126] in the reaction of glycine ethyl ester with <u>27</u>. It has been concluded that k_1 is rate-controlling in all such substitution reactions even though <u>27</u> is highly activated toward nucleophilic attack and contains a "poor" leaving group.

4.3. Peptide Hydrolysis

It has been found that a number of complexes of the type $[CoN_4(OH)(OH_2)]^{2+}$ (N_4 = a system of four nitrogen donor atoms, i.e., 2-en, trien or tren) stoichiometrically cleave the N-terminal amino acid from di- or tripeptides.

Gillard has reported [127] that $[Co(dien)X_3]$ complexes react with glycylglycylglycine to give $[Co(dien)(glygly)]^+$ and free glycine so that, in this case, the two N-terminal amino acids can be removed in a single step.

The peptide cleavage reactions are stoichiometric rather than catalytic, but in spite of this restriction they do provide interesting models for the cobalt-activated peptidases [for example, glycylglycine dipeptidase is activated by cobalt(II)] and for the metal proteinases in general.

A necessary feature for hydrolytic activity in the cobalt complex is the availability of at least two cis sites in the octahedral coordination sphere, the remaining sites being filled by nonlabile ligands.

It was first reported that the cis-β-$[Co(trien)(OH)(H_2O)]^{2+}$ ion [128] rapidly hydrolyzes amino acid esters and amides and acts selectively in the N-terminal hydrolysis of small peptide molecules. Cobalt(III) complexes of triethylenetetramine of the type $[Co(trien)X_2]$ can exist in three geometrical configurations designated cis-α, cis-β, and trans (see structures below). The trans-isomer need not be considered as it does not allow cis-chelation of

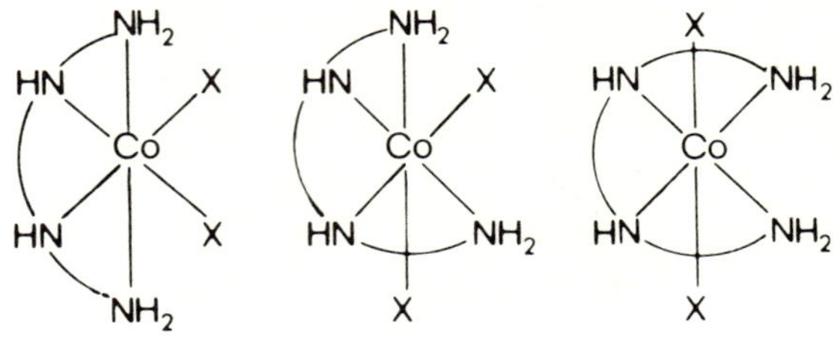

cis-α cis-β trans

$[Co^{III}(trien)X_2]$ isomers

an amino acid. As an amino acid contains two different donor atoms the reaction of an amino acid with the cis-β isomer gives two isomeric complexes. These isomers are designated β_1 where the amino acid nitrogen is trans to a primary amino group of trien, and β_2 where the amino acid nitrogen is trans to a secondary nitrogen of trien [129]. The reaction of cis-β-$[Co(trien)(OH)(H_2O)]^{2+}$ with

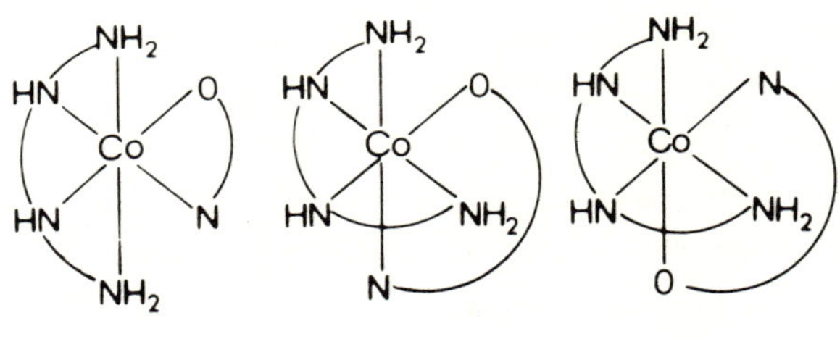

α β_1 β_2

$[Co(trien)gly]^{2+}$ isomers

glycine, L-alanine, L-valine, L-leucine, etc. in the pH range 7.5 to 8.0 and 60°C gives quantitative yields of the corresponding amino acid complex ions, β-$[Co(trien)AA]^{2+}$.

$$\beta\text{-}[Co(trien)(OH)(H_2O)]^{2+} + AA^- \rightarrow \beta\text{-}[Co(trien)AA]^{2+} \qquad (32)$$

where AA^- is the anion $NH_2CHRCO_2^-$. The cis-β-$[Co(trien)(OH)(H_2O)]^{2+}$ also reacts with peptides, amino acid esters, and amino acid amides to give the amino acid complex β-$[Co(trien)AA]^{2+}$ in which the N-terminal amino acid residue resulting from hydrolysis of the amide or ester is coordinated. The requirement for a free N-terminal

$$\beta\text{-}[Co(trien)(OH)(OH_2)]^{2+} + H_2NCHRCONHR' \rightarrow \beta\text{-}[Co(trien)(NH_2CHRCO_2)]^{2+} + R'NH_2 \qquad (33)$$

group was established by the observation that the β-$[Co(trien)(OH)(OH_2)]^{2+}$ ion does not hydrolyze the peptide bonds of such N-protected peptides as N-carbobenzyloxyglycylphenylalanine. Kinetic measurements on the hydrolysis of amino acid esters and glycinamide suggest a rate law, first order in β-$[Co(trien)(OH)(H_2O)]^{2+}$ and first order in the free base form of the amino acid derivative. The proposed mechanisms for the hydrolysis are shown in Scheme 6.

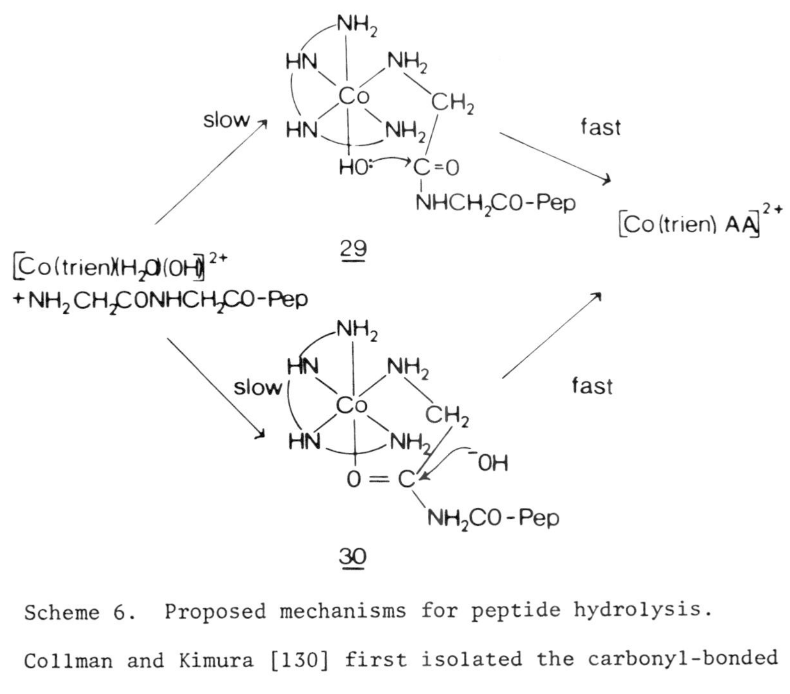

Scheme 6. Proposed mechanisms for peptide hydrolysis.

Collman and Kimura [130] first isolated the carbonyl-bonded

species illustrated in 30, and this complex is considered to be the intermediate involved in the peptide hydrolysis [see also Section 3.7.3]. The preparation of the peptide reagent cis-β-hydroxyaquo (trien)cobalt(III) perchlorate has been described [131].

Attempts to carry out the related reactions with the complex cis-$[Co(en)_2(OH)(H_2O)]^{2+}$ were less successful. A study of the reaction of the above complex with amino acid esters, amino acid amides, and dipeptides has been published [132]. Amide or ester hydrolysis is usually accompanied by ligand disproportionation leading to mixtures of products. The complex hydrolyzes N-terminal amino acids from peptides at pH 7 to 8 and 80°C. However, problems arise due to the equilibrium

$$\text{cis-}[Co(en)_2(OH)(H_2O)]^{2+} \rightleftarrows \text{trans-}[Co(en)_2(OH)(H_2O)]^{2+} \qquad (34)$$

since the cis isomer is only slightly favored over the trans isomer [133].

The cleavage of amino acid esters and peptides by the complex $[Co(tren)(OH)(H_2O)]^{2+}$ has also been studied [134] (tren = 2,2',2''-triaminotriethylamine). The N-terminal amino acid of a peptide is selectively removed as $[Co(tren)AA]^{2+}$ and only a trace of the C-terminal amino acid complex is detected in a few cases. The reactions are, however, much slower than the corresponding peptide cleavage reactions of trien, probably due to slower water exchange rates in the tren complex.

Cyclen complexes [135] and complexes of the diethylenetriamine acetic acids [136] have been prepared as possible peptide cleaving reagents. Cyclen can now be readily synthesized.

A number of N-O chelated glycine amide and peptide complexes of the type $[CoN_4(\text{gly }NR_1R_2)]^{2+}$ have been prepared [137] (N_4 = 2 en or trien). Typical complexes are shown in 31 and 32. The kinetics of base hydrolysis have been studied [137] in order to establish if the C=O chelated species was the reactive intermediate in the $[Co^{(III)}(\text{tetramine})\text{-}(OH)(OH_2)]^{2+}$-promoted reactions. For the ethylenediamine system the kinetics are consistent with the reaction

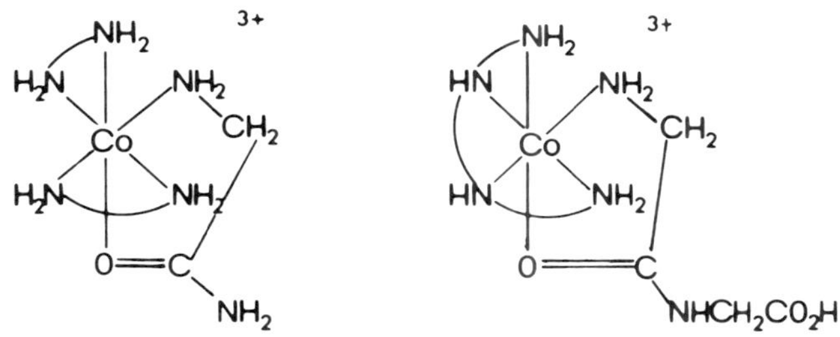

31 32

Scheme 7. At high pH (∿11) deprotonation of the amide or peptide nitrogen occurs. Such deprotonated species do not undergo base hydrolysis. For the carbonyl-bonded species <u>33</u> rate constants for hydroxide ion attack vary from 1 to 25 M^{-1} sec^{-1} at 25°C and I = 1.0 M corresponding to rate enhancements of ca. 10^4 to 10^6 over the uncoordinated substrates. The biological implications of these reactions have been considered by Buckingham et al. [137]. The $Zn^{(II)}$-activated carboxypeptidase A and the $Mn^{(II)}$-activated leucine aminopeptidase have rates some 10 to 10^4 times faster than obtained by $Co^{(III)}$ activation of the carbonyl group. It was suggested that higher rate accelerations could be obtained in the model system if

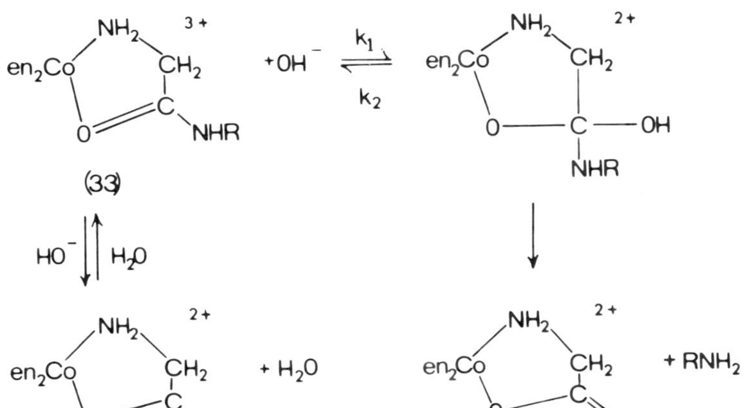

Scheme 7. Base hydrolysis of carbonyl-bonded amides and peptides.

intramolecular effects involving coordinated hydroxide ion could occur.

A comparison of intramolecular and intermolecular hydrolysis has been given [123]. Base hydrolysis of the cis-$[Co(en)_2Br(glyNR_1R_2)]^{2+}$ over the pH range 9 to 14 results in two pathways for the production of $[Co(en)_2(gly)]^{2+}$. Following loss of bromide, competition for the 5-coordinate intermediate by solvent water and the amide carbonyl oxygen results in the cis-$[Co(en)_2OH(glyNR_1R_2)]^{2+}$ and the $[Co(en)_2(glyNR_1R_2)]^{3+}$ (carbonyl chelated) species in the ration 54:46 ($R_1 = R_2 = H$).

Intramolecular hydrolysis in the cis-$[Co(en)_2OH(gly\ NR_1R_2)]^{2+}$ ion (33) is at least 10^7 and possibly 10^{11} times faster than base hydrolysis of uncoordinated glycine amide. Such a mechanism provides a pathway for hydrolysis which is decidedly more efficient

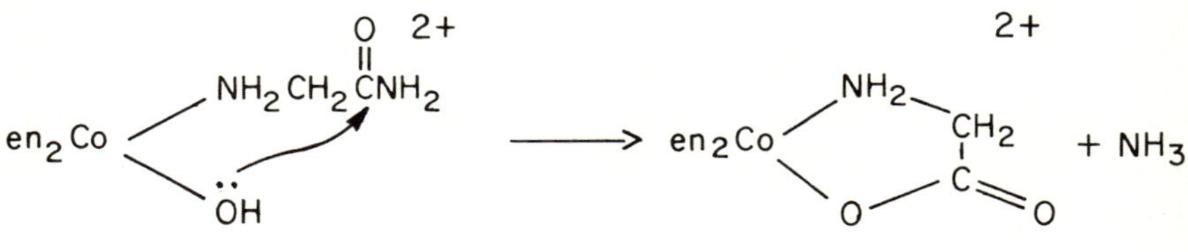

(33)

than that provided by direct metal-ion polarization of the carbonyl function. Divalent metal ions compare favorably with cobalt(III) in promoting hydrolysis in model systems and it seems likely that such mechanisms will also operate in the labile metal ion reactions.

4.4. Peptide Bond Formation

As discussed in Section 4.2, Busch and his coworkers [113] first prepared the series of complexes cis-$[Co(en)_2(glyOR)Cl]^{2+}$ where R = CH_3, C_2H_5, C_3H_7, and t-C_4H_9, by reacting glycine esters with trans-$[Co(en)_2Cl_2]Cl$ in water.

When the reaction of the cobalt(III) complex with a glycine ester is carried out in dimethylsulfoxide (DMSO) or dimethylformamide (DMF), rather than in water, the products are the dipeptide complexes $[Co(en)_2(glyglyOR)]^{3+}$; (R = CH_3, C_2H_5, C_3H_7) [138]. Similar results were obtained [130,139] for the reaction of glycine esters with cis-$[Co(trien)Cl_2]^+$ in water to give cis-$[Co(trien)(glyOr)Cl]^{2+}$, while the corresponding reactions in DMSO or DMF gave the dipeptide ester complexes cis-$[Co(trien)(glyglyOR)]^{3+}$. The structure of the complex cis-$[Co(trien)(glyglyOC_2H_5)](ClO_4)_3 \cdot H_2O$ has been determined by x-ray diffraction [140]. The peptide ester is coordinated through the terminal amino group and the carbonyl oxygen atom of the same amino acid residue (34). The pK_a of the peptide hydrogen is ca. 9.5 in the ester derivative. It has been concluded that peptide bond formation occurs via the chelated ester

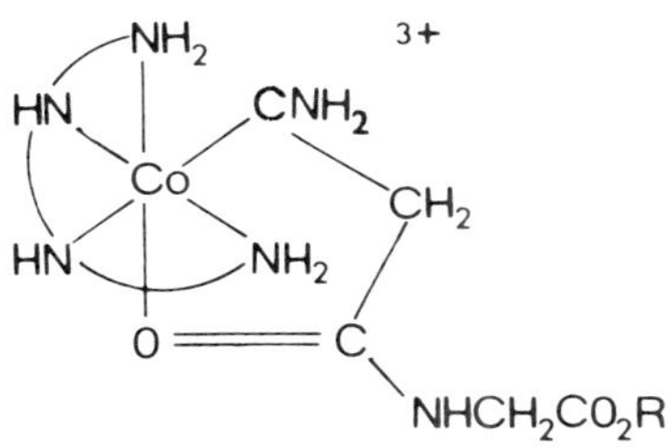

34

complex (35) which is activated toward nucleophilic attack by a further molecule of amino acid ester [reaction (36)]. Wu and Busch

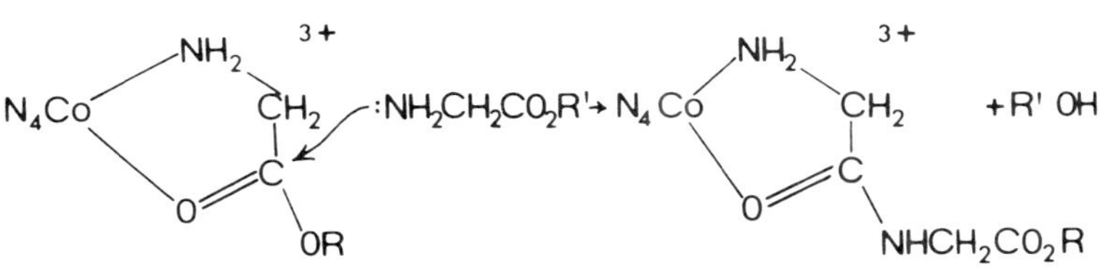

35

(36)

[141] have studied somewhat similar reactions of the $Co(dien)^{3+}$ moiety (a three-site system rather than a two-site system). The reaction of $[Co(dien)X_3]$, (X = Cl, NO_2) with glycine esters and glycylglycine esters gives $[Co(dien)(glyglyOR)X]^{2+}$ and $[Co(dien)(glyglyglyglyOR)]^{2+}$. Base hydrolysis of the nitrodipeptide complexes

at pH ∿10 gives the glycinato derivative

$$[Co(dien)(NH_2CH_2CONHCH_2CO_2R)NO_2]^{2+} + OH^- \rightarrow [Co(dien)(NH_2CH_2CO_2)NO_2]^+ \quad (37)$$

The carbonyl oxygen-bonded amide complexes of cobalt(III) are quite acidic [reaction (38)]. Typical pK_a values are $[Co(dien)(glyglyOR)NO_2]^{2+}$ 9.89; $[Co(dien)(glyNH_2)NO_2]^{2+}$ 10.5, and $[Co(dien)(glyglyOH)NO_2]^{2+}$ 11.1. The deprotonated species are inert to base hydrolysis (Section 4.3). The kinetics of base hydrolysis of some of these complexes has been studied in detail [182].

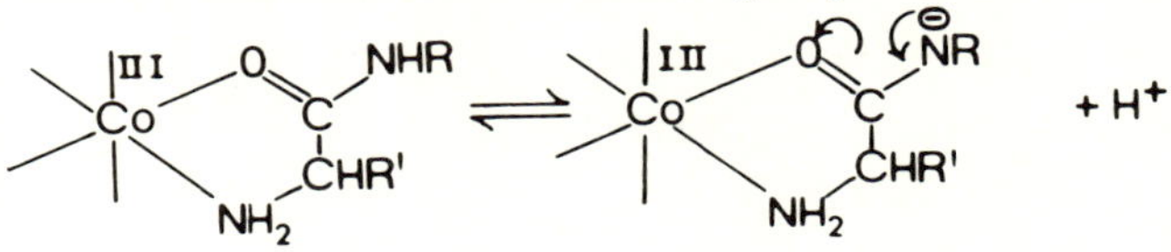

(38)

5. CONCLUSIONS

The remarkable facility of cobalt(III) in inducing the hydrolysis of amino acid esters and amides is sufficiently dramatic to be relevant to similar reactions in biological systems. Divalent metal ions appear to compare favorably with cobalt(III) in promoting hydrolysis in model systems and presumably similar mechanisms occur in the kinetically labile systems. Attack by a coordinated hydroxide ion on a coordinated substrate ("bound nucleophile-bound substrate") can lead to rate accelerations of at least 10^7 and possibly 10^{11}, while external hydroxide ion attack on a species in which the carbonyl group is directly bonded to the metal ion ("bound substrate-free nucleophile") leads to a rate acceleration of $\sim 10^6$.

A number of recent discussions are available on the mechanism of action of carboxypeptidase A [106,142-145]. The enzyme (MW = 34,600) consists of a single polypeptide chain of 307 amino acid residues and contains one atom of zinc per molecule, which is essential for enzymic activity. The enzyme catalyzes the hydrolysis of the peptide or ester bonds of N-acyl α-amino acids and O-acyl α-hydroxy acids adjacent to the free terminal carboxyl group. The zinc ligands have been identified as His-69, Glu-72, and His-196 in addition to a water ligand. The 2.0 Å resolution, x-ray investigation of the carboxypeptidase A-glycyl-L-tyrosine complex has established that the only portions of carboxypeptidase A which are sufficiently close to the peptide bond of the substrate to be directly involved in the catalysis are the Zn(II) ion, Glu-270, and Tyr-248. The carbonyl group of the peptide bond displaces the water ligand from the zinc ion in the complex. The only other group of the protein within 3 Å of a functional group of the substrate is Arg-145 which binds the terminal carboxyl group present in the substrate. As glycyl-L-tyrosine is a very poor substrate for carboxypeptidase A it is difficult to draw firm mechanistic conclusions from the x-ray diffraction data since the complex observed is likely to be nonproductive. However, it is believed that the arrangement of the zinc ion, Glu-270, Tyr-248, and Arg-145 in the complexes of carboxypeptidase A with reactive peptide substrates is essentially the same as in the case of glycyl-L-tyrosine.

A possible mechanism for the action of carboxypeptidase A, given by Lipscombe and coworkers [146], is shown in Scheme 8. Lipscombe has suggested that when the zinc-bound water molecule is replaced by the substrate, the environment around the zinc ion takes on a lower dielectric constant and hence polarization of the carbonyl group by the zinc ion becomes even more effective.

In addition to the "Zn-carbonyl mechanism," the "Zn-hydroxide" mechanism can also be considered, in which a zinc-bound hydroxide ions acts as a nucleophile. Such a mechanism appears to occur for the zinc metalloenzyme carbonic anhydrase. Carbonic anhydrase is not specific for carbon dioxide, in addition the enzyme also catalyzes the hydration of various aldehydes and the hydrolysis of certain esters [147-149]. A possible mechanism for the carbonic anhydrase-catalyzed hydrolysis of p-nitrophenyl acetate is shown in Scheme 9. A recent study of zinc-anion cooperation in aldehyde hydration has been published [150].

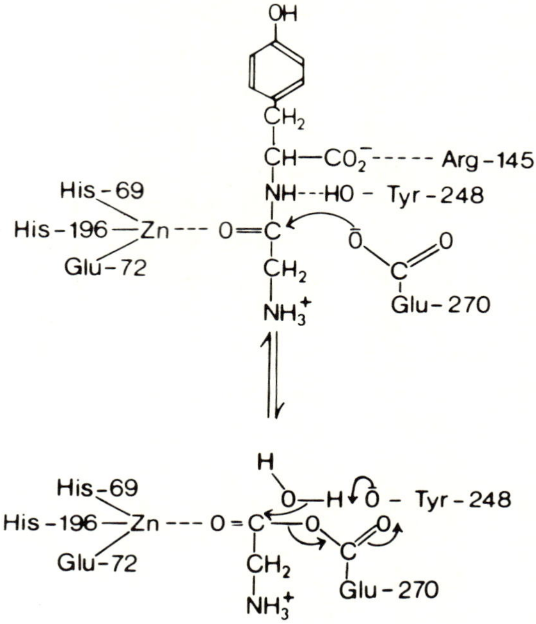

Scheme 8. Possible mechanism for the carboxypeptidase A-catalyzed hydrolysis of glycyl-L-tyrosine suggested by Lipscombe and coworkers [146].

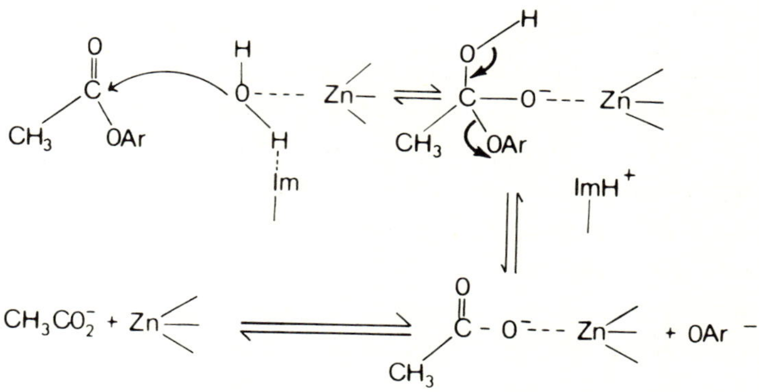

Scheme 9. Hydrolysis of p-nitrophenyl acetate catalyzed by carbonic anhydrase. (After Pocker and Stone [147].)

It is clear that studies of the model systems has done much to clarify the possible mechanistic pathways of the enzymic reactions and further important developments in this area can be expected.

ADDENDUM

Ester Hydrolysis: Labile Systems

A number of additional studies have been made of stereoselective effects in the metal complex-catalyzed hydrolysis of amino acid esters [151-153] with particular reference to histidine methyl ester. Stereoselectivity is observed if the catalyst is the 1:1 complex of nickel(II) with histidine or tryptophan. Stereoselectivity is not observed with the 1:1 complexes of nickel(II) with aspartic acid or methionine. Aspects of the reactivity of amino acids and their derivatives coordinated to metal ions have been reviewed [154] as has the metal ion-promoted hydrolysis of peptides and amino acid derivatives [155]. Catalysis of the hydrolysis of p-nitrophenyl carboalkoxyglycinates by hydroxo complexes of mercury(II) chelates has been the subject of a recent kinetic investigation [156]. This study is probably the first investigation of the hydrolysis of an α-amino acid ester containing a good leaving group; previous work has been concerned with simple alkyl esters.

Rodgers and Jacobson [157] have published the crystal structure of DL-(ethylvalinate-N,N-diaceto)diaquocopper(II), i.e., the copper complex of EVDA. The stereochemistry around the copper is distorted octahedral, the equatorial coordination consisting of the tertiary nitrogen, the two acetate oxygens, and a water molecule. The second water molecule occupies an axial position at a greater distance from the metal, while the ether oxygen of the ester group occupies the opposite axial position lying at 2.84 Å from the metal. The ester carbonyl group is not coordinated.

It is becoming increasingly clear that the rate enhancements observed in metal-promoted hydrolytic reactions are due primarily to more positive (i.e., less negative) entropies of activation. Thus $\Delta(\Delta S^{\ddagger})$ values of as much as +30 cal K^{-1} mol^{-1} have been noted compared with the nonpromoted reactions [158], corresponding to a rate acceleration of 3.2×10^6 at a constant enthalpy of activation. Bimolecular reactions have $\Delta S^{\ddagger}$ values in the range -5 to -15 cal K^{-1} mol^{-1} [159]. The entropy of activation is extremely sensitive to solvent effects. The orientation of solvent molecules around charges or developing charges in the transition state results in a negative entropy change and the effect may be as large, or larger, than the entropy change resulting from the molecularity of the reaction. In a hydrolytic reaction the addition of a nucleophile (H_2O or OH^-) to the free substrate results in a transition state of considerable polarity or hydrogen-bonding ability requiring increased solvent structuring in its vicinity and a corresponding negative contribution to $\Delta S^{\ddagger}$. In ester hydrolysis the metal ion can be regarded as "solvating" the developing negative charge on the carbonyl oxygen thus leading to more positive entropies of activation.

A very useful review on the interaction of histidine and its derivatives with transition metal ions has appeared [160].

Peptides and Peptide Esters

Long and coworkers [161,162] have noted that the rate of glycylglycine hydrolysis is directly proportional to the concentrations of Co(II), Ni(II), and Mn(II) over the pH range 3.8 to 6.0. A simulation analysis by computer of this work has appeared [163]. Nakon and Angelici [164] have carried out a potentiometric and infrared study of copper(II) complexes of glycylglycine methyl ester and glycylsarcosine methyl ester. The copper(II)-catalyzed hydrolysis of the methyl esters of glycylglycine and glycylsarcosine has been studied [165], as has the copper(II)-promoted hydrolysis of ethyl glycylglycinate, ethyl glycyl-β-alaninate, and ethyl glycyl-L-leucinate [166]. Under the conditions of the experiments (pH >7.0) the peptide esters act as tridentate ligands, donation occurring from the terminal amino group and the deprotonated amide nitrogen atom with a weak interaction between the metal ion and the carbonyl group of the ester. Rate accelerations of the order of 10^3 (compared with the unprotonated ligands) were observed. Base (and water) hydrolysis of the 1:1 complex of the tetramethyl ester of EDTA with copper(II) ($[Cu(Me_4EDTA)]^{2+}$) has been investigated. In the complex probably two ester groups of the ligand are bonded to the metal ion in conjunction with the two nitrogen donors. Rate accelerations of the order of 10^5 compared with the unprotonated ligand were observed [167].

Cobalt(III) Systems

The kinetics of the reaction between isopropylglycinate bis-(ethylenediamine)cobalt(III) perchlorate and glycine ethyl ester in DMSO solution to give ethyl glycylglycinate bis(ethylenediamine)-cobalt(III) perchlorate and isopropyl alcohol has been studied in detail and the tetrahedral intermediate observed spectrophotometrically [168]. It has been suggested [169] that high concentrations of ClO_4^- alter the mechanism of hydrolysis of glycine ethyl ester in cis-$(Co(en)_2X(glyOEt)]^{2+}$ ions (X = Cl, Br). Previous workers [114,117,122] have argued that removal of halide as HgX^+ results in the exclusive entry of the carbonyl oxygen into the coordination sphere to form a chelated glycine ester intermediate, but Nomiya and Kobayashi [169] have reported that at high ClO_4^- concentrations (>4 M) a significant amount of the aquo monodentate ester complex $[Co(en)_2(H_2O)(glyOEt)]^{3+}$ is formed and that this compound is relatively inert and only slowly ($t_{1/2}$ ∿23 min) reverts to the chelated ester intermediate by loss of bound water. Buckingham and Wein [170] have claimed that ClO_4^- does not alter the mechanism, but changes the rate and in the presence of mercury(II) the reaction proceeds exclusively via the chelated ester intermediate $[Co(en)_2$-$glyOEt]^{3+}$.

The base hydrolysis of cis-$[Co(en)_2Cl(NH_2(CH_2)_5CO_2Me]^{2+}$ has been studied by pH-stat and stopped-flow techniques [171]. Following loss of chloride a slower base hydrolysis of the ester function occurs. The final product of base hydrolysis is an hydroxypentamine.

A number of additional studies of the sequential hydrolysis of peptides with the cis-β-[Co(trien)(OH)(OH$_2$)]$^{2+}$ ion have appeared [172-176]. Bentley and Creaser [174,175] have shown that the cis-β-hydroxyaquo(triethylenetetramine)cobalt(III) ion is effective in promoting the hydrolysis of amino acid esters, dipeptides, and dipeptide esters at pH 8.0 and 25°C. Their results indicate that the quantitative total hydrolysis of peptides and proteins using cobalt(III)-activated intermediates is a feasible procedure. The hydroxyaquoethylenediaminediacetatocobalt(III) ion has also been shown to be effective in the cleavage of N-terminal peptide bonds [177]. The second-order rate constants for the hydrolysis of L-ala-L-phe with [Co(edda)(OH)(OH$_2$)] and cis-β-[Co(trien)(OH)(OH$_2$)]$^{2+}$ at pH 9.0 are very similar. The results indicate that the dipositive charge carried by the trien complex does not play a large role in promoting the peptide bond hydrolysis.

Girgis and Legg [178] have studied the [Co(trien)(OH)(OH$_2$)]$^{2+}$ promoted hydrolysis of the diethyl ester of L-aspartic acid and dipeptides containing aspartic acid, glutamic acid, and methionine. If the trifunctional amino acid is N-terminal as in α-L-aspartylglycine and L-aspartic acid diethyl ester, essentially no hydrolysis is observed; however, if [Co(dien)(OH)(OH$_2$)$_2$]$^{2+}$ is employed, hydrolysis occurs with the formation of [Co(dien)(L-asp)]$^+$ where the amino acid is coordinated as a tridentate. The crystal structure of [Co(tren)gly]$^{2+}$, a peptide hydrolysis product from [Co(tren)(OH)(OH$_2$)]$^{2+}$, has been reported [179]. Hydrolysis of gly-phe at pH 7.5 by the hydroxoaquotris(2-aminoethyl)aminocobalt(III) ion gives the orange isomer of [Co(tren)(gly)]$^{2+}$. The orange isomer is also obtained when [Co(tren)(glyOR)Cl]$^{2+}$ is hydrolyzed in acid solution in the presence of Hg(II). The crystal structure establishes that the glycine is coordinated with its nitrogen atom trans to the tertiary amino group of tren and that the tren ligand is composed of three ethylene diamine-type chelate rings with two δ and one λ conformations.

REFERENCES

1. See, for example: Q. Fernando, *Advan. Inorg. Chem. Radiochem., 7,* 185 (1965); M. L. Bender, in "Reactions of Coordinated Ligands," Advances in Chemistry Series, No. 37, American Chemical Society, 1963; T. C. Bruice and S. J. Benkovic, "Bioorganic Mechanisms," Vol. 1, Benjamin, New York, 1966; M. L. Bender, "Mechanisms of Homogeneous Catalysis from Protons to Proteins," Wiley-Interscience, 1971.
2. H. Kroll, *J. Amer. Chem. Soc., 74,* 2036 (1952).
3. For a general discussion of metal ions in enzymic catalysis see J. E. Coleman in "Progress in Bioorganic Chemistry," Vol. 1 (E. T. Kaiser and F. J. Kezdy, eds.), Wiley-Interscience, New York, 1971, pp. 159ff.
4. M. P. Springer and C. Curran, *Inorg. Chem., 2,* 1270 (1963).
5. R. W. Hay and L. J. Porter, *Aust. J. Chem., 20,* 675 (1967).
6. H. Shindo and T. L. Brown, *J. Amer. Chem. Soc., 87,* 1904 (1965).
7. M. L. Bender, *Chem. Rev., 60,* 53 (1960).
8. M. L. Bender, *J. Amer. Chem. Soc., 73,* 1626 (1951).
9. M. L. Bender and B. W. Turnquest, *J. Amer. Chem. Soc., 79,* 1889 (1957).
10. W. L. Koltun, M. Fried, and F. R. N. Gurd, *J. Amer. Chem. Soc., 82,* 233 (1960).
11. S. P. Datta and B. R. Rabin, *Biochim. Biophys. Acta, 19,* 572 (1956); *Trans. Faraday Soc., 52,* 1117, 1123, 1130 (1956).
12. R. B. Martin, M. Chamberlin, and J. T. Edsall, *J. Amer. Chem. Soc., 82,* 495 (1960).
13. R. W. Hay and L. J. Porter, *J. Chem. Soc. (B),* 1261 (1967).
14. R. W. Hay and P. J. Morris, *J. Chem. Soc. (B),* 1577 (1970).
15. R. W. Hay, L. J. Porter, and P. J. Morris, *Aust. J. Chem., 19,* 1197 (1966).
16. M. Robson, *Nature, 208,* 265 (1965).
17. M. Robson-Wright, *J. Chem. Soc. (B),* 1265 (1967).
18. R. W. Hay and P. J. Morris, *J. Chem. Soc. (A),* 3562 (1971).
19. R. W. Hay and P. J. Morris, *J. Chem. Soc.* (Perkin II), 1021 (1972).
20. R. W. Hay and P. J. Morris, *J. Chem. Soc. (A),* 1518 (1971).
21. L. J. Porter, D. D. Perrin, and R. W. Hay, *J. Chem. Soc. (A),* 118 (1969).

22. R. W. Hay, P. J. Morris, and D. D. Perrin, *Aust. J. Chem., 21,* 1073 (1968).

23. R. W. Hay and P. J. Morris, *Chem. Commun.,* 732 (1968).

24. W. Schaeg and F. Schneider, *Z. Physiol. Chem., 326,* 40 (1961).

25. J. P. Greenstein, *J. Biol. Chem., 96,* 499 (1932).

26. N. C. Li, E. Doody, and J. M. White, *J. Amer. Chem. Soc., 79,* 5859 (1957).

27. T. R. Kelly, Ph.D. Thesis, University of Glasgow, 1962.

28. A. C. Andrews and D. M. Zebolsky, *J. Chem. Soc.,* 742 (1965).

29. H. L. Conley and R. B. Martin, *J. Phys. Chem., 69,* 2923 (1965).

30. R. W. Hay and P. J. Morris, *Chem. Commun.,* 23 (1967).

31. J. E. Purdie and N. L. Benoiton, *Can. J. Chem., 49,* 3468 (1971).

32. C. Gustaffson, *Ann. Acad. Sci. Fennicae, A II,* No. 15 (1945).

33. J. M. White, R. A. Manning, and N. C. Li, *J. Amer. Chem. Soc., 78,* 2367 (1956).

34. R. P. Bell and B. A. W. Coller, *Trans. Faraday Soc., 60,* 1087 (1964).

35. H. L. Conley and R. B. Martin, *J. Phys. Chem., 69,* 2914 (1965).

36. W. A. Connor, M. M. Jones, and D. L. Tulcen, *Inorg. Chem., 4,* 1129 (1965).

37. J. E. Hix and M. M. Jones, *Inorg. Chem., 5,* 1863 (1966).

38. R. W. Hay and P. J. Morris, *J. Chem. Soc. (A),* 1524 (1971).

39. R. W. Hay and L. J. Porter, *J. Chem. Soc. (A),* 127 (1969).

40. R. W. Hay and L. J. Porter, *Chem. Commun.,* 653 (1967).

41. B. Holmquist and T. C. Bruice, *J. Amer. Chem. Soc., 91,* 2982 (1969).

42. C. K. Ingold, *J. Chem. Soc.,* 1375 (1930); 2170 (1931).

43. F. H. Westheimer and M. W. Shookoff, *J. Amer. Chem. Soc., 62,* 269 (1940).

44. R. P. Bell and G. M. Waind, *J. Chem. Soc.,* 1979 (1950).

R. P. Bell and F. J. Lindars, *J. Chem. Soc.,* 4601 (1954).

R. P. Bell and D. J. Rawlinson, *J. Chem. Soc.,* 4387 (1958).

R. P. Bell and G. A. Wright, *Trans. Faraday Soc., 57,* 1377 (1961).

R. P. Bell and M. Robson, *Trans. Faraday Soc., 60,* 893 (1964).

R. P. Bell and B. A. W. Coller, *Trans. Faraday Soc., 61,* 1445 (1965).

45. G. Aksnes and J. E. Prue, *J. Chem. Soc.*, 103 (1959).

G. Aksnes and P. Froyen, *Acta Chem. Scand.*, *20*, 1451 (1966); 1507 (1967).

46. B. Hansen, *Acta Chem. Scand.*, *12*, 324 (1958); *16*, 1927 (1962); *17*, 1307 (1963); *17*, 1481 (1963).

47. A. Agren, U. Hedsten, and B. Jonsson, *Acta Chem. Scand.*, *15*, 1532 (1961).
A. Agren and G. T. van der Veen, *Acta Pharm. Suec.*, *67*, 437 (1963).

48. R. B. Martin, *J. Amer. Chem. Soc.*, *89*, 2501 (1967).

49. G. W. Wheland, "Advanced Organic Chemistry," 3rd ed., Wiley, New York, 1960.

50. T. C. Bruice and S. J. Benkovic, "Bioorganic Mechanisms," Vol. 1, Benjamin, New York, 1966.
W. P. Jencks, "Catalysis in Chemistry and Enzymology," McGraw-Hill, New York, 1969.
A. J. Kirby and A. R. Fersht in "Progress in Bioorganic Chemistry," Vol. 1 (E. T. Kaiser and F. J. Kézdy, eds.), Wiley-Interscience, 1971, p. 1.

51. R. B. Martin, A. Parcell, and R. I. Hedrick, *J. Amer. Chem. Soc.*, *86*, 2406 (1964).

52. A. Pilbrant, *Acta Pharm. Suec.*, *6*, 469 (1969).

53. "Stability Constants" (L. G. Sillen and A. E. Martell, eds.), Chemical Society Special Publications No. 17 (1964) and 25 (1971).

54. J. E. Hix and M. M. Jones, *J. Amer. Chem. Soc.*, *90*, 1732 (1968).

55. R. D. Gillard, H. M. Irving, R. M. Parkins, N. C. Payne, and L. D. Pettit, *J. Chem. Soc. (A)*, 1159 (1966).

56. C. C. McDonald and W. D. Phillips, *J. Amer. Chem. Soc.*, *85*, 3736 (1963).

57. J. H. Ritsma, G. A. Wiegers, and F. Jellinek, *Rec. Trav. Chim.*, *84*, 1577 (1965).
J. H. Ritsma, J. C. Van de Grampel, and F. Jellinek, *Rec. Trav. Chim.*, *88*, 411 (1969).

58. P. J. Morris and R. B. Martin, *J. Inorg. Nucl. Chem.*, *32*, 2891 (1970).

59. C. Regardh, *Acta Pharm. Suec.*, *3*, 101 (1966).

60. D. L. Leussing and G. S. Alberts, *J. Amer. Chem. Soc.*, *82*, 4458 (1960).
D. L. Leussing, R. E. Laramy, and G. S. Alberts, *J. Amer. Chem. Soc.*, *82*, 4826 (1960).

61. D. D. Perrin and I. G. Sayce, *J. Chem. Soc. (A)*, 82 (1967); 53 (1968).

62. D. S. Barnes and L. D. Pettit, *Chem. Commun.*, 1000 (1970).

63. A. Pilbrant, *Acta Pharm. Suec.*, *6*, 37 (1969).

64. R. W. Hay and P. J. Morris, *J. Chem. Soc. (Dalton)*, 56 (1973); for a preliminary report see *Chem. Commun.*, 732 (1968).

65. C. G. Swain, *J. Amer. Chem. Soc.*, *66*, 1696 (1944).

66. J. M. White, R. A. Manning, and N. C. Li, *J. Amer. Chem. Soc.*, *77*, 5225 (1955).

67. R. Mathur and N. C. Li, *J. Amer. Chem. Soc.*, *86*, 1289 (1964).

68. R. J. Angelici and B. E. Leach, *J. Amer. Chem. Soc.*, *89*, 4605 (1967).

69. R. J. Angelici and B. E. Leach, *J. Amer. Chem. Soc.*, *90*, 2499 (1968).

70. B. E. Leach and R. J. Angelici, *J. Amer. Chem. Soc.*, *90*, *2504* (1968).

71. D. Hopgood and R. J. Angelici, *J. Amer. Chem. Soc.*, *90*, 2508 (1968).

72. R. J. Angelici and D. Hopgood, *J. Amer. Chem. Soc.*, *90*, 2514 (1968).

73. B. E. Leach and R. J. Angelici, *Inorg. Chem.*, *8*, 907 (1969).

74. B. E. Leach and R. J. Angelici, *J. Amer. Chem. Soc.*, *91*, 6296 (1969).

75. R. J. Angelici and J. W. Allison, *Inorg. Chem.*, *10*, 2238 (1971).

76. R. D. Gillard, *Inorg. Chim. Acta Rev.*, *1*, 69 (1967).

77. M. Murakami, H. Itatani, K. Takahashi, J. Kang, and K. Suzuki, *Mem. Inst. Sci. Ind. Res., Osaka Univ.*, *20*, 95 (1963).

78. R. W. Hay and P. J. Morris, *Chem. Commun.*, 18 (1969).

79. R. W. Kretsinger, F. A. Cotton, and R. F. Bryan, *Acta Cryst.*, *16*, 651 (1963).
M. M. Harding and S. J. Cole, *Acta Cryst.*, *16*, 643 (1963).
K. A. Fraser and M. M. Harding, *J. Chem. Soc.* (*A*), 415 (1967).
R. Candlin and M. M. Harding, *J. Chem. Soc.* (*A*), 421 (1967).
M. M. Harding and H. A. Long, *J. Chem. Soc.* (*A*), 2554 (1968).

80. T. Nozawa, Y. Akimoto, and M. Hatano, *Macromol. Chem.*, *158*, 21 (1972).

81. R. P. Houghton and C. S. Williams, *Tetrahedron Lett.*, 5091 (1967).

82. D. S. Sigman and C. T. Jorgenson, *J. Amer. Chem. Soc.*, *94*, 1724 (1972).

83. R. G. Lee, D. A. Long, and T. G. Truscott, *Trans. Faraday Soc.*, *65*, 503 (1969).

D. A. Long and T. G. Truscott, *Trans. Faraday Soc., 64,* 1866 (1968).
D. A. Long and T. G. Truscott, *Trans. Faraday Soc., 64,* 1624 (1968).
D. A. Long and T. G. Truscott, *Trans. Faraday Soc., 59,* 2316 (1963).
D. A. Long and T. G. Truscott, *Trans. Faraday Soc., 59,* 1833 (1963).

84. H. C. Freeman, *Advan. Protein Chem., 22,* 257 (1967).

85. R. D. Gillard, E. D. McKenzie, R. Mason, and G. B. Robertson, *Coord. Chem. Rev., 1,* 263 (1966).

86. A. S. Brill, R. B. Martin, and R. J. P. Williams, in "Electronic Aspects of Biochemistry" (B. Pullman, ed.), Academic Press, New York, 1964.

87. H. C. Freeman, "The Biochemistry of Copper" (J. Peisach, P. Aisen, and W. E. Blumberg, eds.), Academic Press, New York, 1966, pp. 77ff.

88. Specialist Periodical Reports, "Amino-Acids, Peptides and Proteins" (G. T. Young, ed.), The Chemical Society, London (each alternate volume contains a review of metal complexes).

89. For a discussion on this point see, for example, A. P. Brunetti, M. C. Lim, and G. H. Nancollas, *J. Amer. Chem. Soc., 90,* 5120 (1968).

90. B. Strandberg, I. Lindqvist, and R. Rosenstein, *Z. Kristallogr., 116,* 266 (1961).

91. H. C. Freeman, J. M. Guss, and R. L. Sinclair, *Chem. Commun.,* 485 (1968).

92. E. W. Wilson and R. B. Martin, *Inorg. Chem., 9,* 528 (1970).

93. E. D. McKenzie, *J. Chem. Soc. (A),* 1655 (1969).

94. R. D. Gillard, E. D. McKenzie, R. Mason, and G. B. Robertson, *Nature 209,* 1347 (1966).

95. M. T. Barnet, H. C. Freeman, D. A. Buckingham, I-Nan Hsu, and D. van der Helm, *Chem. Commun.,* 367 (1970).

96. D. L. Rabenstein, *Can. J. Chem., 49,* 3767 (1971).

97. E. Bamann, J. G. Haas, and H. Trapmann, *Arch. Pharm., 294,* 569 (1961).

98. E. Bamann, *Angew. Chem., 52,* 186 (1939);

E. Bamann and M. Meisenheimer, *Chem. Ber., 71,* 1711, 1980, 2086, 2233 (1938).

99. E. Bamann, H. Trapmann, and A. Rother, *Chem. Ber., 91,* 1744 (1958); *Naturwissenshaften, 43,* 326 (1956).

100. E. Bamann and H. Trapmann, *Advan. Enzymol., 21,* 169 (1959).

101. L. Meriwether and F. H. Westheimer, *J. Amer. Chem. Soc., 78,* 5119 (1956).

102. I. J. Grant and R. W. Hay, *Aust. J. Chem., 19,* 1189 (1965).

103. M. M. Jones, T. J. Cook, and S. Brammer, *J. Inorg. Nucl. Chem., 28,* 1265 (1966).

104. R. H. Andreatta, H. C. Freeman, A. V. Robertson, and R. L. Sinclair, *Chem. Commun.,* 203 (1967).

105. A. Sugihara, T. Ashida, Y. Sasada, and M. Kakudo, *Acta Cryst., B24,* 203 (1968).

106. E. T. Kaiser and B. L. Kaiser, *Accounts Chem. Res., 5,* 219 (1972).

107. A. Nakahara, K. Hamada, I. Miyachi, and K. Sakurai, *Bull. Chem. Soc. Japan, 40,* 2826 (1967).

108. A. Nakahara, K. Hamada, Y. Nakao, and T. Higashiyama, *Coord. Chem. Rev., 3,* 207 (1968).

109. K. Hamada, H. Ueda, Y. Nakao, and A. Nakahara, *Bull. Chem. Soc. Japan, 42,* 1297 (1969).

110. T. Ueki, T. Ashida, Y. Sasada, and M. Kakudo, *Acta Cryst., B25,* 328 (1969).

111. S. Yamada, M. Wagatsuma, Y. Takeuchi, and S. Terachima, *Chem. Pharm. Bull., 19,* 2380 (1971). See also *Tet. Lett.* 1501 (1970).

112. M. D. Alexander and D. H. Busch, *Inorg. Chem., 5,* 1590 (1966).

113. M. D. Alexander and D. H. Busch, *Inorg. Chem., 5,* 602 (1966).

114. M. D. Alexander and D. H. Busch, *J. Amer. Chem. Soc., 88,* 1130 (1966).

115. Y. Wu and D. H. Busch, *J. Amer. Chem. Soc., 92,* 3326 (1970).

116. D. A. Buckingham, L. G. Marzilli, and A. M. Sargeson, *J. Amer. Chem. Soc., 89,* 4539 (1967).

117. D. A. Buckingham, D. M. Foster, and A. M. Sargeson, *J. Amer. Chem. Soc., 90,* 6032 (1968).

118. See, for example, D. A. Buckingham, I. I. Olsen, and A. M. Sargeson, *J. Amer. Chem. Soc., 89,* 5129 (1967); *Aust. J. Chem., 20,* 597 (1967); *J. Amer. Chem. Soc., 90,* 6654 (1968).

119. D. A. Buckingham, D. M. Foster, and A. M. Sargeson, *J. Amer. Chem. Soc., 91,* 3451 (1969).

120. D. A. Buckingham, C. E. Davis, and A. M. Sargeson, *J. Amer. Chem. Soc., 92,* 6159 (1970).

121. D. A. Buckingham, D. M. Foster, and A. M. Sargeson, *J. Amer. Chem. Soc., 91,* 4102 (1969).

122. D. A. Buckingham, D. M. Foster, L. G. Marzilli, and A. M. Sargeson, *Inorg. Chem., 9,* 11 (1970).

123. D. A. Buckingham, D. M. Foster, and A. M. Sargeson, *J. Amer. Chem. Soc.*, *92*, 6151 (1970).

124. R. W. Hay, R. Bennett, and D. J. Barnes, *J. Chem. Soc. (Dalton)*, 1524 (1972).

125. D. A. Buckingham, J. Dekkers, A. M. Sargeson, and M. Wein, *J. Amer. Chem. Soc.*, *94* 4032 (1972).

126. D. A. Buckingham, J. Dekkers, and A. M. Sargeson, *J. Amer. Chem. Soc.*, *95*, 4173 (1973).

127. R. D. Gillard and D. A. Phipps, *Chem. Commun.*, 800 (1970).

128. D. A. Buckingham, J. P. Collman, D. A. R. Happer, and L. G. Marzilli, *J. Amer. Chem. Soc.*, *89*, 1052 (1967).
For a preliminary report see J. P. Collman and D. A. Buckingham, *J. Amer. Chem. Soc.*, *85*, 3039 (1963).

129. L. G. Marzilli and D. A. Buckingham, *Inorg. Chem.*, *7*, 1042 (1967).

130. J. P. Collman and E. Kimura, *J. Amer. Chem. Soc.*, *89*, 6096 (1967).

131. C. B. Storm and P. Ellsworth, *Biochem. Preps.*, *13*, 59 (1971).

132. D. A. Buckingham and J. P. Collman, *Inorg. Chem.*, *6*, 1803 (1967).

133. J. Bjerrum and E. Rasmussen, *Acta Chem. Scand.*, *6*, 1265 (1952).

134. E. Kimura, S. Young, and J. P. Collman, *Inorg. Chem.*, *9*, 1183 (1970).

135. J. P. Collman and P. W. Schneider, *Inorg. Chem.*, *5*, 1380 (1966).

136. P. W. Schneider and J. P. Collman, *Inorg. Chem.*, *7*, 2010 (1968).

137. D. A. Buckingham, C. E. Davis, D. M. Foster, and A. M. Sargeson, *J. Amer. Chem. Soc.*, *92*, 5571 (1970).

138. D. A. Buckingham, L. G. Marzilli, and A. M. Sargeson, *J. Amer. Chem. Soc.*, *89*, 4538 (1967).

139. D. A. Buckingham, L. G. Marzilli, and A. M. Sargeson, *J. Amer. Chem. Soc.*, *89*, 2772 (1967).

140. D. A. Buckingham, P. A. Marzilli, I. E. Maxwell, and A. M. Sargeson, *Chem. Commun.*, 488 (1968).

141. Y. Wu and D. H. Busch, *J. Amer. Chem. Soc.*, *94*, 4115 (1972).

142. W. N. Lipscombe, *Accounts Chem. Res.*, *3*, 81 (1970).

143. W. N. Lipscombe, *Chem. Soc. Rev.*, *1*, 319 (1972).

144. J. A. Hartsuck and W. N. Lipscombe, in "The Enzymes," Vol. 3, 3rd ed. (P. D. Boyer, ed.), Academic Press, New York, p. 1.

145. F. A. Quiocho and W. N. Lipscombe, *Advan. Protein Chem., 25,* 1 (1971).

146. W. N. Lipscombe, J. A. Hartsuck, G. N. Reeke, F. A. Quiocho, P. H. Bethge, M. L. Ludwig, T. A. Steitz, H. Muirhead, and J. C. Coppola, *Brookhaven Symp. Biol., 21,* 24 (1968).

147. Y. Pocker and J. T. Stone, *Biochemistry, 6,* 668 (1967).

148. J. A. Veerpoorte, S. Mehta, and J. T. Edsall, *J. Biol. Chem., 242,* 4221 (1967).

149. K.-W. Lo and E. T. Kaiser, *Chem. Commun.,* 834 (1966).

150. R. H. Prince and P. R. Woolley, *J. Chem. Soc. (Dalton),* 1548 (1972).

151. J. R. Blackburn and M. M. Jones, *J. Inorg. Nucl. Chem., 35,* 1597 (1973).

152. J. R. Blackburn and M. M. Jones, *J. Inorg. Nucl. Chem., 35,* 1605 (1973).

153. J. R. Blackburn and M. M. Jones, *J. Inorg. Nucl. Chem., 35,* 2421 (1973).

154. A. Passini and L. Casella, *J. Inorg. Nucl. Chem., 36,* 2133 (1974).

155. E. Kimura, *Yuki Gosei Kagaku Kyokai Shi, 29,* 1096 (1971).

156. M. M. Werber and Y. Shalitin, *Bioinorg. Chem., 2,* 275 (1973).

157. J. Rodgers and R. A. Jacobson, *Inorg. Chim. Acta, 13,* 163 (1975).

158. R. W. Hay and C. R. Clark, to be published.

159. L. L. Schaleger and F. A. Long, *Advan. Phys. Org. Chem., 1,* 1 (1963).

160. R. J. Sundberg and R. B. Martin, *Chem. Rev., 74,* 471 (1974).

161. D. A. Long, T. G. Truscott, J. R. Cronin, and R. G. Lee, *Trans. Faraday Soc., 67,* 1094 (1971).

162. J. R. Cronin, D. A. Long, and T. G. Truscott, *Trans. Faraday Soc., 67,* 2096 (1971).

163. K. Ohkubo and H. Sakamoto, *Bull. Chem. Soc. Japan, 46,* 2579 (1973).

164. R. Nakon and R. J. Angelici, *Inorg. Chem., 12,* 1269 (1973).

165. R. Nakon and R. J. Angelici, *J. Amer. Chem. Soc., 95,* 3170 (1973).

166. R. W. Hay and K. B. Nolan, *J. Chem. Soc. (Dalton),* 2542 (1974).

167. R. W. Hay and K. B. Nolan, *J. Chem. Soc. (Dalton)* in press.

168. D. A. Buckingham, J. Dekkers, and A. M. Sargeson, *J. Amer. Chem. Soc., 95,* 4173 (1973).

169. K. Nomiya and H. Kobayashi, *Inorg. Chem., 13,* 409 (1974).

170. D. A. Buckingham and M. Wein, *Inorg. Chem., 13,* 3027 (1974).

171. K. B. Nolan, B. R. Coles, and R. W. Hay, *J. Chem. Soc., (Dalton),* 2503 (1973).

172. M. D. Fenn and J. H. Bradbury, *Anal. Biochem., 49,* 498 (1972).

173. E. Kimura, *Inorg. Chem., 13,* 951 (1974).

174. K. W. Bentley and E. H. Creaser, *Inorg. Chem., 13,* 115 (1974).

175. K. W. Bentley and E. H. Creaser, *Biochem. J., 135,* 507 (1973).

176. G. R. Stark, *Advan. Protein Chem., 24,* 261 (1970).

177. S. K. Oh and C. B. Storm, *Bioinorg. Chem., 3,* 89 (1973).

178. A. Y. Girgus and J. I. Legg, *J. Amer. Chem. Soc., 94,* 8420 (1972).

179. Y. Mitsui, J. Watanabe, Y. Iitaka, and E. Kimura, *J. Chem. Soc. Chem. Comm.,* 280 (1975).

180. T. Nakata, M. Tasumi, and T. Miyazawa, *Bull. Chem. Soc. Japan, 48,* 1599 (1975).

181. J. E. Richman and T. J. Atkins, *J. Amer. Chem. Soc., 96,* 2268 (1974).

182. R. W. Hay and D. P. Piplani, *Kémiai Kozlemények* (in press).

Chapter 5

CATALYTIC ACTIVITY OF POLY-L,α-AMINO ACID-METAL ION COMPLEXES: NEW APPROACHES TO ENZYME MODELS

Masahiro Hatano and Tsunenori Nozawa
Chemical Research Institute of Non-Aqueous Solutions,
Tohoku University, 2-1-1 Katahira,
Sendai, Japan

1. INTRODUCTION

The significance of metal ions in biological systems is currently obvious. For example, the function of many enzymes is metal ion-dependent [1-4]. Recently Sigel and McCormick discussed the discriminating behavior of metal ions and ligands with regard to their biological significance [5] and attempted to answer the question: what are the control mechanisms that determine the coordination and coordination tendency of metal ions?

During the past two decades, considerable progress in understanding the role of metals in biological processes has resulted from the discovery and purification of a large number of metalloproteins and from efforts to ascertain their function, e.g., in catalysis, electron transport, or gas-exchange reaction [6].

Experiments on biological systems continue to progress admirably; meanwhile, the study of biochemical mechanisms may be approached from a new angle, i.e., by synthesizing and studying model molecules or model molecular systems having similar biochemical activity [7].

Breslow developed extensive models for enzyme catalysis [8]. Thus, model enzymes which bind the substrate into an oriented mixed complex and then attack the substrate with appropriately placed functional groups have been produced. Metal complexing produces such systems, with accelerations of hydrolysis by up to 10^9 times [8]. However, as a first approximation Breslow denies the idea that only a macromolecule can have enzyme-like properties; it is an open question whether or not a macromolecule is required.

It is the purpose of this chapter to answer whether or not a macromolecule is required. First we shall review the related studies historically in several branches.

1.1. The Environmental Effect of the Polyelectrolyte

As catalysts for metabolic processes, most enzymes possess amazing activity and specificity. Although complete understanding of enzyme action has not been established, certain pertinent facts are known. The principal constituent of an enzyme is protein, a naturally occurring, high molecular weight polyamide of polypeptide [9]. Some enzymes are metal ion chelates.

Synthetic polyamino acids have been shown to be instructive protein models, and it is reasonable that their combination with metal ions should bear some resemblance to metalloproteins. This possibility was examined by measuring the catalytic activity of complexes of copper(II) ion with polyamino acids [10-12]. The cupric ion catalyzes oxidations of ascorbic acid and hydroquinone with molecular oxygen, and the cupric complexes have been suggested as a model for copper-containing oxidases. The Cu(II)-catalyzed oxidations by molecular oxygen of ascorbate and homogentisate anions, hydroquinone, and tetramethyl-p-phenylenediamineammonium cation were studied in the presence of poly-L-histidine (PLH). It was found that the addition of PLH enhances the catalytic efficiency of Cu(II) toward negatively charged and neutral substrates, but inhibits the oxidation of positively charged substrates. The Cu(II)-PLH-catalyzed reaction, unlike the Cu(II)-aquo complex-catalyzed reaction, becomes zero order in substrate concentration at relatively low concentration of substrate and exhibits the Michaelis-Menten-type kinetics with increase of concentration of substrate. The specific effect of PLH on the Cu(II)-catalyzed reactions may be eliminated by the presence of a moderate concentration of inert electrolytes, if the environmental effect on the catalytic activity of Cu(II)-PLH originates from its polyelectrolyte nature. In this case, the addition of neutral salt to the reaction system results in decrease in the specific effect of PLH on the Cu(II)-catalyzed reactions. Thus, the enhancement of catalytic activity is interpreted as an electrostatic effect of the polyelectrolyte on the concentration of the reactant involved in the close vicinity of the bound Cu(II). It may be stated that, formally, from the standpoint of reaction kinetics, the polyelectrolyte behaves as a second phase in which the concentration of reactants is different from those in the bulk of solution. The distribution coefficients which determine the relative concentrations of the reactants in the two phases depend primarily on electrostatic parameters. In other words, owing to electrostatic factors, one encounters a feature like heterogeneous kinetics in a homogeneous medium in this polyelectrolyte system. The kinetic behavior of the PLH-Cu(II)-catalyzed oxidation reactions are reminiscent of some enzymic oxidase processes. The similarity between the two types of catalysts, i.e., enzyme catalyst and the PLH-Cu(II) system catalyst, is even greater if one considers the fact that Cu(II) bound to imidazole groups is likely to help part of active site of some oxidases. This similarity is, however, rather apparent, as the products of reduction of oxygen are different for the two types of reactions; H_2O is produced in the oxidase reactions, whereas PLH-Cu(II) yields H_2O_2 in stoichiometric yield. Although the mechanisms of reaction of enzymic oxidase and of PLH-Cu(II) are evidently different, the specific effect as a polyelectrolyte on the pattern of catalytic reactions involving charged species, demonstrated in Pecht's studies [10-12], may perhaps contribute to resolve enzymic reactions.

1.2. Polymer Effect on Complex Stability

In the preceding section the enhancement of the catalytic activity in the presence of the polyelectrolyte was ascribed to an environmental effect of the polymer. An enhancement of the catalytic activity of polymer-metal complexes may be explained on the basis of the complex formation.

Many polymer-metal complexes were found to catalyze the decomposition of hydrogen peroxide [13-16]. In order to clarify the mechanism of the catalytic decomposition reaction of hydrogen peroxide by polyacrylhydroxamatocopper(II) complexes, the stability constants of these complexes were measured by using a modified Bjerrum's method, and the fraction of each of the chemical species was estimated [17]. The complex formation proceeds through the following two steps:

$$Cu^{2+} + HL \xrightarrow{K_1} CuL^+ + H^+ \quad (1)$$

$$CuL^+ + HL \xrightarrow{K_2} CuL_2 + H^+ \tag{2}$$

where HL is the ligand group, and K_1 and K_2 are the apparent stability constants of CuL^+ and CuL_2, respectively. The value of K_1 decreases with an increase in the ratio of the total concentration of Cu^{2+} to the total concentration of the ligand residues, T_{Cu}/T_{HL}, and also decreases with an increase in the molecular weight of the ligand polymer. The variation of K_1 is larger than that of K_2. Therefore, it was confirmed that an increase of T_{Cu}/T_{HL} and an increase of the molecular weight of the ligand polymer result in an increase in the fraction of CuL^+ species in the polymer-Cu(II) complexes.

The initial rate of the decomposition reaction of hydrogen peroxide, which depends upon the initial concentration of the substrate, increases with an increase in the ratio T_{Cu}/T_{HL} [16]. The relationship between the initial concentration of hydrogen peroxide and the initial rate of the reaction may be obtained from the Michaelis-Menten-type rate expression. The initial decomposition rate was discussed on the basis of the complex formation constants obtained [16].

The chemical species which contribute to the catalytic activity was presumed to be CuL^+, i.e., the chemical species derived from a cupric ion and one hydroxamato group. Therefore, further addition of this ligand over the formation of CuL^+ causes a competitive inhibition of the hydrogen peroxide decomposition reaction. These experimental results are consistent with the results in Sigel's paper [18].

The assumption of the CuL^+-active species was supported by the studies on the kinetics of the hydrogen peroxide decomposition reaction using a polymethacroylacetonato-Cu(II) complexes [19]. The catalytic activity of the complex is more pronounced than that of a low molecular weight analog, i.e., bis(2,4-pentanedionato)-Cu(II) complex. The effect of the molecular weight of polymeric ligands on the catalytic activity was examined using polymeric ligands designated P-1 and P-2, whose molecular weights were 6000 and 10,000, respectively. The catalytic activity of Cu(II) complexes was greater with P-2 than with P-1. In order to estimate the chemical species in the complexes on the basis of the formation constants of these systems, pH titrations were carried out, and the results were analyzed by a modified Bjerrum's method. From the relationship between the catalytic activity and the amount of the complex species, it was shown that the catalytically active species is CuL^+, and that the catalytic activity of CuL^+ is enhanced by an increase in the molecular weight of the ligand polymer. The similar experimental results were obtained in the various sorts of the reactions catalyzed with other metal complexes of the ligand polymer [20].

In enzyme systems, the existence of similar chemical species to those of CuL^+ type may be plausible, and these species may behave as active sites in enzymic reaction. The variation of the formation constant of the complexes with variation of the ligand moiety in polymer-metal complexes may be one of the environmental effects originating from the polyelectrolyte character of the ligand polymers.

1.3. Role of Ternary Complexes in Enzyme Reactions

The fact that the assumption of the CuL^+-active species was supported by studies on the kinetics of the hydrogen peroxide decomposition reaction using polymer-Cu(II) complexes, which is consistent with the experimental results on the formation constants of the polymer complexes, suggests the important role of ternary complexes in the metalloenzymes.

Recently, Gillard [21] summarized the possible reactions at an active site of a metalloenzyme in the following way: (a) The metal ion may induce, by coordination, a specific "lock" geometry of the apoprotein metal binding site so that only certain substrates are able to attach to the framework produced. (b) The metal ion may activate a bond or bonds of the substrate (or the protein) through coordination. This is quite feasible, since it may well be that the metal binding site of the apoprotein has a constant geometry, whether or not the metal ion is present. (c) The metal ion may induce by coordination a specific "key" geometry of the substrate so that it will fit specifically the "lock" of the protein.

In enzymes, the endgroups such as amino and carboxylate groups may coordinate with metal ions, and for transition metal ions such as Cu^{2+} the amide group in peptides may act as ligand. More important for the active sites of enzymes which coordinate metal ions are the binding sites created within an apoprotein by the occurrence of trivalent amino acids. Histidine, lysine, and glutamic acid offer a strong binding site, while methionine, serine, and threonine offer a weak one. The strong binding sites have often been suggested as essential to the active site of an enzyme. In this case, ternary complexes can be considered as models for enzyme-metal ion-substrate complexes. With regard to these complexes, it is informative to compare the stability constant of the binary complex with that of the ternary complex. The difference in stability between the binary and ternary complexes is a way to characterize the tendency toward formation of ternary complexes. Sigel discussed the difference in stability between the binary and ternary complexes [22-25]. Since more coordination positions are available for bonding of the first ligand to a given multivalent metal ion than for the second ligand, the stability constant for the formation of the 1:1 complex is usually greater than that for the 1:2, i.e., $\log K^{M}_{ML} > \log K^{ML}_{ML_2}$, and negative values are expected for $\Delta \log K$, according to equation (3).

$$\Delta \log K = \log K^{MA}_{MAB} - \log K^{M}_{MB} \tag{3}$$

Sigel found that the $\Delta \log K$ values for several Cu(II) complexes are positive. This is true for all systems which contain 2,2'-bipyridyl and a ligand with oxygen atoms as donors.

Investigation of the ethylenediamine-Cu(II)-pyrocatechol system shows that not each ternary N-ligand-Cu(II)-O-ligand system has a positive Δ log K value. An essential condition is obviously the π system of the nitrogen ligand. Since one might expect that the imidazole group has bipyridyl-like qualities as a ligand and it often has been suggested as a binding site for metal ions in proteins, it is of interest to compare the results for some complexes containing the imidazole group as a binding site.

The data of some ternary Cu(II) systems containing histamine as an imidazole ligand show that the 1:1 Cu(II)-histamine complex forms stabler ternary complexes with O donors than with N donors. This observation suggests that the same might be possible in ternary Cu(II) complexes of Cu(II)-proteins containing an imidazole group as the binding site. That two different ligands in the coordination sphere of a single metal ion influence each other was emphasized by the investigations of the stability of ternary complexes. The same is evident from kinetic studies. The Cu(II)-catalyzed decomposition of hydrogen peroxide is strongly dependent on the kind of ligand which is bound to the metal ion [23,24]. These examples show that a protein can exert a direct influence on the rate of a reaction, which is not only dependent on the metal ion it binds but also dependent on the kind of donor group it provides for the binding of the metal ion.

Recently, Davankov and Rogozkin [26,27] and Angelici and coworkers [28,29] found that a polystyrene resin containing an optically active Cu(II) complex can be used chromatographically to resolve partially or completely several optically active amino acids. Davankov reported the complete resolution of D,L-proline on the polymer containing an L-proline complex unit. On the other hand, Angelici used the polymer containing (N-carboxymethyl-L-valine)copper(II) units as an asymmetric sorbent. In these cases, the complex units have larger stability constants for the L-amino acidates than those for the D enantiomers. It is essential to form the ternary complexes of the complex unit in the asymmetric sorbent with amino acidate. Although the structures of the ternary complexes on the resin are unknown, these results suggest that there is steric repulsion between the R group on the amino acid and the bulky side chain of the complex unit in the resin. The asymmetrically selective complex formation in the resins containing the optically active copper(II) complex is instructive to study the mechanism of enzymic reactions and the role of the ternary complexes in the reactions.

2. POLY-L-LYSINE-COPPER(II) COMPLEXES AS COPPER-CONTAINING ENZYME MODELS

Many polymer-metal complexes have been found to exhibit high catalytic efficiencies in comparison with lower molecular weight analogs, and the catalytic activities of these complexes were compared with those of metalloenzymes. No polymer-metal complex exhibiting an asymmetrically selective catalysis was reported, however, until the poly-L-lysine-copper(II) complex was found to have a large catalytic activity and to be an asymmetrically selective catalyst for the oxidation of D- and L-(3,4-dihydroxyphenyl)alanine (DOPA).

2.1. Structures of Poly-L-Lysine-Copper(II) Complexes

Recently, metal-complex formations of poly-α-amino acids, such as poly-L-lysine, poly-L-glutamic acid, poly-L-histidine, and the catalytic effects of these metal complexes have been studied by several authors [30-33]. No conformational studies of these complexes have been reported, although in metalloenzymes, a knowledge of the conformational effects of apoenzymes on enzymic activities is important for the clarification of the mechanisms of reaction. The conformational changes of poly-α-amino acids were investigated in their metal complexes in order to correlate the conformations of poly-α-amino acid-metal complexes with catalytic activities. In this investigation, a poly-α-amino acid with amino groups in side chains, poly-L-lysine (PLL), was selected as a polymer ligand, since PLL has a more stable helical conformation than poly-L-ornithine (PLO) or poly-L-α,γ-diaminobutyric acid (PLB) at a pH greater than 10 [34,35]. The amide groups in the main chain or the protonated amino groups in the side chains of PLL are considered to be the ligand groups which release hydrogen ions through complex formations with cupric ions. The pH titration curve of the poly-L-lysine-copper(II) complex [PLL-Cu(II)] at $T_{Cu}/T_N = 0.13$ has two different buffer regions, before and after $\alpha = 0.5$, where T_{Cu} means the total molar concentration of cupric ion, and T_N means the total molar concentration of lysine residue. From the pH region in which the complex formation takes place, it was assumed that the complexes are formed between cupric ions and amino groups in the lower pH region ($\alpha = 0.5$), where α means the degree of neutralization. This assumption was verified by ORD, CD, and electronic spectral measurements, as described later.

In order to evaluate the average number of amino groups bound to one cupric ion ($\bar{n}$), it is necessary to calculate the total molar concentration of the amino groups bound to cupric ions and to divide it by the total molar concentration of the cupric ions present in the system. The total molar concentration of the amino groups bound is calculated from the following equations, derived from the total amino group concentration and electroneutrality [16].

$$T_N = [HL^+] + [L] + [CuL^{2+}] + 2[CuL_2{}^{2+}] + 3[CuL_3{}^{2+}] + 4[CuL_4{}^{2+}] \quad (4)$$

$$[Na^+] + [H^+] + [HL^+] + 2T_{Cu} = [OH^-] + [NO_3{}^-] + 2T_{.Cu} + T_N \quad (5)$$

The brackets specify the equilibrium concentrations, which are identical with the concentrations of the added reactants in the cases of $[Na^+]$ and $[NO_3{}^-]$. From equations (4) and (5), the following equation was obtained.

$$\text{Total bound amino groups} = (1 + K_a/[H^+])([H^+] + [Na^+] - [NO_3^-] - [OH^-]) - (K_a/[H^+])T_N \quad (6)$$

Here K_a is the apparent acid dissociation constant for PLL defined as in equation (7) and the values may be estimated from the modified titration curve at each value of α.

$$K_a = [L][H^+]/[HL^+] \quad (7)$$

The values of $\bar{n}$ calculated from equation (6) and T_{Cu} are plotted against $-\log([HL^+]/[H^+])$, which may be determined by use of equation (7). The formation curves for the PLL and PDLL systems are at larger $-\log(HL^+]/[H^+])$ values than that for the tetraammine complex system, which indicates that the formation constants of the cupric complexes of PLL and PDLL are much larger than that of the ammine complex. Furthermore, the separating factors between successive formation constants are much smaller in PLL and PDLL complexes than in the ammine complex, since the slopes of the curves for the former are steeper than that of the latter. This means that PLL-Cu(II) releases the amino groups as the pH decreases in the system.

Since Bjerrum's average coordination number concerning amino group, $\bar{n}$, amounts to 4 at pH 8.5, the PLL-Cu(II) complexes consist of mainly a cupric ion coordinated with four amino nitrogens at pH 8.5. This is shown in Figure 1 [36].

Figure 2 shows the visible absorption curves for PLL-Cu(II) at various degrees of neutralization (α). The visible spectra show that the wavelengths of the absorption maxima remain almost constant for α from 0 to 0.5, although the absorbances vary significantly. This fact indicates that only one complex species is formed in this region and the amount of the species vary with a variation of the α value. Since $\bar{n}$ has a maximum of 4, one cupric ion may bind four amino groups. Figure 2 shows also that the wavelengths and the extinction coefficients of the bands change remarkably with a

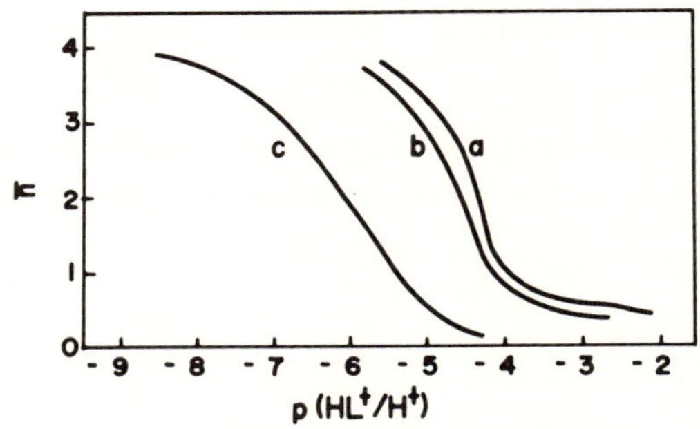

FIG. 1. Modified Bjerrum's plot for (a) PLL-Cu(II), (b) PDLL-Cu(II), and (c) tetrammine-Cu(II), at T_N = 0.01 M, T_{Cu}/T_N = 0.13, and 25°C.

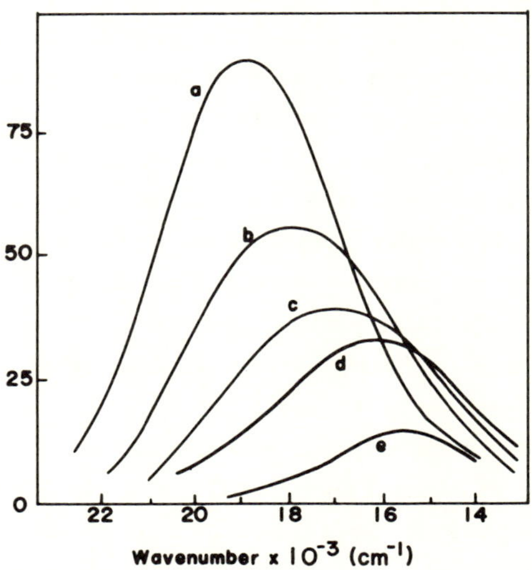

FIG. 2. Visible spectra of PLL-Cu(II) at various degrees of neutralization (a); a: 1.19, b: 0.98, c: 0.87, d: 0.80, and e: 0.55. (T_{Cu}/T_N = 0.14, 25°C).

variation of the α values (for α values greater than 0.5). Thus, in the higher pH region (pH = 9.0), the formation of a new chemical species is expected. As one possibility, an amide group in the polypeptide main chain may coordinate to the tetramino complex and induce an asymmetric character due to the asymmetric carbon atom in the polypeptide main chain.

In order to analyze the asymmetrical structures of PLL-Cu(II) its CD and AB spectra were measured at various temperatures, pH values, and conditions of preparation. The wavelengths and molar ellipticities at the extrema in the CD curves of PLL-Cu(II) depend remarkably on the pH of the solution (Table 1). The magnitudes of the extrema also depend upon the temperature and the ratio of T_{Cu}/T_N (Table 2). The CD spectrum of PLL-Cu(II) at pH 10.5 was resolved into several components by approximate Gaussian analysis (Figs. 3 and 4). It may be reasonably considered to consist of three groups of bands, i.e., four peaks in the 350 to 650 nm range. three peaks in the 240 to 350 nm range, and one peak at around

TABLE 1

Circular Dichroism of PLL-Cu(II) with pH Variation

(T_{Cu}/T_N = 0.075, T_N = 5 x 10^{-3} M, 22°C)

pH	α	λ_{min} or λ_{max} (nm) ([θ] [deg·cm^2/dmole])				
8.93	0.427	230 (-100)	255 (780)	280 (-880)	315 (590)	545 (-780)
9.53	0.508	228 (-200)	255 (2640)	285 (-590)	320 (660)	530 (-1170)
10.01	0.698	225 (-4150)	255 (3760)	284 (-1000)	318 (1000)	515 (-1900)
10.60	0.942	222 (-17,400)	255 (5300)	286 (-1900)	318 (1900)	513 (-3300)

TABLE 2

Circular Dichroism of PLL-Cu(II) at pH 10.5[a]

T_{Cu}/T_N	Temp. (°C)	$[\theta]^b$ at λ nm (x 10^{-3} deg cm^2/dmole)				
		222	255	290	320	510
0.05	23	-18.4	9.0	-2.8	2.8	-4.9
0.075	23	-12.1	7.7	-2.6	2.6	-4.1
0.10	23	- 8.2	4.3	-2.2	1.8	-3.5
0.13	25	- 5.0	3.7	-2.7	1.9	-3.9
0.13	20	- 5.5	3.7	-2.3	1.8	-3.8
0.13	15	- 7.7	3.6	-2.4	1.6	-3.5
0.13	10	- 8.8	3.5	-2.1	1.7	-3.6
0.15	23	- 2.0	4.4	-2.3	1.6	-3.7

[a]The measurements were carried out at T_{Cu} = 5 x 10^{-3} M.

[b]The values of [θ] refer to the residue concentration of PLL (the peak at 222 nm) and the molar concentration of cupric ion (the peaks at 255-510 nm).

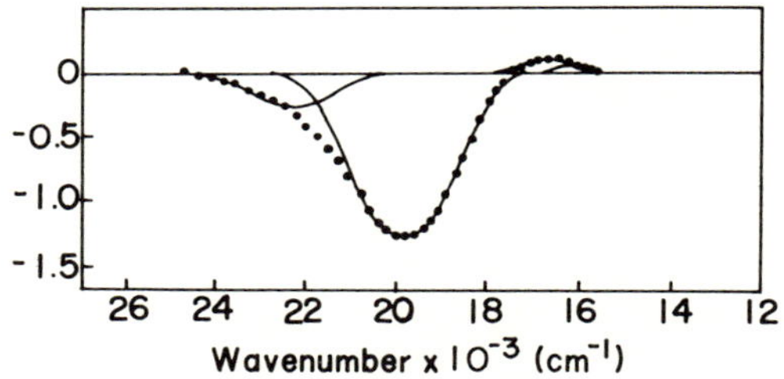

FIG. 3. Circular dichroism spectrum of PLL-Cu(II), d-d transition band. T_{Cu}/T_N = 0.13, T_N = 5·10^{-} M, pH = 10.5. ····· experimental curve, ——— resolved curves.

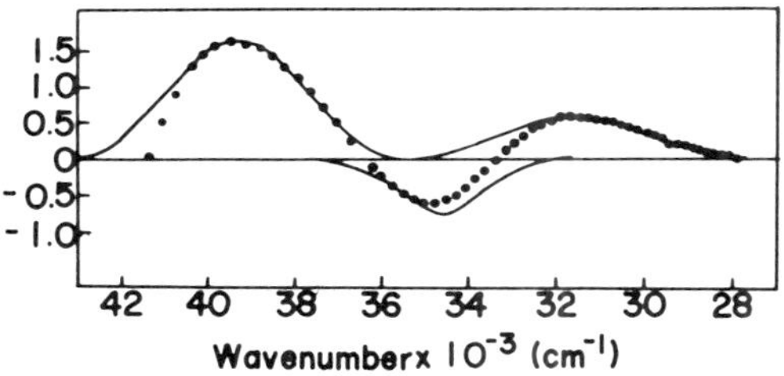

FIG. 4. Circular dichroism spectrum of PLL-Cu(II) charge transfer band. $T_{Cu}/T_N = 0.075$, $T_N = 5 \cdot 10^{-}$ M, pH = 10.5. •••• experimental curve, —— resolved curves.

222 nm. From a comparison of the CD spectra with the AB spectra, these three groups of bands may be attributed to d-d transitions of the cupric ion, charge-transfer transitions of the complex, and an $n\text{-}\pi^*$ carbonyl group transition of the amide in the helical PLL, respectively. The CD bands in the region 240 to 650 nm, therefore, could be ascribable to an asymmetric coordination structure of the complex.

In the discussion of the viscosity measurements [36] it was suggested that PLL-Cu(II) takes a helical or rodlike structure as a whole. The results in Table 2 support this assumption. The whole structure at $T_{Cu}/T_N = 0.05$ may be similar to that at $T_{Cu}/T_N = 0.15$, because the molar residue ellipticity in the wavelength region of 240 to 650 nm does not differ greatly when T_{Cu}/T_N is varied from 0.05 to 0.15. At a value of 0.05, the overall structure of PLL-Cu(II) is probably a helical rodlike one, since the molar residue ellipticity at 222 nm is as high as -18,400 deg•cm^2/dmol (Table 2).

The relative amount of asymmetric coordination structure decreases with decrease of pH and reaches zero for pH values less than 8.5. In the buffer region of higher pH, 2 more equivalents of alkali are consumed than the amount necessary for the deprotonation of the protonated amino groups in the side chain. It is inferred that the coordination of a peptide nitrogen to the cupric ion takes place with the displacement of a proton, and the consumption of another equivalent of alkali can be accounted for by the release of a proton from the water coordinated apically to the cupric ion.

In order to test for the coordination of the nitrogen from the peptide linkages to Cu^{2+}, the IR spectra of PLL and PLL-Cu(II) in D_2O solutions were taken. The IR spectra of PLL-Cu(II) have a band around 1610 cm^{-1}, which is not seen in the IR spectra of PLL, and which develops with an increase in pH. The band around 1610 cm^{-1} in PLL-Cu(II) is ascribable to the peptide carbonyl groups where the nitrogen atoms have coordinated to cupric ion with the displacement of proton [37-39].

From the CD bands in the wavelength region of 240 to 650 nm certain inferences may be made. Wellman has suggested a relationship between the signs and the magnitudes of the Cotton effects and the chelate ring conformations of the amino acidatocopper(II) complexes [40,41]. This suggestion has been somewhat modified by Martin et al. [42]. Simple octant rules have been proposed to account for the results of Cotton effect signs in the visible absorption region of transition metal complexes [42,43]. Yasui et al. [44] have ascertained experimentally the possibility of resolving the ligand field bands of copper complexes in solution into their split components by means of the CD. Copper(II) complexes of L-amino acids and L-amino acid amides exhibit four CD components in the ligand field region [44,45]. The CD spectrum of PLL-Cu(II) resembles that of L-amino acid amides in that both have four split components with the same signs in the ligand field region. If the main coordination plane with four amino nitrogens remains unchanged at pH 9.0 and 10.5, and the peptide amide nitrogen coordinates apically to cupric ion at greater distance when pH = 10.5, the λ_{max} in the d-d transition region should exhibit a red shift with an increase in pH [46]. Contrary to this expectation, the λ_{max} in the CD and AB spectra has a blue shift with increasing pH (Table 1). This fact suggests that the amide nitrogen may coordinate to cupric ion more closely than the amino nitrogens in the plane.

In comparison with the spectrum of PLL-Cu(II), the CD and AB spectra of the cupric complexes of L-alanineamide with ethylenediamine or diethylenetriamine were examined [47]. In this study it was found that the copper(II) complex of L-alanineamide with diethylenetriamine has the amide nitrogen coordinated to the Cu^{2+} in the apical position, and that the copper(II) complex of L-alanineamide with ethylenediamine has the amide nitrogen coordinated to the Cu^{2+} in the planar position.

Accordingly, in PLL-Cu(II) the peptide nitrogen and the hydroxy ion are considered to coordinate in the buffer region of high pH. Hence, the asymmetry in PLL-Cu(II) appears with increasing pH. Therefore, the asymmetric structure, other than the α helix, is attributable to the coordination of the peptide amide nitrogen to the cupric ion. In other words, the asymmetric structures in PLL-Cu(II) are induced by a vicinal effect of the asymmetric carbon of the peptide linkage. The addition of phenylalanine decreases $[\theta]_{240-650}$ and increases $[\theta]_{222}$. Therefore, the unit structure of PLL-Cu(II) cannot be a biuret-type complex. Accordingly, the complex unit might be considered to be an octahedral cupric complex coordinated with four amino groups, and having a hydroxyl ion and a peptide nitrogen at the axial positions, where the length of the Cu-N(amide) bound may be shorter than that of the Cu-N(amino) bond. From the above discussion, a plausible structure of PLL-Cu(II) is a main chain of helical or rodlike shape whose unit structure of PLL-Cu(II) may consist of a cupric ion coordinated with four amino groups in a plane, which is held nearly parallel to the axis of the helix, and with an amide nitrogen and a hydroxyl group axially to this plane (Fig. 5) [39]. (See also the Addendum on p. 273.)

2.2. Asymmetrically Selective Oxidation of 3,4-Dihydroxyphenylalanine

Shibata and Shibata [48] reported that $(-)$-$[Co(en)_2(NH_3)Cl]Br_2$ oxidizes the L enantiomer of racemic 3,4-dihydroxyphenylalanine (DOPA) more easily than the D enantiomer. Therefore, the oxidation of DOPA to its quinone catalyzed by optically active metal complexes may offer some stereochemical interest. On the other hand, DOPA oxidase in nature includes Cu(II) complexes bound to the protein in the enzyme.

Therefore, the oxidation of DOPA catalyzed by PLL-Cu(II) was

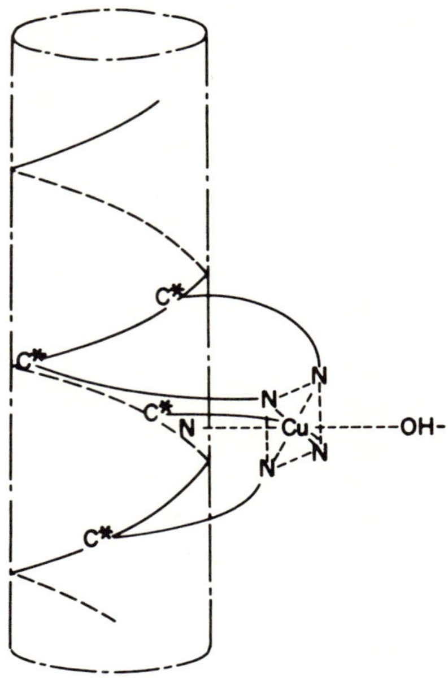

FIG. 5. A schematic structure of PLL-Cu(II).

examined kinetically in order to ascertain the correlation between the asymmetric structures of the polymer complex and its catalytic activity, and the results were analyzed on the basis of the structural analyses of the complex [49]. The kinetic results show that the helical PLL-Cu(II) complex accelerates the oxidation of D-DOPA more than that of L-DOPA. In order to clarify the cause for the asymmetrically selective catalysis, the oxidation of DOPA was followed by using PLL-Cu(II) at various temperatures and ratios of T_{Cu}/T_N [49]. Table 3 shows the effect of temperature variation on the oxidation rate when the catalyst is PLL-Cu(II). The ratio of asymmetric selectivity in the oxidation of DOPA, V_D/V_L (the ratio of the initial oxidation rate of D-DOPA to that of L-DOPA), depends remarkably on the reaction temperature. Table 3 shows the results of DOPA oxidation at various temperatures when $T_{Cu}/T_N = 0.13$. The oxidation rates of D-DOPA are greater than those of L-DOPA, and the

TABLE 3

Oxidation Rate of DOPA Catalyzed by PLL-Cu(II) at pH 10.5[a]

T_{Cu}/T_N	Temp.	V_D[b] x 10^5	V_L[b] mole/liters min	Selectivity ratio (V_D/V_L)	α-Helix content (%)
0.13	25	2.9_3	2.7_4	1.0_7	15_6
0.13	20	2.2_0	1.9_0	1.1_6	17_2
0.13	15	1.8_5	1.3_0	1.4_2	24_0
0.13	10	1.4_3	0.9_3	1.5_3	27_6
0.15	20	2.8_2	2.7_9	1.0_2	6_3
0.13	20	3.5_0	2.7_9	1.2_5	17_2
0.10	20	2.9_5	2.7_6	1.0_7	25_6
0.05	20	2.7_7	2.6_8	1.0_3	57_5

[a]The kinetic experiments were carried out at T_{Cu} = 2.0 x 10^{-4} M. $[DOPA]_0$ was 1.0 x 10^{-4} M in the upper four runs and 2.1 x 10^{-4} M in the lower four runs.

[b]The rates V_D and V_L represent the initial oxidation rates of D-DOPA and of L-DOPA, respectively.

differences between both rates are more remarkable at low temperature. The selectivity ratios increase with a decrease in the reaction temperature.

It is evident from Table 3 that, at a constant T_{Cu}/T_N (0.13), the asymmetric selectivity ratio in the reaction increases with an increase in α-helix content in the catalyst, but when T_{Cu}/T_N is varied, the selectivity does not increase with the amount of α-helix content. That is, the catalyst prepared at T_{Cu}/T_N = 0.05 exhibits almost no selectivity, though it has a much higher α-helix fraction than the catalyst prepared at T_{Cu}/T_N = 0.13. These results indicate that the α-helical structure in the catalyst may be necessary, but not sufficient, to show an appreciable selectivity. The important difference, except for the α-helix content, between the catalyst prepared at T_{Cu}/T_N = 0.05 and that prepared at T_{Cu}/T_N = 0.13 may be the distances between the cupric complex unit along the polymer chain. The asymmetric structure in the complex unit itself does not differ much with a variation in the T_{Cu}/T_N ratio, judging from the molar ellipticity data which change little for wavelengths between 240 and 650 nm (Table 4). Therefore, the selectivity in the catalyst cannot be correlated with individual cupric complex units, but two or more units may act cooperatively in the selective catalysis. The experimental results may be explained by the assumption that the helical structure in the catalyst serves to hold the two or more cupric complex units in a spatial configuration, thus forming sites which can bind D-DOPA selectively.

TABLE 4

Circular Dichroism of PLL-Cu(II) at pH 10.5[a]

T_{Cu}/T_N	Temp. (°C)	$[\theta]^b$ at nm (10^{-3} deg·cm²/dmole)				
		222	255	290	320	510
0.05	23	-18.4	9.0	-2.8	2.8	-4.9
0.075	23	-12.1	7.7	-2.6	2.6	-4.1
0.10	23	- 8.2	4.3	-2.2	1.8	-3.5
0.13	25	- 5.0	3.7	-2.7	1.9	-3.9
0.13	20	- 5.5	3.7	-2.3	1.8	-3.8
0.13	15	- 7.7	3.6	-2.4	1.6	-3.5
0.13	10	- 8.8	3.5	-2.1	1.7	-3.6
0.15	23	- 2.0	4.4	-2.3	1.6	-3.7

[a]The measurements were carried out at $T_{Cu} = 5 \times 10^{-3}$ M.

[b]The values of $[\theta]$ refer to the residue concentration of PLL (the peak at 222 nm) and the molar concentration of cupric ion (the peaks at 255-510 nm).

The oxidation of DOPA exhibits a kinetic behavior of the Michaelis-Menten type, and equations (8) and (9) are applicable. Equation (9) may be rewritten as equation (10).

$$E + S \underset{k_2}{\overset{k_1}{\rightleftharpoons}} ES \xrightarrow{k_3} E + P \tag{8}$$

$$V_0 = k_3[ES] = k_3[E]_0[S]_0/(k_2+k_3)/k_1 + [S]_0 \tag{9}$$

$$V_0 = k_1k_3[S]_0[E]_0/(k_2 + k_3 + k_1[S]_0) \tag{10}$$

$[E]_0$ and $[S]_0$ represent the initial concentrations of PLL-Cu(II) and DOPA, respectively, and [ES] is the concentration of an intermediate complex formed in the reaction between the catalyst and substrate. Since k_3 can be assumed to be small, the oxidation rate is expressed as in equation (11) when $k_1[S]_0$ is small compared to k_2.

$$V_0 = (k_1k_3/k_2)[S]_0[E]_0 \tag{11}$$

Taking logarithms of both sides of equation (11) and substituting for the rate constants in terms of activation energies, equation (12) may be obtained.

$$\ln V_0 = (-E_3 + E_2 - E_1)/RT + I_c \tag{12}$$

$$E_A = E_3 - E_2 + E_1 \tag{13}$$

$$\ln V_0 = (-E_A/R)(1/T) + I_c \tag{14}$$

where E_1, E_2, and E_3 represent the apparent activation energies in the elementary reactions whose rate constants are k_1, k_2, and k_3, respectively. Therefore, a plot of log V_0 vs 1/T should yield a straight line, and the apparent activation energy can be obtained from the slope. The apparent energy values were calculated to be 7.9 and 12.0 kcal/mole for D-DOPA and L-DOPA, respectively. Since

k_3 for D-DOPA is the same as that of L-DOPA, the difference of E_A values must be ascribable to the difference of the Michaelis constants (K_m) for D-DOPA and L-DOPA (Table 5). The equilibrium

TABLE 5

Michaelis Constant (K_m) and Initial Oxidation Rates (V_0) in the Oxidation of DOPA

(T_{Cu}/T_N = 0.13, T_N = 2.0 x 10^{-4} M, $[DOPA]_0$ = 0.5 x 10^{-4}, 20°C)

Catalyst	Substrate	pH	$V_0 \times 10^5$ (mole/liter min)	$K_m \times 10^4$ (mole/liter)
PLL-Cu(II)	L-DOPA	10.5	1.04	2.25
PLL-Cu(II)	DL-DOPA	10.5	1.40	1.60
PDLL-Cu(II)	L-DOPA	10.5	1.60	1.68
PDLL-Cu(II)	DL-DOPA	10.5	1.60	1.68
PLL-Cu(II)	L-DOPA	6.9	0.067	0.34
PLL-Cu(II)	DL-DOPA	6.9	0.067	0.34

constant K in the reaction between PLL-Cu(II) and DOPA corresponds nearly to $1/K_m$. From the temperature dependence of K, thermodynamic parameters were determined. The ΔH for D-DOPA is more negative than that for L-DOPA, which indicates stronger binding of the catalyst [PLL-Cu(II)] to D-DPA than to L-DOPA in the complex. The fact that the TΔS terms for D-DOPA are more negative than those for L-DOPA suggests that the intermediate complex for L-DOPA has a more random structure than that for D-DOPA (Table 6). These results agree with the CD measurements of PLL-Cu(II) on addition of DOPA [37,49].

When PLL-Cu(II) and D-DOPA are used, a bifunctional catalysis may operate in the oxidation of DOPA. It is assumed that PLL-Cu(II) activates DOPA in two different ways. One is an ordinary activation, i.e., an activation of DOPA by the coordination of its catechol endgroup to a cupric ion, and the other is a so-called

TABLE 6

Activation Parameters for the PLL-Cu(II)-Catalyzed Oxidation of DOPA

(T_{Cu}/T_N = 0.13, T_{Cu} = 2.0 x 10^{-4} M)

Substrate	Temp. (°C)	K_m x 10^{-5} (M)	ΔG	ΔH (kcal/mole)	TΔS
D-DOPA	20	15.3	-5.10	-15.7	-10.6
D-DOPA	15	8.93	-5.42	-15.7	-10.3
D-DOPA	13	7.95	-5.49	-15.7	-10.2
L-DOPA	20	22.5	-4.88	- 7.8	- 2.9
L-DOPA	15	16.2	-5.07	- 7.8	- 2.7
L-DOPA	13	16.0	-5.08	- 7.8	- 2.7

bifunctional activation. In the latter, DOPA may coordinate to PLL-Cu(II) with both endgroups, the amino acid end and the catechol end, at the same time. The former type of activation may be the same for both L- and D-DOPA, whereas the bifunctional activation must favor the D-DOPA. This bridge-type intermediate complex agrees with the discussion of the TΔS term, in which the more negative value of the D-DOPA compared with that of L-DOPA implies greater order. The CD spectral change shows that more α-helical structure is recovered on addition of D-DOPA than on the addition of L-DOPA. Thus, the TΔS term reflects not only a complex unit but also the intermediate bifunctional complex.

2.3. Asymmetrically Selective Hydrolysis of Amino Acid Esters

In general, metalloenzymes exhibit a high degree of specificity toward substrates of a particular optical configuration [50]. Since, apparently, the ligands surrounding the metal ion in these enzymes are largely responsible for the overall specificity of the enzymes, numerous attempts have been made to design simple metal complexes which also have some specificity. Recently, various studies have shown that several metal ions and their complexes are effective catalysts for the hydrolysis of amino acid esters [51-55]. In these instances, coordination of the ester to a metal ion makes it more susceptible to attack by nucleophiles, e.g., water and hydroxide ion. More recent works have indicated stereoselectivity in related processes involving some metal complexes [56-59]. Hix and Jones [59] examined in detail the hydrolysis of (R)-(-)- and (S)-(+)-histidine methyl esters in the presence of the catalytically active complexes Ni[(R)-(-)-histidinate]$^+$ and Ni[(S)-(+)-histidinate]$^+$. In this study, it was found that the rate of hydrolysis of the (R)-(-) ester is greater in the presence of the Ni[(S)-(+)-histidinate]$^+$ ion than in the presence of Ni[(R)-(+)-histidinate]$^+$, and that these rate differences are mirrored in the presence of the (S)-(+) ester. The system with opposite ester-histidine configurations has an observed rate constant approximately 40% greater than that found with identical ester-histidine configurations. Since no detectable differences in the stability constants of the histidinato(methyl histidinate)nickel(II) complexes could be observed experimentally, it must be concluded that the main source of stereoselectivity results from differences in the hydrolytic activities of the isomeric complexes once formed. It has been shown that these differences in activities can result from the greater ease of interaction of the metal and ester group in the proposed complex when the ester and histidine are of opposite configuration. This interaction may facilitate a greater polarization of the carbonyl group in the intermediate complex, thereby making nucleophilic attack on the ester more favorable.

In the course of studies on some polymer-metal complexes which exhibit enzyme-like catalytic activity, PLL-Cu(II) was found to exhibit asymmetrically selective catalysis in the hydrolysis of phenylalanine esters at pH 7. The reaction rates are summarized in Table 7. The data for the initial portion of the reaction produce a straight line when plotted as first-order reaction. It is

TABLE 7

Pseudo-First-Order Rate Constants (k) for the Hydrolysis of D- and L-Phenylalanine Esters at Ester Concentrations of 1.0×10^{-2} M and 24.8°C

Catalyst species	Concentration of copper(II) ($\times 10^{-3}$ M)	Enantiomorphic forms[a]	pH	k ($\times 10^{-3}$ min^{-1})
Non		L	7	<0.01
		L	8	<0.01
		L	9	<0.01
		L	10.5	26.1
Cu[b]	0.5	L	7	<0.01
	5.0	L	7	<0.01
	5.0	L	7	0.01
[Cu(II) (bipyridyl)$_2$][b]	5.0	L	9	1.69
	5.0	L	10.5	5.99
PLL-Cu(II)[b]	0.52	D	7	13.3
	0.52	L	7	5.52

[a] L-Phenylalanine ethyl ester was used in the experiments, and D-phenylalanine methyl ester was also used with PLL-Cu(II).

[b] See text.

readily seen by the data in Table 7 that the rate of hydrolysis of phenylalanine esters catalyzed by PLL-Cu(II) is higher for the D-amino acid ester than that for the L-amino acid ester. The variation of the PLL-Cu(II)-catalyzed reaction rates with the ester concentrations was determined at 24.8°C, and the data are summarized in Table 8. Table 8 also indicates that PLL-Cu(II) derived from the higher molecular weight PLL (P-14; degree of polymerization, 920) catalyzed the hydrolysis of phenylalanine ester about four times faster than the PLL-Cu(II) from the lower molecular weight PLL (P-13; degree of polymerization, 470).

TABLE 8

Pseudo-First-Order Rate Constants (k) for the Hydrolysis of D- and L-Phenylalanine Methyl Ester Catalyzed by PLL-Cu(II) at 24.8°C and pH 7.0

($T_{Cu} = 5.2 \times 10^{-4}$ M, $T_{Cu}/T_N = 0.13$)

PLL	Enantio-morphic forms	Concentrations ($\times 10^{-3}$ M)	k ($\times 10^{-3}$ min^{-1})
P-13[a]	D	8.04	6.13
	L	8.04	1.84
	D	8.51	22.1
P-14[b]	L	9.22	9.12
	D	3.76	3.80
	L	4.26	0.92

[a]Degree of polymerization, 470.
[b]Degree of polymerization, 920.

As discussed elsewhere [60,61], catalytic hydrolysis of an amino acid ester (Es) depends upon coordination of Es to the metal complex; this is followed by hydrolysis of the coordinated Es. The general mechanism assumed here requires a rapid coordination equilibrium [eq. (15)], followed by a rate-determining attack of OH^- on the ester in the complex Cu(PLL)(Es).

$$\text{Cu(PLL)} + \text{HEs}^+ \underset{}{\overset{K}{\rightleftharpoons}} \text{Cu(PLL)(Es)} + \text{H}^+ \tag{15}$$

The nonmetal complex-catalyzed base hydrolysis of phenylalanine methyl ester occurs with a rate constant of less than 1.0×10^{-5} min^{-1} at pH 7.0 as compared to that of the hydrolysis of the ester in Cu(PLL)(Es) for which the value is 5.5×10^{-3} min^{-1}. Thus, the first step in the mechanism of the reaction in equation (16) is probably the chelation of the ester group to the metal ion.

$$Cu(PLL)(Es) + OH^- \xrightarrow{k_{OH}} Cu(PLL)(A^-) + \text{alcohol} \quad (16)$$

In terms of this mechanism, the calculated rate constant k is Kk_{OH}, where K and k_{OH} are the equilibrium constant in equation (15) and the rate constant in equation (16). Hence the difference in k values for PLL-Cu(II) with D ester from those with L ester may be due to a difference in K and/or k_{OH}. It is possible to propose a coordinated ester species in which the L ester would coordinate less strongly than the D ester. Thus, the K value for the D ester would be larger than that for the L ester [62].

The activation parameters for the hydrolysis of phenylalanine methyl ester catalyzed by PLL-Cu(II) can be calculated in an usual way and are summarized in Table 9. Table 9 shows that the apparent activation energies for the hydrolysis of D- and L-phenylalanine methyl esters are 9.31 and 32.8 kcal/mole, respectively, and that the activation enthalpy (ΔH) and the activation entropy (ΔS) are smaller for D-phenylalanine methyl ester than for L-phenylalanine methyl ester. The activation parameters indicate that the hydrolysis of D-phenylalanine methyl ester is more accelerated by the ΔH part than L-phenylalanine methyl ester is. Moreover, the ΔS part rather retards the hydrolysis of D-phenylalanine methyl ester more than of L-phenylalanine methyl ester.

TABLE 9

The Activation Parameters for the Hydrolysis of Phenylalanine Methyl Esters Catalyzed by the Poly-L-lysine-Copper(II) Complex at $T_{Cu}/T_N = 0.13$ and $T_{Cu} = 5.2 \times 10^{-4}$ M

Substrate	E_{exp}	ΔH (kcal/mole)	ΔG	TΔS
D-MePhe	9.31	8.72	19.7	-11.0
L-MePhe	32.8	32.2	20.5	11.7

The CD data described above suggest that PLL-Cu(II) forms more of the intermediate complex with D-phenylalanine methyl ester than with L-phenylalanine methyl ester [63]. If this is kept in mind, the difference in the activation parameters tells us that the ester in the intermediate complex will be more labile (with small ΔH) for the D enantiomer than for the L enantiomer, and that the entropy for PLL-Cu(II) is higher than that for the intermediate complex.

3. CONCLUSIONS

What are "polymer effects" in the polymer metal complexes with regard to their catalytic activities and/or their catalytic specificities? This problem is not clear yet because the structures are too complex to analyze them in terms of polymer chemistry, metal complex chemistry, and the related chemical fields.

However, it is true that some polymer-metal complexes have been found to exhibit high catalytic activity and specificity in comparison with low molecular weight analogs, and the catalytic activities of these complexes were discussed in comparison with those of metalloenzymes. In addition, some polymer effects on their catalyses were proposed in order to elucidate the elaborating chemical behavior of the polymer-metal complexes, that is:

1. Environmental effects, such as electrostatic field effect originated from the polymer ligand as a polyelectrolyte
2. Variation of the stability constant affected by the electrostatic field in the polymer ligand moiety
3. Ternary complex formation, which is performed around the defined ligand residues, such as histidyl residue
4. Formation of chiral structures surrounding the metal ion, which are held by ordered structures in the polymer ligands (in one case an α-helix in polypeptides, in another case, a locally chiral structure can be considered)

For catalytic activity enhancement of the polymer-metal complexes, the first three factors are essential. Furthermore, the last is necessary to aim at finding the asymmetrically selective catalysts like a metalloenzyme. The formation of the chiral structures surrounding metal ion would be performed in the complex formation of metal ion with poly-L,α-amino acids including ordered structures such as α-helix. On the other hand, the polymer ligands including local chiral structure, such as "Davankov's adsorbent" [26,27] would be used as asymmetrically selective catalysts.

ADDENDUM

Recently, Dr. Sigel and we discussed the following point. In our manuscript one reads on page 258 "this fact suggests that the amide nitrogen (in an apical position) may coordinate to cupric ion more closely than amino nitrogens in the plane." According to Sigel's statement, this assumption is in contrast to the observations commonly made in the chemistry of Cu(II): usually the atoms coordinated in the square plane are more closely at Cu(II) than the (one or) two axially coordinated ligand atom(s) [64]. Hence, Dr. Sigel's paper [65] concerning the ternary 2,2'-bipyridyl-Cu(II)-oligoglycine complexes, indicates also that the terminal amino group and the first three deprotonated amide nitrogens of an oligoglycine are bound to a Cu(II) ion, in accord with the finding by Freeman; additionally, he can obtain no hint for the binding of a deprotonated amide group to an apical position. From the above-mentioned points, Dr. Sigel concluded that the structure of PLL-Cu(II) could not be as proposed by us in Figure 5; he suggested that (at pH 9 and 10.5) an amide and OH^- group should replace two amino groups in the square plane of the Cu(II) complex. Accordingly, we examined the structural analyses of PLL-Cu(II) at pH 7.5, where PLL holds no helical structure, according to the CD measurements. At this low pH, an appreciable magnitude of a CD spectrum could be observed in the range of d-d transitions. This means that the existence of a helical portion in PLL-Cu(II) may not be needed for the appearance of the CD bands in the range of d-d transitions, and suggests that the ligation of the amide nitrogen to Cu(II) ion occurs appreciably. Now, further studies are proceeding using model complexes. We are greatly indebted to Dr. Sigel for stimulating discussions.

ABBREVIATIONS

AB: Absorption

α: Degree of neutralization of the protonated ligand

CD: Circular dichroism

DOPA: 3,4-Dihydroxyphenylalanine

K: Equilibrium constant in the reaction between PLL-Cu(II) and DOPA

K_1: Apparent stability constant of CuL^+

K_2: Apparent stability constant of CuL_2

K_a: Apparent acid dissociation constant

K_m: Michaelis constant

L: Deprotonated ligand

LH: Protonated ligand

λ_{max}: Wavelength of absorption maximum

MePhe: Phenylalanine methyl ester

$\bar{n}$: Average number of amino groups bound to one copper(II)

ORD: Optical rotatory dispersion

PDLL: Poly-D,L-lysine

PDLL-Cu(II): Copper(II) complex of poly-D,L-lysine

PLH: Poly-L-histidine

PLL: Poly-L-lysine

PLL-Cu(II): Copper(II) complex of poly-L-lysine

(R)-(-)- or (S)-(+)-histidine methyl ester: Histidine methyl ester having R or S configuration

T_{HL}: Total concentration of ligand residues

T_{Cu}: Total concentration of Cu^{2+}

$[\theta]$: Molar ellipticity of CD band, where the suffix numerical figure shows the wavelength of peak or trough of the CD band

V_D and V_L: Initial oxidation rates of D-DOPA and of L-DOPA

REFERENCES

1. B. G. Malmström and A. Rosenberg, *Advan. Enzymol., 21,* 131 (1959).

2. B. L. Vallee and J. E. Coleman, *Comp. Biochem., 12,* 165 (1964).

3. R. D. Gillard, *Inorg. Chim. Acta Rev., 1,* 69 (1967).

4. M. T. Beck, *Coord. Chem. Rev., 3,* 91 (1968); D. D. Ulmer and B. L. Vallee, *Advan. Chem. Ser.,* (1971) 187.

5. H. Sigel and D. B. McCormick, *Accounts Chem. Res., 3,* 201 (1970).

6. A. S. Brill, R. B. Martin, and R. J. P. Williams, "Electronic Aspects of Biochemistry," Academic Press, New York (1964).

7. J. H. Wang, *Accounts Chem. Res., 3,* 90 (1970).

8. R. Breslow, *Advan. Chem. Ser., 100,* 21 (1971).

9. E. Katchalsky, M. Sera, H. I. Silman, and A. Berger, "The Proteins; Composition, Structure, and Function," 2nd ed., Vol. 2, Academic Press, New York, 1964, p. 406.

10. A. Levitzki, I. Pecht, and M. Anbar, *Nature, 207,* 1356 (1965).

11. I. Pecht, A. Levitzki, and M. Anbar, *J. Amer. Chem. Soc., 89,* 1587 (1967).

12. A. Levitzki, I. Pecht, and A. Berger, *J. Amer. Chem. Soc., 94,* 6844 (1972).

13. R. C. W. Welch and H. F. Rase, *Ind. Eng. Chem. Fundamentals, 8,* 389 (1969).

14. Y. Nose, M. Hatano, and S. Kambara, *Makromol. Chem., 98,* 136 (1966).

15. T. Nozawa, Y. Nose, M. Hatano, and S. Kambara, *Makromol. Chem., 112,* 73 (1968).

16. M. Hatano, T. Nozawa, T. Yamamoto, and S. Kambara, *Makromol. Chem., 115,* 1 (1968).

17. M. Hatano, T. Nozawa, T. Yamamoto, and S. Kambara, *Makromol. Chem., 115,* 10 (1968).

18. H. Sigel and G. Blauer, *Helv. Chim. Acta, 51,*1246 (1968).

19. T. Nozawa, M. Hatano, and S. Kambara, *Kogyo Kagaku Zasshi* (in Japanese), *72,* 373 (1969).

20. T. Nozawa, M. Hatano, and S. Kambara, *Kogyo Kagaku Zasshi, 72,* 369 (1969).

21. R. D. Gillard, *Inorg. Chim. Acta Rev., 1,* 69 (1967).

22. H. Sigel and R. Griesser, *Helv. Chim. Acta, 50,* 1842 (1967).

23. H. Sigel and B. Prijs, *Helv. Chim. Acta, 50,* 2357 (1967).

24. H. Sigel, *Angew. Chem., 80,* 124 (1968).

25. R. F. Pasternack and H. Sigel, *J. Amer. Chem. Soc., 92,* 6146 (1970).

26. V. A. Davankov and S. V. Rogozkin, *J. Chromatogr., 60,* 280 (1971).

27. S. V. Rogozkin and V. A. Davankov, *Chem. Commun.,* 490 (1971).

28. B. E. Leach and R. J. Angelici, *J. Amer. Chem. Soc., 91,* 6296 (1969); *Inorg. Chem., 8,* 907 (1969).

29. R. V. Snyder, R. J. Angelici, and R. B. Meck, *J. Amer. Chem. Soc., 94,* 2660 (1972).

30. Y. Moriguchi, *Bull. Chem. Soc. Japan, 39,* 2656 (1966).

31. H. Takesada, Y. Yamaguchi, and A. Wada, *Biopolymers, 4,* 713 (1966).

32. R. C. W. Welch and H. F. Rase, *Ind. Eng. Chem., Fundamentals, 8,*389 (1969).

33. A. Levitzki, I. Pecht, and M. Anbar, *Nature, 207,* 1386 (1965).

34. M. Hatano, M. Yoneyama, I. Ito, T. Nozawa, and M. Nakai, *J. Amer. Chem. Soc., 91,* 2165 (1969).

35. M. Hatano and M. Yoneyama, *J. Amer. Chem. Soc., 92,* 1392 (1970).

36. M. Hatano, T. Nozawa, S. Ikeda, and T. Yamamoto, *Makromol. Chem., 141,* 1 (1971).

37. T. Nozawa and M. Hatano, *Makromol. Chem., 141,* 21 (1971).

38. M. Hatano, T. Nozawa, and M. Yoneyama, *Bull. Chem. Soc. Japan, 43,* 295 (1970).

39. M. Nakai, M. Yoneyama, and M. Hatano, *Bull. Chem. Soc. Japan, 44,* 874 (1971).

40. K. M. Wellman, T. G. Mecca, W. Mugall, and C. R. Hare, *J. Amer. Chem. Soc., 89,* 3646 (1967).

41. K. M. Wellman, T. G. Mecca, W. Mugall, and C. R. Hare, *J. Amer. Chem. Soc., 90,* 805 (1968).

42. R. B. Martin, J. M. Tsangaris, and J. W. Chang, *J. Amer. Chem. Soc., 90,* 821 (1968).

43. C. J. Hawkins and E. Lavsen, *Acta Chim. Scand., 19,* 185 (1965).

44. T. Yasui, J. Hidaka, and Y. Shimura, *J. Amer. Chem. Soc., 87,* 2762 (1965).

45. T. Yasui, *Bull. Chem. Soc. Japan, 38,* 1746 (1965).

46. R. L. Belford, M. Calvin, and G. Belford, *J. Chem. Phys., 26,* 1165 (1957).

47. T. Murakami, T. Nozawa, and M. Hatano, *J. Amer. Chem. Soc., 92,* 5768 (1970).

48. Y. Shibata and K. Shibata, *Nippon Kagaku Zasshi* (in Japanese), *41,* 35 (1920).

49. M. Hatano, T. Nozawa, S. Ikeda, and T. Yamamoto, *Makromol. Chem., 141,* 11 (1971); *141,* 31 (1971).

50. A. E. Dennard and R. J. P. Williams, "Transition Metal Chemistry," Vol. 2 (R. L. Carlin, ed.), Dekker, New York, 1966, p. 116.

51. M. L. Bender, *Advan. Chem. Ser., 37,* 22 (1963).

52. H. L. Conley and R. B. Martin, Jr., *J. Phys. Chem., 69,* 2923 (1965).

53. M. D. Alexander and D. H. Busch, *J. Amer. Chem. Soc., 88,* 1130 (1966).

54. J. E. Hix, Jr. and M. M. Jones, *Inorg. Chem., 5,* 1863 (1966).

55. R. J. Angelici and B. E. Leach, *J. Amer. Chem. Soc., 89,* 4605 (1967).

56. M. Murakami, H. Itatani, K. Takahashi, J. Kang, and K. Suzuki, *Mem. Inst. Sci. Ind. Res. Osaka Univ., 20,* 95 (1963).

57. R. G. Asperger and C. F. Lia, *Inorg. Chem., 6,* 796 (1967).

58. B. E. Leach and R. J. Angelici, *J. Amer. Chem. Soc., 91,* 6296 (1969).

59. J. E. Hix, Jr. and M. M. Jones, *J. Amer. Chem. Soc., 90,* 1723 (1968).

60. B. E. Leach and R. J. Angelici, *Inorg. Chem., 8,* 907 (1969).

61. R. J. Angelici and D. Hopgood, *J. Amer. Chem. Soc., 90,* 2514 (1968).

62. T. Nozawa, Y. Akimoto, and M. Hatano, *Makromol. Chem., 158,* 21 (1972).

63. T. Nozawa, Y. Akimoto, and M. Hatano, *Makromol. Chem., 161,* 289 (1972).

64. J. F. Blount, H. C. Freeman, R. V. Holland, and G. H. W. Milburn, *J. Biol. Chem., 245,* 5177 (1970).

65. H. Sigel, R. Griesser, and B. Prijs, *Z. Naturforsch., 27b,* 353 (1972).

Chapter 6

REACTIONS OF MOLYBDENUM COORDINATION COMPOUNDS: MODELS FOR BIOLOGICAL SYSTEMS

Jack T. Spence
Department of Chemistry and Biochemistry
Utah State University
Logan, Utah

1. INTRODUCTION: BIOLOGICAL IMPORTANCE OF MOLYBDENUM

There is currently considerable interest in the coordination chemistry of molybdenum. This is mainly due to the fact that molybdenum is a trace element in biological systems, participating in a number of important enzymic redox reactions, among which are the oxidation of purines and the reduction of nitrate and atmospheric nitrogen.

Biological molybdenum is always bound to an enzyme and, in a number of cases, appears to be associated with the coenzyme flavin adenine dinucleotide (FAD). Despite the obvious importance of the reactions of these molybdenum enzymes, the nature of the binding site and the functions of the metal in the redox systems are not well understood.

The purpose of this chapter is to bring together the results of studies concerned with the reactions of model complexes of molybdenum and determine what importance they have for our understanding of the biochemical role of the metal.

2. BIOCHEMICAL REACTIONS OF MOLYBDENUM ENZYMES

2.1. Enzymic Systems

Molybdenum is known to be a necessary cofactor for five enzymically catalyzed redox reactions: nitrogen and nitrate reduction, and purine, aldehyde, and sulfite oxidation. There appear to be a number of varieties of some of these enzymes, differing in electron donors or acceptors, other cofactors, and molecular weights. This is summarized in Table 1. A review concerning these enzymes, particularly detailed with respect to xanthine oxidase, has recently been published by Bray and Swann [1].

It should be noted that for all the molybdenum enzymes except the nitrogenases, evidence, mainly from ESR studies, indicates that the metal participates in the electron transfer process, undergoing cyclic changes in oxidation state [1].

TABLE 1

MOLYBDENUM ENZYMES

Enzyme	Substrate	Electron donor or acceptor	g-Atoms Mo/mole enzyme	Other cofactors g-atoms or moles/mole enzyme[f]	Ref.
Nitrogenase[a,b]	N_2	Ferredoxin	1	Fe[d] 30	69
Nitrogenase[a,c]	N_2	Ferredoxin	2	Fe[d] 72	69
Nitrate reductase	NO_3^-	NADPH	1-2	Fe-Cyt[e], FAD	18
Nitrate reductase	NO_3^-	NADPH	?	FAD	70
Nitrate reductase	NO_3^-	NADPH	?	Fe[d]	68
Nitrate reductase	NO_3^-	NADPH	?	Fe-Cyt[e]	71
Sulfite oxidase	SO_3^{2-}	O_2	2	Fe-Cyt[e]	8
Aldehyde oxidase	Aldehydes	O_2	2	FAD 2, Fe 8	72
Xanthine dehydrogenase	Purines	NAD	2	FAD 2, Fe 8	73
Xanthine dehydrogenase	Purines	Ferredoxin	2	FAD 2, Fe 8	2
Xanthine oxidase	Purines	O_2	2	FAD 2, Fe 8	1

[a] Enzyme containing Mo and Fe only.

[b] From *Clostridium*.

[c] From *Azotobacter*.

[d] Fe = nonheme iron.

[e] Fe-Cyt = heme iron.

[f] Where no number is given, the number of g-atoms or moles of cofactor/mole enzyme is unknown.

2.2. Current Problems

2.2.1. Oxidation States

It seems quite probable that in the oxidized form of all the molybdenum enzymes, except xanthine dehydrogenase from *Micrococcus lactilyticus* [2], the metal is present as the +6 state [1]. With xanthine oxidase, which has been most extensively studied, Mo(VI) is first reduced to an ESR-active Mo(V) species. In the presence of excess substrate, there is evidence to indicate it may be reduced further to Mo(IV). This conclusion is based on the subsequent disappearance of the ESR signal due to Mo(V) with time [3] and on inhibition studies [4]. The main difficulty with this hypothesis is that Mo(IV) is known to be thermodynamically unstable in aqueous environments, disproportionating to Mo(V) and Mo(III) species. In fact, with the exception of the octacyano comples, $Mo(CN)_8^{4-}$, which has unusual symmetry, and one or two species that have never been well characterized, no Mo(IV) complexes stable in water are known. There are, however, a number of Mo(IV) complexes that can be prepared in nonaqueous solvents [5] and the environment about the metal in the enzyme could be somewhat hydrophobic [6]. Alternatively the disappearance of the Mo(V) ESR signal might be due to dimer formation, since most Mo(V) complexes exist as ESR-inactive dimers in water, and most of the molybdenum enzymes contain two Mo atoms per molecule [7]. In addition, the loss of signal might be attributed to reduction to Mo(III), which is ESR-inactive in solution, probably because of unfavorable relaxation properties. There are, however, other difficulties associated with these interpretations [1] and the question of the oxidation state in the completely reduced enzyme remains open.

In the case of the nitrate reductases [1] and sulfite oxidase [8], the same oxidation states (+6, +5) appear to be involved, while with the nitrogenases no ESR signals attributed to Mo have been detected [9]. In the case of the *Micrococcus* xanthine dehydrogenase, a large Mo(V) signal is observed with the resting enzyme, making lower oxidation states mandatory if this ESR-detectable molybdenum is involved in subsequent electron transfer [2].

2.2.2. Binding Sites

Considerable evidence, again mainly due to ESR studies, indicates the molybdenum is most likely bound to one or more sulfurs (cysteine side chains) at the active site [1]. This is based primarily on the finding that in the molybdenum enzymes, ESR signals having high g values and small hyperfine splittings (A) are observed, as is the case with a number of Mo(V) complexes with thiol ligands [10]. The properties and reactions of Mo(V) complexes with cysteine [11], glutathione [12], and similar ligands are consequently of interest and have been studied more than other Mo(V) complexes.

A major difficulty with the binding site being a cysteine side chain of the protein is that at the pH for maximum enzymic activity of xanthine oxidase ($\sim$8), Mo(VI) complexes only weakly, if at all, with mercapto groups [13]. In fact, it forms very few stable complexes at this pH, even with such strong chelating ligands as EDTA [14] and 8-hydroxyquinoline [15]. It is difficult to imagine what group, or combination of groups, present in the enzyme would bind Mo(VI) strongly enough to explain the apparent stability of the metal-enzyme complex. Of course, the pH at the binding site might be much lower than 8, or steric factors may greatly reduce the rates of ligand exchange, as is the case with proton exchange in certain macromolecules [16]. Work concerning both oxidation state and binding sites has recently been reviewed by Spence [17].

2.2.3. Reactions

As has been mentioned, evidence exists for most of the molybdenum enzymes which supports an active redox role for the metal [1]. In the case of xanthine oxidase, the most recent work indicates that after reduction of Mo(VI) by substrate, the reduced molybdenum species (+5 or +4) subsequently transfers electrons either to the iron-sulfur or to flavins, or both [1]:

$$\text{Substrate} \rightarrow \text{Mo(VI)} \begin{array}{l} \nearrow \\ \searrow \end{array} \begin{array}{c} \text{FAD} \rightarrow O_2 \\ \uparrow\downarrow \\ \text{Fe-S} \end{array} \tag{1}$$

With the nitrate reductases, Mo(VI) appears to be reduced by FAD. It then transfers electrons either to a cytochrome, a nonheme iron, or directly to NO_3^-, depending on the specific enzyme [18,19]:

$$\text{NADPH} \rightarrow \text{FAD} \rightarrow \text{Mo(VI)} \rightarrow \text{Fe(III)} \rightarrow NO_3^- \tag{2}$$

(with a second arrow leading directly from Mo(VI) to NO_3^-)

With nitrogenase, however, no direct evidence for the involvement of molybdenum in the redox reaction has been obtained (no ESR signals attributable to Mo(V) are observed during reduction of N_2 [9]). At present, therefore, almost nothing can be said concerning the role the metal plays in the sequence of reactions leading to ammonia production in the nitrogenase system.

3. MODEL REACTIONS

3.1. Redox Reactions

3.1.1. Nitrate Reduction

The reduction of nitrate in acid solution by Sn(II) in the presence of Mo(VI) has been studied by Haight et al. [20,21]. Both the products and the rate of reduction were found to depend on the acid used. In 3 M HCl:

$$NO_3^- + 4\text{Sn(II)} + 10H^+ \xrightarrow{\text{Mo(VI)}} 4\text{Sn(IV)} + NH_4^+ + 3\ H_2O \tag{3}$$

In 3 M H_2SO_4:

$$2NO_3^- + 4\text{Sn(II)} + 10H^+ \xrightarrow{\text{Mo(VI)}} N_2O + 4\text{Sn(IV)} + 5H_2O \tag{4}$$

A mechanism for these reactions involving the formation of Mo(IV) as reactive intermediate was developed, consistent with the observed kinetics. The Mo(IV) is presumably formed by a fast reaction with Sn(II):

$$Sn(II) + Mo(VI) \rightarrow Mo(IV) + Sn(IV) \qquad (5)$$

In H_2SO_4, the Mo(IV) then proceeds to reduce NO_3^-, regenerating Mo(VI), while in HCl it forms a complex with NO_3^-, rendering the nitrate susceptible to reduction by Sn(II). Although the kinetics are complicated, it would appear they are equally compatible with a monomeric Mo(V) species as the catalytically active intermediate. In both cases, the form of Mo(VI) or Mo(IV) is unknown; Mo(VI), however, is most likely a polymeric anionic species under these conditions.

Haight also studied the Mo(VI)-catalyzed reduction of NO_3^- at the dropping mercury electrode and again concluded that Mo(IV) is the active species [22]. This conclusion, however, was questioned by Kolthoff and Hodara [23], who proposed a monomeric form of Mo(V) as the active intermediate.

Detailed kinetic studies of the reduction of nitrate by Mo(V) in this laboratory found only traces of nitrate are reduced in phosphate buffer, while in tartrate or citrate the reaction is quantitative and proceeds with a reasonable rate [24]. This was explained by the observation of a small Mo(V) ESR signal in tartrate and citrate, which was not found in phosphate. Furthermore, the rate of reduction was shown to be dependent on the total Mo(V) concentration to the one-half power and the nitrate concentration to the first power, giving rise to the mechanism:

$Mo(V)_2$ ESR inactive dimer	$\rightleftarrows$	$2Mo(V)$ ESR active monomer	fast	(6)
$Mo(V) + NO_3^-$	$\rightarrow$	$Mo(VI) + NO_2$	slow	(7)
$2NO_2$	$\rightarrow$	$NO_3^- + NO^+$	fast	(8)
$Mo(V) + NO^+$	$\rightarrow$	$NO + Mo(VI)$	fast	(9)
$H_2O + NO^+$	$\rightleftarrows$	$HNO_2 + H^+$	fast	(10)

The function of the tartrate or citrate, confirmed by further ESR studies [25], is to stabilize the monomer to some extent.

The structures of the $Mo(V)_2$, Mo(V), and Mo(VI) tartrate and citrate complexes are uncertain. Work in this laboratory indicated the $Mo(V)_2$ is present as a 2:1 [tartrate:Mo(V)] complex [25], while other work by Brown and MacPherson reports the reverse molar ratio for this complex [26].

A more complete model system, using reduced flavin mononucleotide ($FMNH_2$), Mo(VI) and NO_3^- in tartrate buffer was recently studied in this laboratory [27]. Although the reduction of NO_3^- by $FMNH_2$ is thermodynamically favorable, the reaction does not proceed in the absence of Mo(VI) or $Mo(V)_2$. When a small amount of either MoO_4^{2-} or $MoOCl_5^{2-}$ (which immediately dimerizes) is added to the system, however, the reduction proceeds quantitatively, the molybdenum compound acting as a catalyst, similar to nitrate reductase:

$$2NO_3^- + 3FMNH_2 \xrightarrow[\text{tartrate}]{\text{Mo(V,VI)}} 2NO + 3\ FMN \qquad (11)$$

Kinetic studies indicated $FMNH_2$ first reduces Mo(VI) to Mo(V), which then proceeds to reduce NO_3^-, as described previously in reactions (6)-(10) regenerating Mo(VI):

$$FMNH_2 + 2Mo(VI) \rightarrow Mo(V)_2 + FMN \qquad (12)$$

In this model system, the rate-controlling step is the same as in the reduction of NO_3^- by $Mo(V)_2$ [reaction (7)] since the rate of disappearance of $FMNH_2$ was found to be zero order in $FMNH_2$, first order in NO_3^-, and one-half order in $Mo(V)_2$.

It should be noted that the product of NO_3^- reduction in these model systems is NO, rather than NO_2^- (or HNO_2) as in the enzymic reaction. Although some nitrite is formed in the model systems, it does not accumulate since it reacts with $Mo(V)_2$ faster than NO_3^- [28]. In the biological system, nitrite is reduced by a nonheme iron enzyme.

3.1.2. Nitrogen Reduction

There are now a rather large number of transition metal complexes known, including some molybdenum species [29], which bind molecular nitrogen as a ligand [30]. The majority of these are organometallic in nature, stable only in nonaqueous environments. There are some, however, such as $Ru(NH_3)_5N_2^{2+}$, which are easily obtained in water [31]. A number of systems involving low oxidation state transition metal species have been shown to catalyze the reduction of N_2 to NH_3 under relatively mild, but nonbiological conditions. In particular, those models based on organometallic compounds of Ti(II) have been extensively studied in this regard, and detailed mechanisms for their reactions have been developed [32,33].

The reduction of ligating nitrogen in a molybdenum complex to N_2H_2 (diimide) was recently reported by Chatt et al. [34]. It was found that when trans-$Mo(N_2)_2(dppe)_2$ and trans-$Mo(N_2)_2(depe)_2$ are treated with HBr in tetrahydrofuran at 0°C they give high yields of $MoBr_2(N_2H_2)(dppe)$ and $MoBr_2(N_2H_2)(depe)_2$ with a loss of one N_2 molecule (dppe = $Ph_2PCH_2CH_2PPh_2$, depe = $Et_2PCH_2CH_2PEt_2$).

The ability of a molybdenum compound to reduce molecular nitrogen to NH_3 was first reported by Vol'pin and Shur in 1965 [35]. They found that when N_2 at 150 atm and 20°C is added to an ether solution of $MoCl_5$ and ethylmagnesium bromide, a small amount of NH_3 is formed [36].

More recently, it was found by Russian workers that a number of systems using molybdenum compounds as catalysts and Ti^{3+}, V^{2+}, or Cr^{2+} as reducing agents in aqueous solutions reduce N_2 to N_2H_4 and NH_3 in reasonable yields [37,38]. It was reported, e.g., that in a system using Ti^{3+} and $MoOCl_3$ in aqueous alcohol, small amounts of N_2H_4 and NH_3 were produced at atmospheric pressure and room temperature. These results were subsequently confirmed by Hill and Richards using $^{15}N_2$ [39]. Better yields were obtainable at higher N_2 pressures (20 to 100 atm) and elevated temperatures (85°C), and the best system was found to consist of $MoOCl_3$ or K_2MoO_4, Mg^{2+} and KOH in aqueous methanol:

$$N_2 + 4Ti^{3+} \xrightarrow[\substack{KOH,\ 85°C \\ H_2O,\ CH_3OH}]{MoOCl_3,\ Mg^{2+}} N_2H_4 + 4Ti(OH)_4 \qquad (13)$$

These workers proposed that the active catalyst is a Mo(III) species, formed by reduction with Ti^{3+}. Evidence for this was their finding that when CO was added to the system (in place of N_2), a Mo(III)-carbonyl compound was formed, as determined by spectral analysis of the solution. It was also found, however, that Mo(III) compounds alone do not reduce N_2, but are only effective in the presence of reducing agents and Mg^{2+}. A mechanism to explain these results was proposed involving the intermediates [38] [reaction (14)]. Further evidence in favor of Mo(III) was the finding that

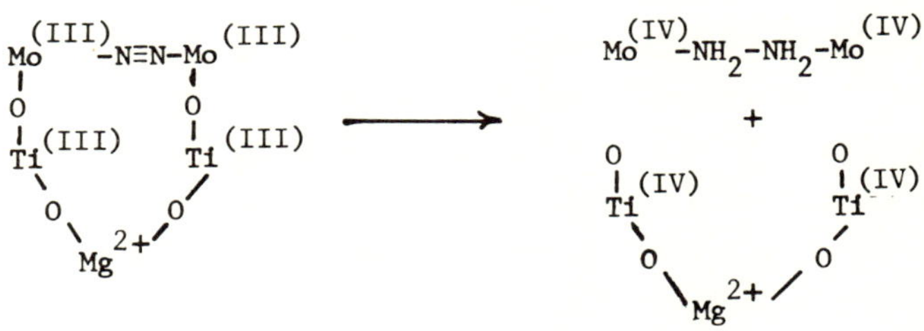

(14)

V^{2+}, which is a d_3 ion as is Mo(III), will reduce N_2 in the absence of any Mo compound (this was not observed when Ti^{3+} or Cr^{2+} were the reducing agents). In fact, a quantitative yield of N_2H_4 was obtained using V^{2+} and N_2 at 10 atm. In this regard, it is of considerable interest to note that V can replace Mo in nitrogenase, although the V enzyme is not as active in reducing N_2 as the Mo enzyme [40].

Although these results certainly support Mo(III) as the active species in N_2 reduction, it should be pointed out that other interpretations are possible. In the presence of strong reducing agents, it is possible that both Mo(IV) and even Mo(II) species are present

[41], perhaps in equilibrium. If so, the formation of a Mo(III)-carbonyl upon addition of CO might simply represent the shift of the equilibrium by formation of this stable Mo(III) compound. In any case, considerably more evidence is necessary to firmly establish Mo(III) as the active state.

A recent series of papers by Schrauzer et al. has reported the development of a model nitrogen-reducing system based on molybdenum-thiol complexes [42-46]. Nitrogenase, in addition to reducing N_2, also reduces a large number of other compounds, such as acetylene, 1-alkynes, cyanide, isocyanides, nitriles, and nitrous oxide. Because of its simplicity and convenience, the reduction of acetylene to ethylene is often used as a measure of nitrogenase activity in enzymic reactions, and Schrauzer's group used this reaction as a model for N_2 reduction in much of their work.

Because of the presence of relatively large amounts of cysteine in nitrogenase, and other evidence supporting Mo-sulfur binding at the active site, these workers mainly used systems consisting of Na_2MoO_4 and cysteine or the Mo(V)-cysteine complex, $Na_2Mo_2O_4$-$(cyst)_2$, as the molybdenum species and $NaBH_4$ or $Na_2S_2O_4$ as the reducing agents in aqueous solution at pH 7 to 12. In addition, other thiol compounds as well as NaSH and Na_2MoS_4 were tested, but, in general, their activities were lower than those with cysteine.

In the reduction of acetylene, optimum conditions were found to be pH 9.6, with $Na_2Mo_2O_4(cyst)_2$ as the catalyst and $NaBH_4$ as the reducing agent [43]. In addition to ethylene, some ethane is also produced by this system, the relative amounts of the two products depending on conditions and age of the catalyst solutions.

A mechanism for the reduction of acetylene to ethylene involving Mo(IV) as the active form of the catalyst was developed to explain the results [43] and is shown in Scheme 1. The rate-controlling step in this mechanism was presumed to be the formation of the substrate-catalyst complex (2), in agreement with the observed rate law for acetylene uptake. Neither this complex, nor the proposed Mo(IV)-catalytic species (1), however, were detected during the reaction. This was explained as being due to the short lifetime of

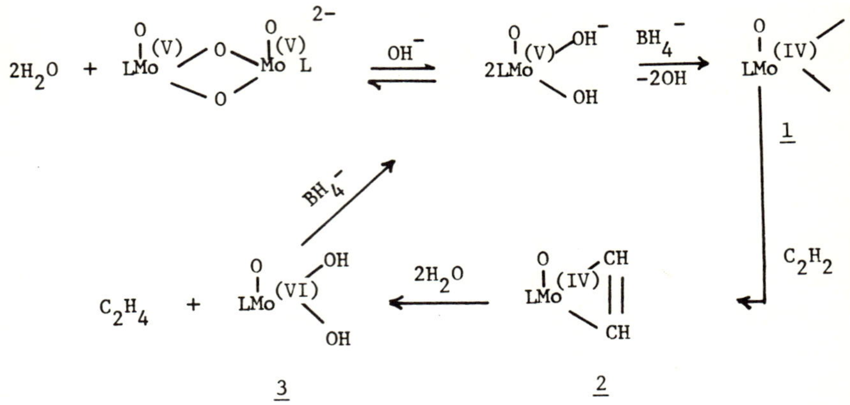

(L = cysteine)

Scheme 1

both species. The oxidized form of the catalyst was assumed to be an Mo(VI) species (3), since after exhaustion of BH_4^-, the solution turned blue and this was attributed to molybdenum blue [a mixed Mo(V)-Mo(VI) species]. Finally, to explain the production of some ethane, a dimeric form of the Mo(IV) catalyst (4) was invoked (Scheme 2). This dimeric form was also postulated to form slowly

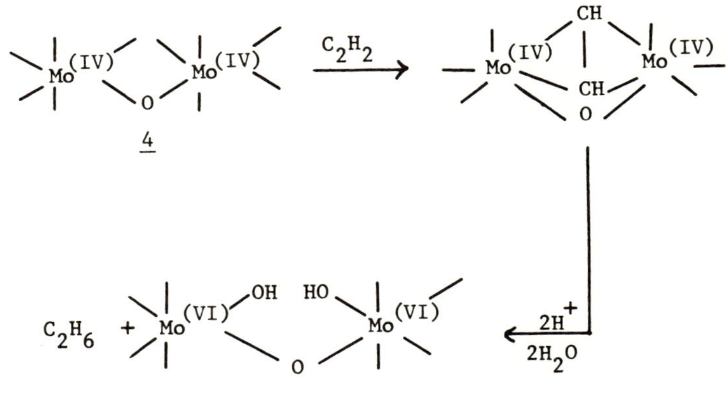

Scheme 2

from the monomer (1), since in aged solutions the ratio of ethane/ethylene in the products increases.

There appear to be several difficulties with this mechanism. First, when BH_4^- is added to $Mo_2O_4(cyst)_2^{2-}$ in the absence of substrate, extensive reduction takes place, since the yellow color of this complex (λ_{max} = 306 nm) is replaced by a brown-red color [43]. Now, if the proposed Mo(IV) catalyst [either the monomer (1) or dimer (4)] has such a short lifetime to be undetectable even in the absence of substrate (as stated by the authors) then it can hardly account for the color change on addition of BH_4^-. Therefore, the majority of the Mo must be present in some other form not accounted for by the mechanism. Since Mo(IV) species are known to be unstable in H_2O, disproportionating into Mo(III) and Mo(V), it seems more likely that the majority of the Mo is in the Mo(III) state. Also, in view of some recent work by Taube [41], it is possible BH_4^- reduces the original Mo(V) to a Mo(II) state. It is possible, of course, that the postulated Mo(IV) species is indeed present in small amounts, in equilibrium with other species, and may be the catalytically active form. Second, the observation that the solution turns blue after exhaustion of BH_4^- does not substantiate the claim that the oxidized form of the catalyst is Mo(VI), since it is well known that molybdenum blue does not form in basic solution [47]. Furthermore, as first reported by Huang and Haight [12] and further studied in this laboratory, the original yellow $Mo_2O_4(cyst)_2^{2-}$ changes to blue upon standing in basic solution, without the production of any Mo(VI) [48]. Since, however, little data concerning the properties of the system have been published (in five papers concerning the work, e.g., not a single visible absorption spectrum is reported) no conclusions concerning the active Mo species can be drawn and the proposed mechanism must be regarded as highly speculative.

In view of the fact that nitrogenase requires ATP in order to reduce N_2, it is of considerable interest that this model system is also stimulated by ATP [43]. This effect was proposed to be due to a phosphorylation reaction which improves the leaving-group properties of the OH^- groups of the Mo(IV) species, thus facilitating the formation of the substrate-catalyst complex [reaction (15)].

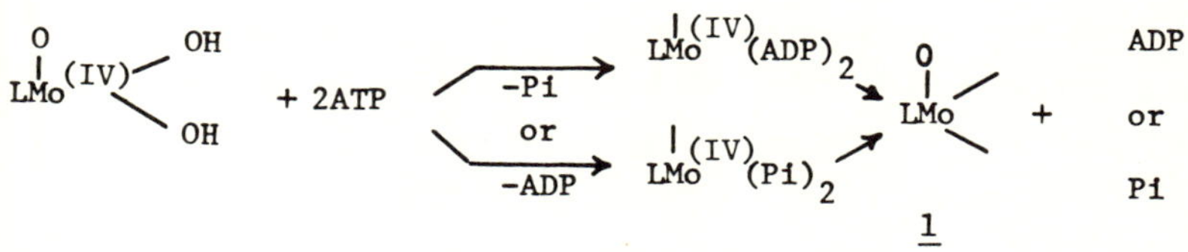

(15)

Additionally, it was found that $FeSO_4$ also stimulates the reduction of C_2H_2 to C_2H_4. Since nitrogenase contains nonheme iron, and must function in concert with a second enzyme that contains only nonheme iron as a cofactor, this finding is most interesting. No mechanism explaining the effect of $FeSO_4$ was proposed, except to suggest it aids in the original transfer of electrons from BH_4^- to Mo(V) or Mo(VI) [43].

These same workers have studied the reduction of isonitriles [45] and nitriles [46] by their model system and have found it functions similarly to nitrogenase. A mechanism, based on the same Mo(IV)-cysteine complex (1) as proposed for C_2H_2 reduction was developed, and a stimulation of the reactions by ATP and $FeSO_4$ was also observed.

Of greater importance with respect to a model for nitrogenase is the report by Schrauzer et al. that their system [45], and a similar one based on Na_2MoO_4, 1-thioglycerol, and $FeSO_4$ [44], reduces molecular N_2 to NH_3. The Na_2MoO_4-thioglycerol system functioned at 2000 psi of N_2, while the system described above for C_2H_2 reduction reduces N_2 at 1 atm and room temperature. This was confirmed independently by Hill and Richards, with N_2^{15} using a system consisting of 2-aminoethanethiol, Na_2MoO_4, $FeSO_4$, and $NaBH_4$, which gave good yields of $N^{15}H_3$ at 1 atm and 25°C [39].

Thus, model systems similar in properties to nitrogenase have been successfully developed. At the present, however, the mechanistic details of the models are almost as obscure as those of nitrogenase itself. Further work in the near future promises to reveal these details and provide us with a greatly improved understanding of biological nitrogen fixation.

3.1.3. Reactions with Flavins

As can be seen from Table 1, many of the molybdenum enzymes also require a flavin, usually FAD, as an additional cofactor. Furthermore, in the cases of xanthine oxidase and aldehyde oxidase there is evidence from ESR studies that the flavin is involved in electron transfer, receiving electrons from reduced molybdenum [eq. (1)] [1]. Additionally, it is generally believed that in certain nitrate reductases the path of electron transfer is from FAD to Mo(VI) and thence to NO_3^- [eq. (2)] [1]. Therefore, model reactions between flavins and Mo(V) and (VI) complexes are of considerable interest with regard to information they might give concerning Mo-flavin interaction in enzymes.

Preliminary studies of the reactions between flavins and Mo(V) and (VI) were reported in 1963 by Mitchell and Williams [49]. They found that $MoOCl_5^{2-}$ partially reduces FMN to $FMNH_2$ at basic pH, while MoO_4^{2-} partially oxidizes $FMNH_2$ under similar conditions. Additionally, increased visible absorption due to flavosemiquinone (FMNH•) was observed in partially reduced solutions of FMN in the presence of Mo(VI), suggesting the presence of a Mo(V)- or Mo(VI)-flavosemiquinone complex.

More detailed studies of the reactions between the Mo(V, VI) redox couple and flavins were carried out by Hemmerich and Spence [50] and Spence et al. [51]. They found, using lumiflavin-3-acetic acid, that a pH-dependent equilibrium is established in phosphate buffer, pH 6 to 8:

$$FH_2 + 2Mo(VI) \rightleftarrows F + Mo(V)_2 \qquad (16)$$

where F = lumiflavin-3-acetic acid; FH_2 = reduced lumiflavin-3-acetic acid; Mo(VI) and $Mo(V)_2$ are undoubtedly present as phosphate complexes of undetermined structure. At pH 6.0, the reaction proceeds essentially completely to the right. As the pH is raised, the equilibrium shifts to the left, approaching completion in that direction at pH 8.0.

During the course of the forward reaction, before equilibrium is reached, a deep red transient color was observed. By working at 0°C, it was possible, using a combination of polarography and spectrophotometry, to obtain a spectrum of this intermediate (Fig. 1) [50]. Under optimum conditions, ∿45% of total flavin is present as red intermediate. Further work using ESR spectrometry established that the red species is ESR-inactive. By analogy with other flavosemiquinone-metal ion complexes which have similar spectra [52], the red intermediate was proposed to be 5. The

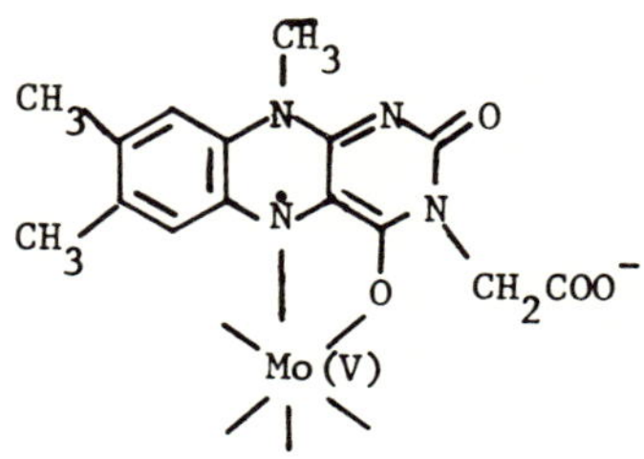

5

complex is ESR-inactive because of spin-pairing, most likely between the flavosemiquinone radical and Mo(V).

No such intermediate was observed during the reverse reaction, although an increase in flavosemiquinone radical ESR signal was detected. This was interpreted as evidence for the presence of an ESR-active Mo(VI)-flavosemiquinone complex; this now seems doubtful, however, since no other evidence has been obtained, e.g., a visible spectrum for such a species.

The results were interpreted by the following mechanism:

F + Mo(VI)	→	F•-Mo(V) complex 5	slow	(17)
F•-Mo(V)	→	F• + Mo(V)	slow	(18)
2Mo(V)	⇄	$Mo(V)_2$	fast	(19)

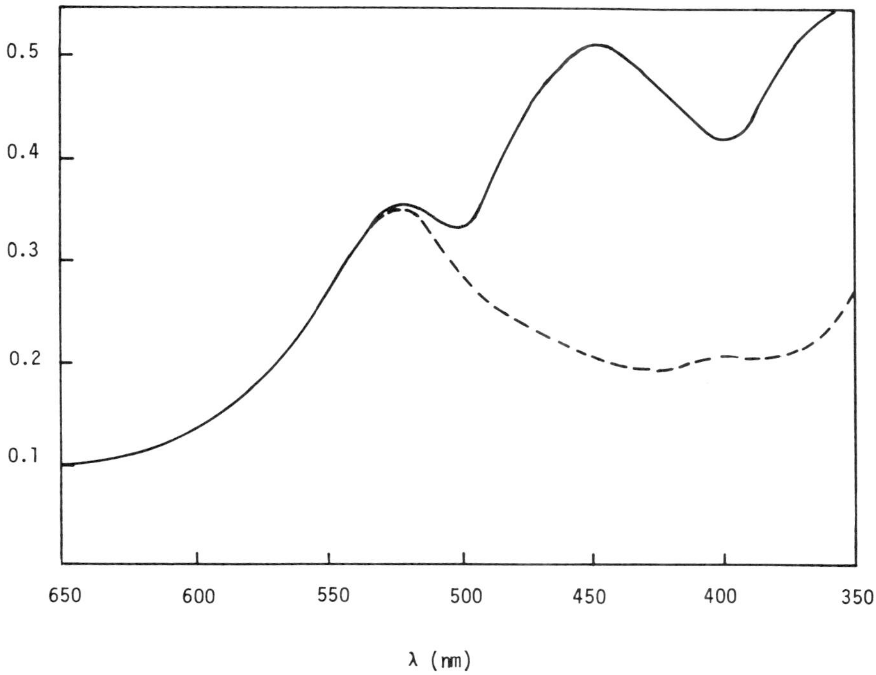

FIG. 1. Spectrum of Mo(V)-flavosemiquinone complex (5): measured spectrum; (- - -) spectrum after subtraction of spectra of uncomplexed flavin present. Total lumiflavin-3-acetic acid concentration = 1.00×10^{-3} M, pH 7.5, 1 M phosphate buffer, 0°C.

$$2F\cdot \rightleftarrows F + FH_2 \quad \text{(disproportionation)} \quad \text{fast} \tag{20}$$

$$Mo(V) + F \rightarrow F\cdot + Mo(VI) \quad \text{(reverse reaction)} \quad \text{fast} \tag{21}$$

Thus, the mechanism indicates the reaction in both directions proceeds via a one-electron transfer involving monomeric Mo(V) and flavosemiquinone as intermediates. At pH 6.0, where there is little reverse reaction, the rate was found to obey second-order kinetics, with a rate constant $k_2 = 1.73 \pm 0.23\ M^{-1}\ sec^{-1}$ (25°C) [52]. At higher pH, due to the mathematical complexity of the rate expression, no test of the mechanism was obtained.

Recently a detailed kinetic study of this reaction in tartrate buffer using FMN at pH 2.50 to 5.00 was completed in this laboratory [53]. Since Mo(V) monomer was implicated in the previous studies of the reaction between flavins and Mo(V, VI), tartrate was chosen because it is known to stabilize the monomer somewhat [25]. Again, a pH-dependent equilibrium was observed, with high pH (5.0) favoring reactants and low pH (2.5) favoring products:

$$2\mathrm{Mo(VI)} + \mathrm{FMNH_2} \rightleftarrows \mathrm{Mo(V)_2} + \mathrm{FMN} \tag{22}$$

Because of experimental considerations, the reverse reaction was chosen for kinetic studies, and its rate was investigated over a wide range of conditions. Contrary to previous studies, the kinetics of the reaction in tartrate [53] are not compatible with a one-electron transfer mechanism and could only be rationalized in terms of a two-electron transfer involving Mo(IV) as a reactive intermediate:

$$\mathrm{Mo(V)_2} \underset{k_{-1}}{\overset{k_1}{\rightleftarrows}} \mathrm{Mo(IV)} + \mathrm{Mo(VI)} \tag{23}$$

$$\mathrm{Mo(IV)} + \mathrm{FMN} \underset{k_{-2}}{\overset{k_2}{\rightleftarrows}} \mathrm{Mo(VI)} + \mathrm{FMNH_2} \tag{24}$$

Applying the steady-state assumption to Mo(IV), this gives the rate expression:

$$\frac{-d[\mathrm{FMN}]}{dt} = \frac{k_1[\mathrm{Mo(V)_2}][\mathrm{FMN}] - [\mathrm{Mo(VI)}]^2[\mathrm{FMNH_2}]/Ke}{k_{-1}/k_2\ [\mathrm{Mo(VI)}] + [\mathrm{FMN}]} \tag{25}$$

Since Ke [equilibrium constant for reaction (22)] could be measured, this expression can be integrated when both $\mathrm{Mo(V)_2}$ and Mo(VI) are present in large excess over FMN. Using a computer curve-fitting program to find the best value of k_{-1}/k_2, the integrated expression then allowed both k_1 and k_{-2} to be evaluated. The integrated expression was found to fit the data to 90 to 95% reaction, and consistent values of k_1 and k_{-1}/k_2 were obtained.

If the reaction mixture containing a large excess of Mo(VI) was allowed to stand for considerable time after reaching equilibrium, a second, slow change was observed. The solution gradually became a deep wine red and the total amount of free FMN and $FMNH_2$ (as measured by the combined polarographic cathodic and anodic wave heights) decreased. This same phenomenon was observed in the previous studies in the phosphate buffer where it was attributed to the formation of a Mo(V)-flavosemiquinone complex as an intermediate in the reaction. This color is due to the same complex in this case, since its visible absorption spectrum was found to be identical to that previously reported (Fig. 1). It cannot, however, be an intermediate, since it does not form until long after the reaction between $Mo(V)_2$ and FMN has reached equilibrium. It therefore must be a competing reaction, proceeding through a one-electron transfer. Apparently, in tartrate buffer this one-electron reaction does not compete favorably with the two-electron reaction. This may be due to the formation of a Mo(VI)-tartrate complex which inhibits the formation of the red Mo(V)-flavosemiquinone complex:

$$\mathrm{Mo(VI)} + \mathrm{FMNH_2} \longrightarrow \mathrm{FMNH{\cdot}\text{-}Mo(V)} \tag{26}$$

favorable in phosphate

unfavorable in tartrate

It would seem that which of the two paths is followed depends on the ligands present in the Mo complexes involved.

More recently as part of a program concerning models for Mo enzymes we have been investigating the chemistry of the Mo(V)-cysteine complex, $Na_2Mo_2O_4(cyst)_2 \cdot 5H_2O$, the structure of which has been determined by x-ray crystallography [54] (<u>6</u>). When <u>6</u> is allowed to stand in basic solution (pH $\geqq$ 7.5) its color slowly changes from the original yellow (λ_{max} = 306 nm) to blue (λ_{max} = 580 to 630 nm). Investigation of this reaction has shown that the

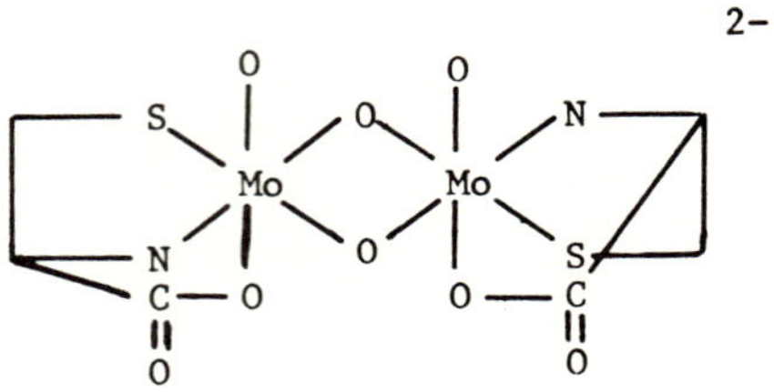

6

dioxobridged 6 is transformed to monobridged 7 under these conditions [48].

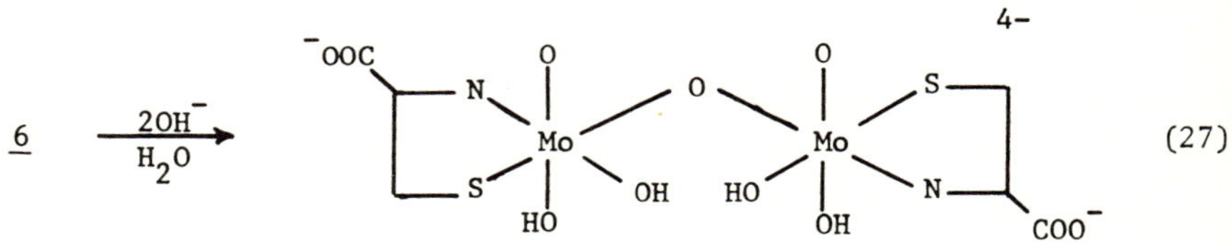

7

Complex 7 was found to rapidly reduce FMN at pH 9 to 11:

$$7 + FMN \rightarrow FMNH_2 + 2Mo(VI) + 2Cyst \tag{28}$$

The rate of the reaction is at least an order of magnitude faster than that observed in the previous study of the reduction of FMN by $Mo(V)_2$ in tartrate. Furthermore, complex 6, the dioxobridged species, reacts only slowly, if at all, with FMN under these conditions. Other complexes, such as $Mo_2O_4(hist)_2$ or the disulfur-bridged analog of 6, $Na_2Mo_2O_2S_2(cyst)_2$, do not undergo bridge opening and therefore do not react with FMN. Clearly, the monobridged structure is much more reactive than the dibridged.

Although the investigation of this reaction is not yet complete, a number of interesting facts have been discovered [48].

First, as 6 is converted to 7 a small ESR signal, very similar to that reported for xanthine oxidase, develops. This was first reported by Huang and Haight [11], who erroneously concluded the ESR signal and the blue color are due to the same species. Although the conversion of 6 to 7 is essentially quantitative at pH 10, the ESR signal never accounts for more than ∿1% of the total species. Second, the kinetics of the reaction of 7 with FMN are second order, first order in FMN and first order in 7, thus eliminating both monomeric Mo(V) and Mo(IV) as being the reactive species in the electron transfer. The kinetics are compatible with a two-electron transfer step involving simultaneous oxidation of both Mo(V) ions or, perhaps more probably, a one-electron transfer step followed by decomposition of the Mo(V)-Mo(VI) mixed valence species (8):

$\underline{7}$ + FMN $\xrightarrow{k}$ FMNH· + 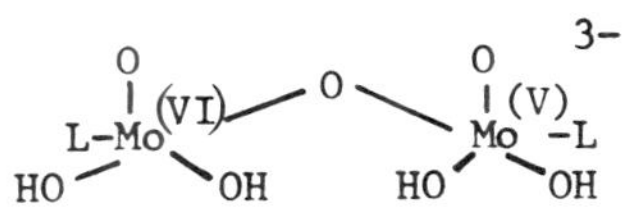 (29)

8

$\underline{8}$ ⟶ Mo(VI) + L + 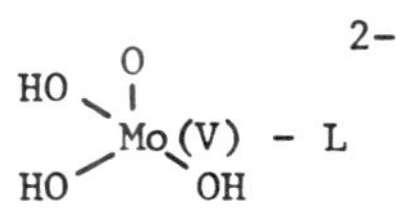 (30)

9

$$2\,\underline{9} \rightleftarrows \underline{7} \tag{31}$$

$$2\text{FMNH}\cdot \rightleftarrows \text{FMN} + \text{FMNH}_2 \tag{32}$$

(L = cysteine)

Such mixed valence intermediates (8) have recently been postulated in the electrochemical reduction of the dimeric Mo(V)-8-hydroxyquinoline complex [55]. Third, the rate of reduction of FMN by 7 at pH > 9.0 is increased when cysteine is added to the system [56]. Cysteine by itself does not reduce FMN under these conditions. This suggests an intermediate complex involving the addition of another cysteine to <u>7</u> might be involved:

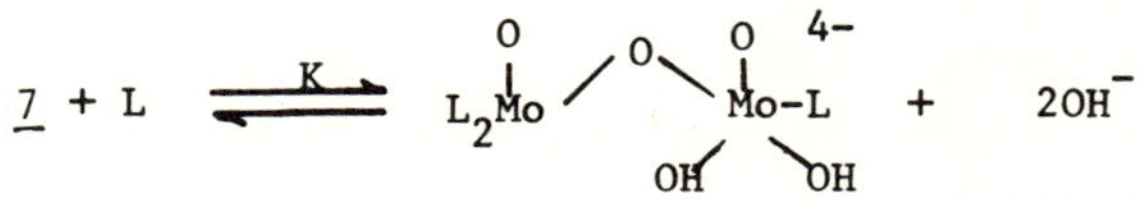

(33)

<u>10</u>

$$\underline{10} + \text{FMN} \xrightarrow{k'} 2\text{Mo(VI)} + 3\text{L} + \text{FMNH}_2 \tag{34}$$

This would lead to a complete rate law [reactions (29) and (34)] of the form:

$$k_{obs} = k + k'K[\text{cyst}]$$

Since cysteine concentration remains constant during the reaction a plot of k_{obs} vs [cysteine] for different initial concentrations of cysteine should be a straight line, with slope = k'K, and intercept = k. As seen in Figure 2 this is in fact the case. Furthermore, the value of k obtained from the intercept is in good agreement with the value of k obtained in the absence of added cysteine. If this mechanism is correct, K [reaction (33)] must be small. The value of k'K obtained from the slope in Figure 2 is almost 2 orders of magnitude larger than k. This means k' must be very large indeed, indicating complex <u>10</u> is highly reactive toward FMN (and probably other oxidants). This suggests thiol ligands activate Mo(V)-dimers toward redox reactions. The 2:1 [cyst:Mo(V)] complex should therefore react extremely rapidly with FMN. Unfortunately, this complex, which can be prepared in nonaqueous solvents [57], loses two cysteine ligands in the presence of H_2O, giving complex 7.

It seems then that there are a number of pathways of electron

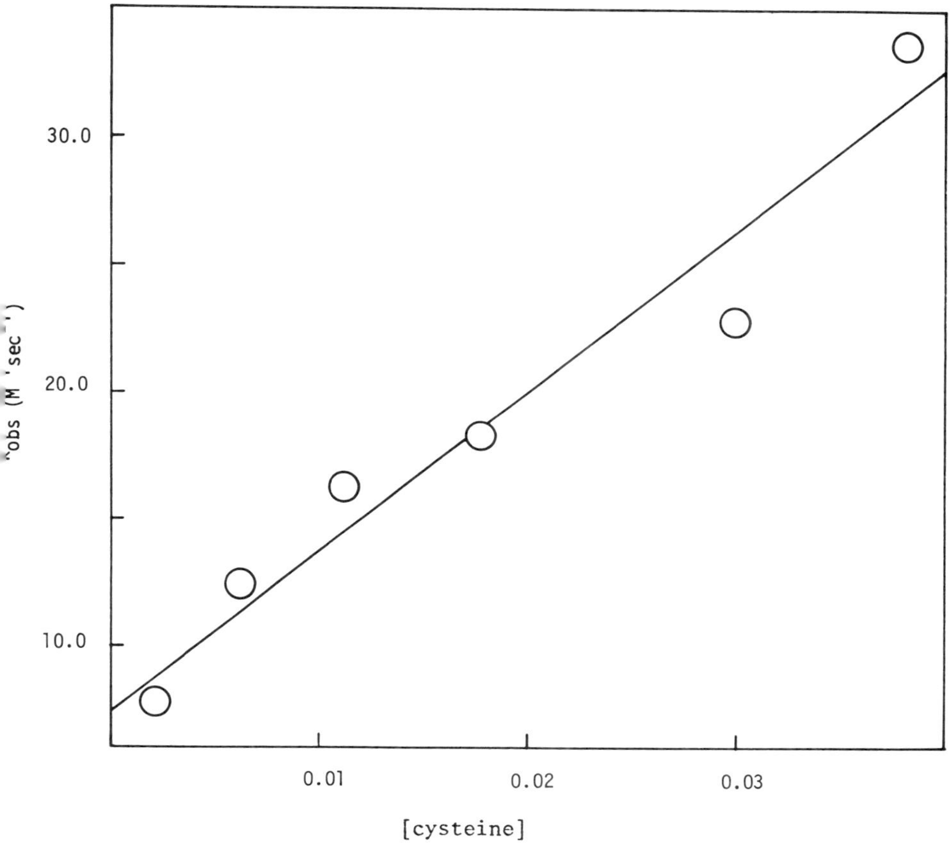

FIG. 2. Plot of k_{obs} vs [cysteine] for reduction of FMN by $Mo_2O_3(OH)_4(Cyst)_2^{4-}$ (7) in presence of added cysteine. pH 10.0, 0.5 M borate, 25°C.

transfer between Mo(V, VI) and flavins. The preferred pathway apparently depends on the nature of the ligands of the metal ion, which suggests a critical role for the binding groups present at the active site of the enzyme. The results also suggest the possibility that the molybdenum at the binding site may function catalytically in both one- and two-electron reactions, perhaps accounting for its biological versatility.

As a model for the redox behavior of molybdenum-flavin

complexes, the electrochemistry of Mo(V, VI)-8-quinolinol complexes in dimethylsulfoxide has been studied by Isbell and Sawyer [55]. It was necessary to work in a nonaqueous solvent because of solubility problems; they reasoned, however, that the active site of enzymes may provide a nonaqueous environment [6]. Their choice of 8-hydroxyquinoline as ligand was based on its similarity to flavosemiquinone, the form of flavin with the greatest metal ion affinity (see structures below). They studied the electrochemical be-

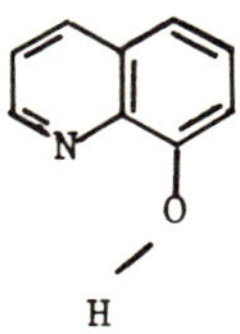

8-OH-Quinoline

11

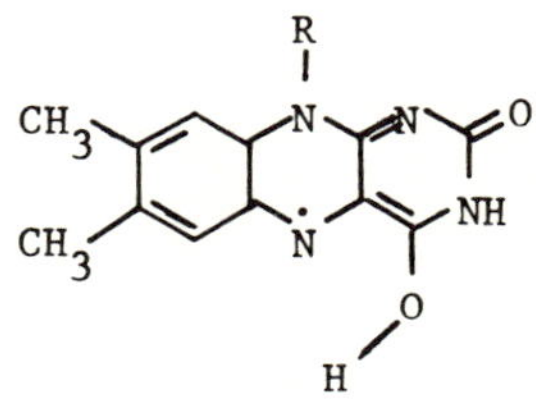

Flavosemiquinone

havior of both the Mo(VI) and Mo(V) complexes of this ligand, using cyclic voltammetry and coulometric techniques and employing ESR to detect species having unpaired spins. Their results revealed a complex set of redox reactions and chemical equilibria. Scheme 3 summarizes their major findings.

In addition to the species shown in Scheme 3, a number of others involving hydrolysis and dissociation of the ligand were detected. The only species in this scheme for which the structures are known with certainty are the $Mo^{(VI)}O_2Q_2$ (12) and $Mo^{(V)}{}_2O_3Q_4$ (13) complexes.

The results indicate all the electrochemical reductions proceed via one-electron steps, giving rise in some cases to mixed valence intermediates. Furthermore, a consideration of the potentials involved indicates the Mo(VI) complex is a poor oxidizing agent, while all lower states are good reducing agents, increasing in reducing power from $Mo^{(V)}{}_2O_3Q_4$ to $Mo^{(III)}$-Q.

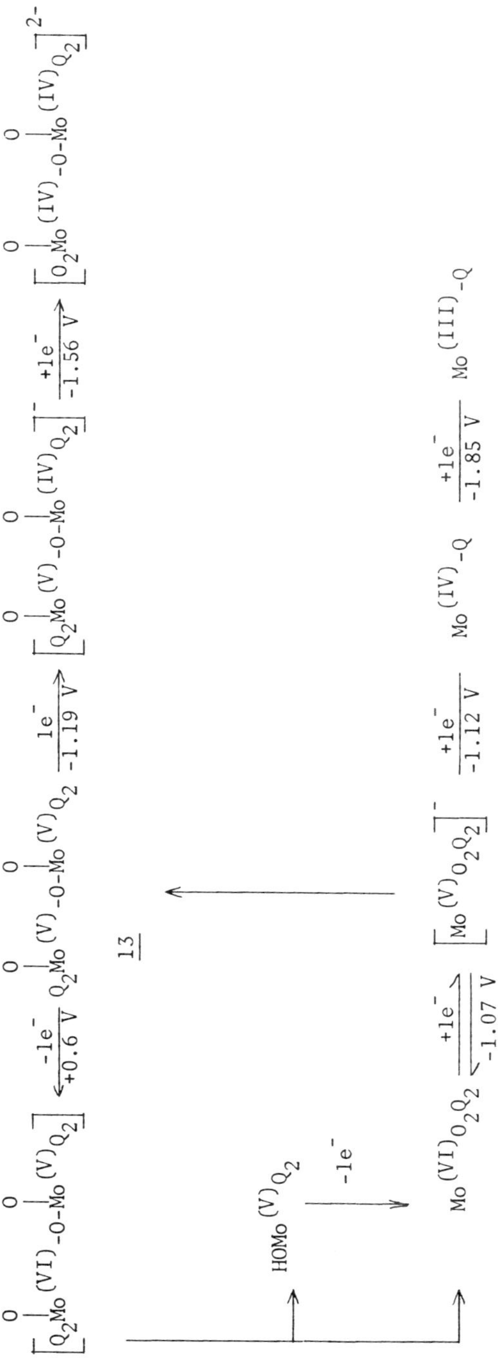

Scheme 3

Q = 8-hydroxyquinoline. Species in brackets indicate unstable intermediates. Voltages are vs SCE.

Although the results are of interest, their applicability to biological molybdenum-flavin reactions is somewhat questionable. In the enzymic reactions, electrons appear to be transferred between flavin and molybdenum, although intermediate complexes may also be formed. The main interest of this work lies in the demonstration of the existence of a number of 8-hydroxyquinoline complexes in which the molybdenum is present in different oxidation states, and particularly, in the apparent formation of intermediate mixed-valence complexes [Mo(VI)-Mo(V) and Mo(V)-Mo(IV)] which may offer an alternative explanation for the appearance of an ESR signal during the enzymic reactions. Whether similar complexes exist, even as intermediates, with flavosemiquinone as ligand remains to be seen.

3.1.4. Reactions with Thiols

Redox reactions between the Mo(V, VI) couple and the thiol-disulfide system, 2RSH/RSSR, are of interest as models for such reactions in molybdenum enzymes, where the metal ion appears to be bound to one or more cysteine side chains. Many of the molybdenum enzymes have a high cysteine content and in some cases, considerable "labile sulfide" [1]. Recent work concerning xanthine oxidase suggests the possibility of redox reactions between a persulfide group and the cysteine-bound Mo, although no positive evidence for such a reaction exists [58,59].

The oxidation of thioglycolic acid (TGA), cysteine, and glutathione by Mo(VI) have been investigated in this laboratory [60,61].

The reaction of Mo(VI) with TGA proceeds in two steps: Mo(VI) is first reduced to a dimeric Mo(V)-TGA complex and then to what appears to be a Mo(IV) species of unknown structure [60]. In both steps TGA is oxidized to dithiodiglycolic acid, DTDGA. At pH 6.0 in phosphate buffer, Mo(VI) and Mo(V) form complexes with TGA, both of which oxidize excess TGA to DTDGA. If excess TGA is added to a solution of Mo(VI) (Na_2MoO_4) at this pH, Mo(V) first increases and then decreases with time (Fig. 3). Using an NMR technique, the amount of DTDGA formed in the overall reaction was estimated to be equal to the amount of Mo(VI) reacted:

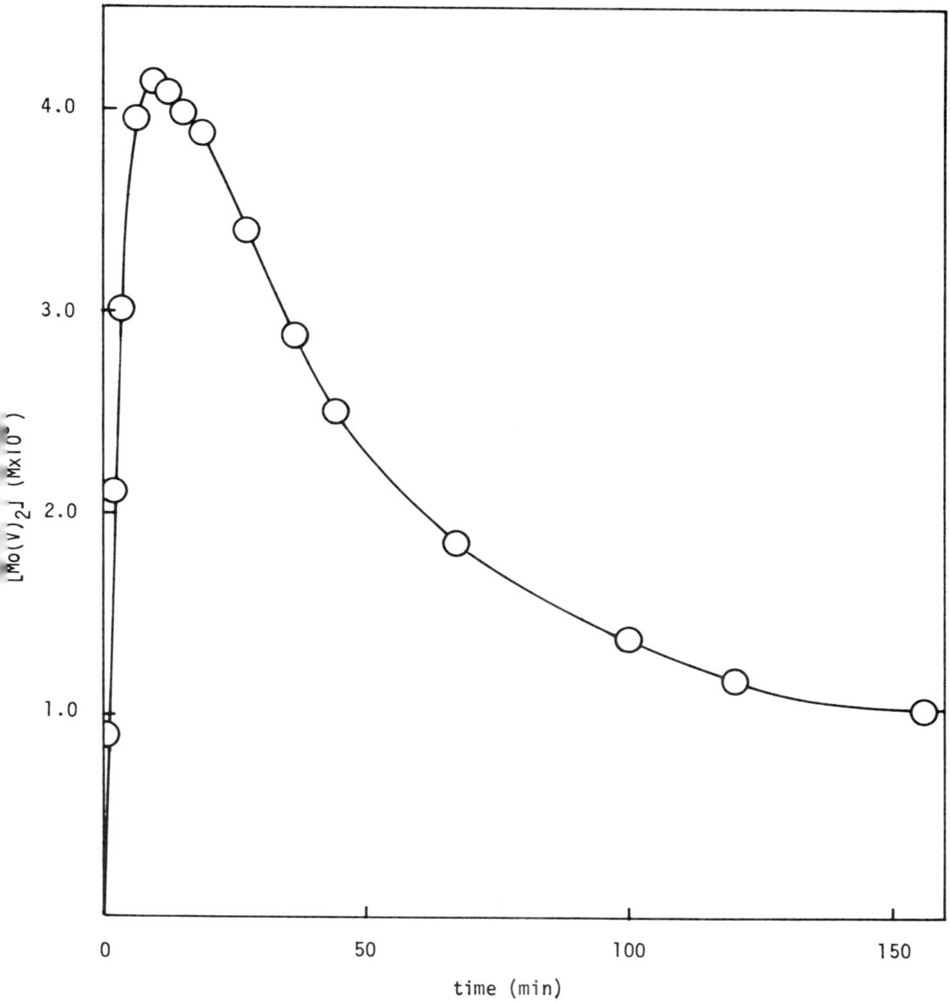

FIG. 3. Change of concentration of $Mo(V)_2$ with time during oxidation of TGA by Mo(VI). $[Mo(VI)]_0 = 1.00 \times 10^{-2}$ M; $[TGA]_0 = 5.20 \times 10^{-1}$ M; pH = 6.00, 1 M phosphate, 60°C.

$$2TGA + Mo^{(VI)}O_2(TGA)_2^{2-} \rightarrow Mo(IV) + DTDGA \qquad (35)$$

When $Mo^{(V)}{}_2O_3(TGA)_4^{4-}$ was added to excess TGA, DTDGA equal in amount to the Mo(V) complex reacted is formed:

$$2TGA + Mo^{(V)}{}_2O_3(TGA)_4{}^{4-} \rightarrow 2Mo^{(IV)} + DTDGA \qquad (36)$$

The structure of the Mo(IV) species formed is unknown, although it is most likely complexed with TGA.

The apparent formation of a Mo(IV) complex that is stable in H_2O is indeed surprising, since, with the exception of $Mo(CN)_8{}^{4-}$, no water-stable Mo(IV) complexes are known. The NMR method for determining the amount of DTDGA formed, however, is relatively imprecise, and it is possible the product is in fact an Mo(III)-TGA complex. Attempts to precipitate pure solids for analysis were unsuccessful, giving only intractable tars. If in fact the product is an Mo(IV) complex, it would lend weight to recent suggestions that a cysteine-bound Mo(IV) is formed during reactions catalyzed by xanthine oxidase [3,4].

The mechanism of oxidation of TGA by $Mo^{(VI)}O_2(TGA)_2{}^{2-}$ was found to be a straightforward one-electron reaction:

$$TGA + Mo^{(VI)}O_2(TGA)_2{}^{2-} \rightarrow Mo^{(V)}O_2(TGA)_2{}^{3-} + TGA\cdot \quad \text{slow} \qquad (37)$$

$$2TGA\cdot \rightarrow DTDGA \quad \text{fast} \qquad (38)$$

$$H_2O + 2Mo^{(V)}O_2(TGA)_2{}^{3-} \rightarrow Mo^{(V)}{}_2O_3(TGA)_4{}^{4-} + 2OH^- \quad \text{fast} \qquad (39)$$

In the oxidation of TGA by $Mo^{(V)}{}_2O_3(TGA)_4{}^{4-}$, the rate was found to be second order in complex and first order in TGA. No satisfactory mechanism to explain these kinetics was developed.

In the oxidation of cysteine, Mo(VI) was found to undergo only a one-electron reduction to $Mo^{(V)}{}_2O_4(cyst)_2{}^{2-}$ [61]. At pH 7.50, at which the reaction was studied, Mo(VI) gives no spectral evidence of complex formation with cysteine. The rate of the reaction, however, was found to be second order in cysteine and first order in Mo(VI), suggesting a 1:1 Mo(VI)-cysteine complex, present in small amounts, as the reactive species:

$$H_2O + MoO_4^{2-} + L \overset{K}{\rightleftharpoons} Mo^{(VI)}O_3L^{2-} + 2OH^- \tag{40}$$

$$Mo^{(VI)}O_3L^{2-} + L \xrightarrow{k_2} Mo^{(V)}O_3L^{3-} + L\bullet \tag{41}$$

$$2H_2O + 2Mo^{(V)}O_3L^{3-} \longrightarrow Mo_2O_4(L)_2^{2-} + 4OH^- \tag{42}$$

$$2L\bullet \longrightarrow L\text{-}L \tag{43}$$

(L = cysteine, L• = cysteine radical, L-L = cystine)

This mechanism leads to the observed rate expression:

$$-d[Mo(VI)]/dt = k_2K[Cyst]^2[Mo(VI)] \tag{44}$$

The rate of oxidation of glutathione by Mo(VI) at pH 7.50 was found to be independent of Mo(VI) concentration at high Mo(VI) concentrations ($\geq 2 \times 10^{-2}$ M) and dependent to the first power on Mo(VI) concentration at low Mo(VI) concentrations ($\leq 5 \times 10^{-4}$ M). In order to explain these results, it was necessary to postulate the existence of an unreactive form of glutathione (GSH') which is in equilibrium with an active form, GSH (glutathione is known to exist in a number of cyclic forms in acid solution [62], but little is known of its structure in neutral or basic solution).

$$GSH' \underset{k_{-1}}{\overset{k_1}{\rightleftharpoons}} GSH \tag{45}$$

$$MoO_4^{2-} + GSH \xrightarrow{k_2} Mo(V) + GS\bullet \tag{46}$$

$$2Mo(V) \rightleftharpoons Mo(V)_2 \tag{47}$$

$$2GS\bullet \rightleftharpoons GSSG \tag{48}$$

(GSH' = inactive, possibly cyclic, form; GSH = active form of glutathione; GS• = radical; GSSG = oxidized glutathione.) Applying the steady state assumption to GSH leads to the following rate expression:

$$-d[Mo(VI)]/dt = \frac{k_1 k_2 [Mo(VI)][GSH']}{k_{-1} + k_2[Mo(VI)]} \qquad (49)$$

In the presence of high Mo(VI) concentration, $k_2[Mo(VI)] >> k_{-1}$, which gives the experimental first-order rate law observed under these conditions:

$$-d[Mo(VI)]/dt = k_1(GSH'] \qquad (50)$$

At low Mo(VI) concentrations, $k_2[Mo(VI)] << k_{-1}$ and the expression reduces to the observed second-order rate law:

$$-d[Mo(VI)]/dt = k_1 k_2/k_{-1}[Mo(VI)][GSH'] \qquad (51)$$

No spectral evidence was found for the existence of either Mo(VI) or Mo(V) complexes with glutathione at this pH, although subsequent work by Haight and Huang [12] indicates a $Mo(V)_2$-glutathione complex is formed, possibly in small amounts, under these conditions.

These results indicate electron-transfer reactions, most likely from thiol groups to Mo(VI) complexes, are a possibility in molybdenum enzymes.

3.2. Ligand Displacement Reactions

3.2.1. Complex Formation and Ligand Exchange

Enzymic molybdenum is undoubtedly complexed with metal-binding groups of the protein. As already discussed, one or more of these groups appears to be a mercapto function of cysteine [1]. Other possible sites include histidine, serine, and tyrosine. It also seems quite likely that substrate binds to the metal ion, either through displacement of protein ligands or bound OH or water. This is especially probable with xanthine oxidase, aldehyde oxidase, and nitrogenase. Therefore, ligand displacement reactions of model molybdenum complexes are of importance for the information they might give concerning such processes in the enzymes. Unfortunately, few such reactions have been studied in any detail.

As a model for substrate binding in xanthine oxidase, the

formation of the Mo(VI)-8-hydroxyquinoline complex was studied by Knowles [63]. The assumptions of the work are that xanthine (and other purines) are first bound to a Mo(VI) binding site, and the kinetics and mechanism of complex formation of 8-hydroxyquinoline (which resembles the substrate) with Mo(VI) should be similar to those of enzyme-bound Mo(VI)-xanthine complex formation (see structures below). The second-order rate constant for the formation of

8-OH-Quinoline   Xanthine

11

enzyme-substrate complex with xanthine oxidase has been estimated to be $5 \times 10^{7}\ M^{-1}\ sec^{-1}$ [64]. The formation of the Mo(VI) complex at pH 8.0 to 9.0 was investigated using the temperature-jump technique. At this pH, Mo(VI) is present as MoO_4^{2-}, while the ligand is only partially ionized ($pK_a = 9.81$). In order to make certain only the 1:1 complex was formed, a large excess of MoO_4^{2-} was used in the studies.

The results indicate the reaction between MoO_4^{2-} and ligand is several orders of magnitude too slow to be a model for enzyme-substrate binding. This was attributed to the necessity for the tetrahedral MoO_4^{2-} to assume octahedral geometry in the complex given in reaction (52).

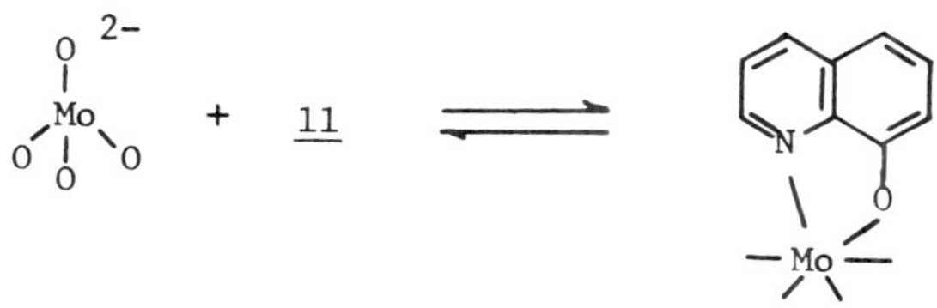

(52)

On the other hand, evidence obtained from the kinetic studies indicates a second pathway for complex formation [reaction (53)] involving a protonated form of Mo(VI), MoO_3OH^-, has a rate constant within 2 orders of magnitude of that for enzyme-substrate binding. This was rationalized on the basis that protonation of the tetrahedral oxyanion results in a weakened and lengthened metal-oxygen bond, thereby causing distortion of the tetrahedral geometry toward an octahedral structure. Presumably then, this partially octahedral species reacts much faster than the tetrahedral MoO_4^{2-} to form the complex, since less rearrangement of the coordination sphere is necessary [reaction (53)]. It was concluded by the author that this second pathway would be a much better model for enzyme-

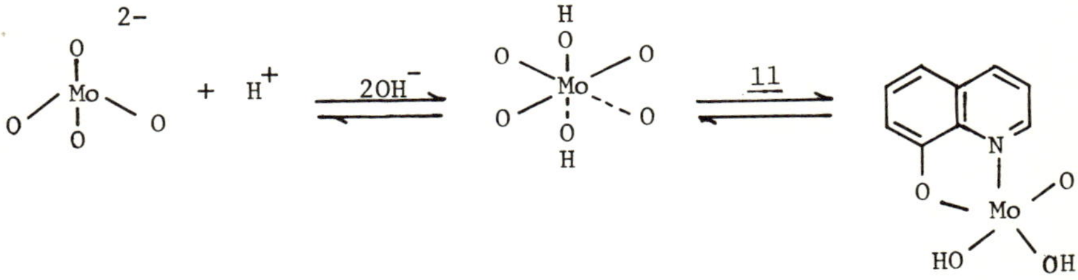

(53)

substrate binding, since Mo(VI) in the enzyme is most likely present in an octahedral complex. The results therefore support the role of Mo(VI) as being the initial binding site for substrate in xanthine oxidase.

Mitchell has investigated ligand exchange reactions of the dioxobridged Mo(V)-oxalate complex at pH 4.0 [65] $[Mo^{(V)}{}_2O_4(C_2O_4)_2\text{-}(H_2O)_2{}^{2-}]$. This complex was found to undergo displacement of both coordinated H_2O and oxalate, as well as conversion from a dibridged to a monobridged structure, in the presence of a variety of ligands. Typical substitution reactions of this complex are found in Scheme 4.

Solid products were obtained with all ligands except EDTA and the structures are based on analysis, infrared and electronic spectra and magnetic susceptibility measurements. All complexes

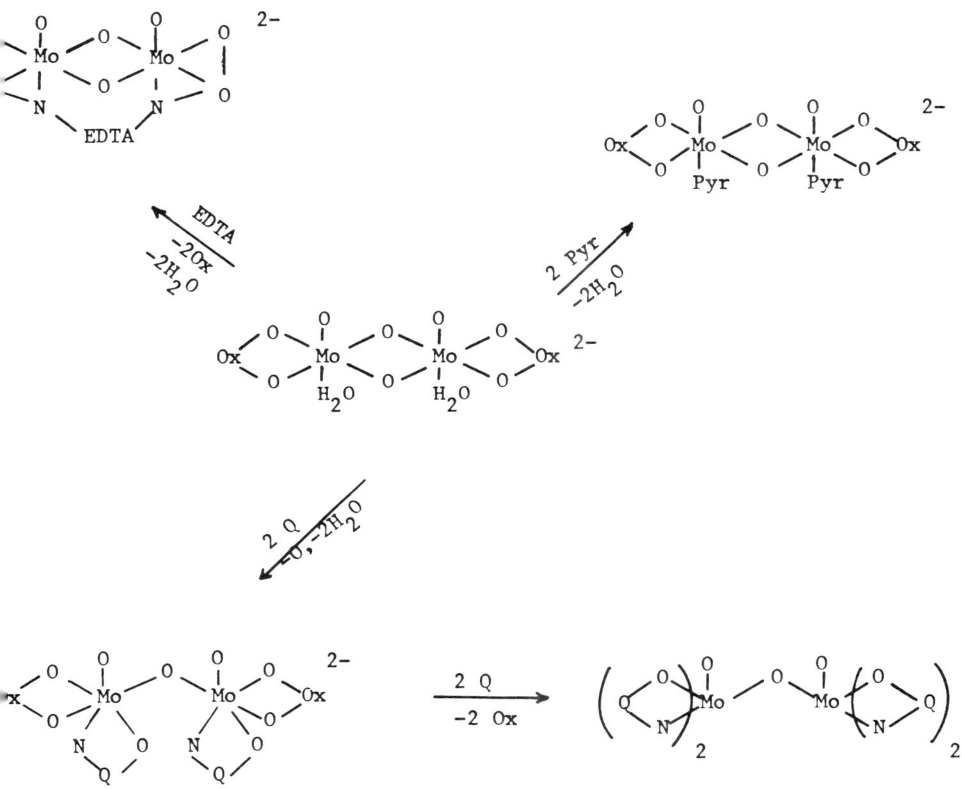

(Ox = oxalate; Q = 8-OH-quinoline; pyr = pyridine)

are essentially diamagnetic (0.4 to 0.5 BM, as compared with 1.73 BM expected for one unpaired electron). The ease of displacement was observed to be: H_2O > one bridging O > oxalate >> terminal O, while the ease of substitution is in the order 8-OH-quinoline >> EDTA >> pyr. It appears there are a number of pathways by which ligand substitution can occur in Mo(V) dimers, producing a variety of products. One might anticipate that in the enzyme, one coordination position on each Mo is occupied by H_2O, which could easily be displaced by substrate. In this regard, however, it should be noted that no displacement of H_2O was found when xanthine was used as ligand in this study [65]. This might be taken as evidence in favor of the view that in xanthine oxidase, the substrate is preferentially bound at an Mo(VI) site; subsequent oxidation of substrate and reduction of the enzyme to Mo(V) would then result in the dissociation of the product.

3.2.2. Bridge-Breaking Reactions of Mo(V) Dimers

At present, most evidence indicates that the Mo(V) oxidation state is of direct significance in most molybdenum enzymes [1]. Although there is no positive evidence for Mo-Mo interaction in any of the enzymes, the occurrence of two atoms of molybdenum per enzyme molecule in many cases (Table 1), and the great tendency of Mo(V) complexes to exist as dimers in aqueous environments suggests this structure may be of biological importance. It is expected that if Mo(V) dimers are present in the enzymes, they will involve either bridging oxygen or sulfur. Furthermore, recent work with complexes indicates monooxobridged species are more reactive in redox reactions than dioxobridged complexes [48]. A number of dimeric Mo(V) complexes containing two sulfur bridges are known, but no monosulfur-bridged species have as yet been reported [66].

Both mono- and dioxobridged Mo(V) complexes with thiol ligands react with H_2S to replace the oxo bridges with sulfur bridges [66].

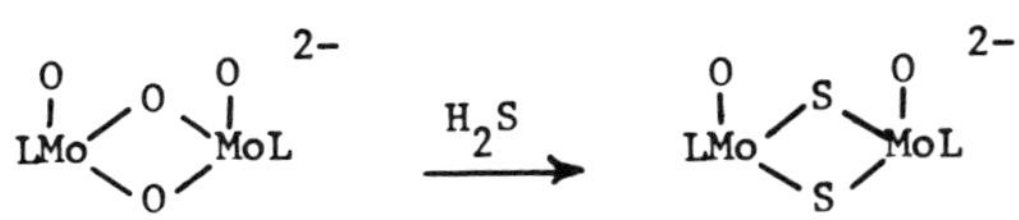

(54)

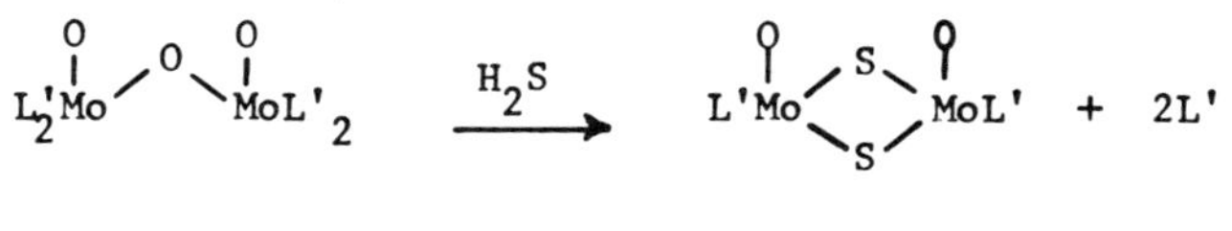

(55)

L = cysteine

L' = $HO-CH_2CH_2-SH$

It was reported by Kay and Mitchell that Mo(V) complexes which do not contain coordinated sulfur do not undergo this reaction [66]. Spivack and Dori, however, were able to prepare a Mo(V)-EDTA complex containing both an oxo and a sulfur bridge, and a disulfur-bridged histidine species [67].

Although no mechanistic studies were reported for these reactions with H_2S, work in this laboratory indicates the reaction most likely involves initial breaking of the oxobridge, followed by displacement with S [56] [reaction (56)].

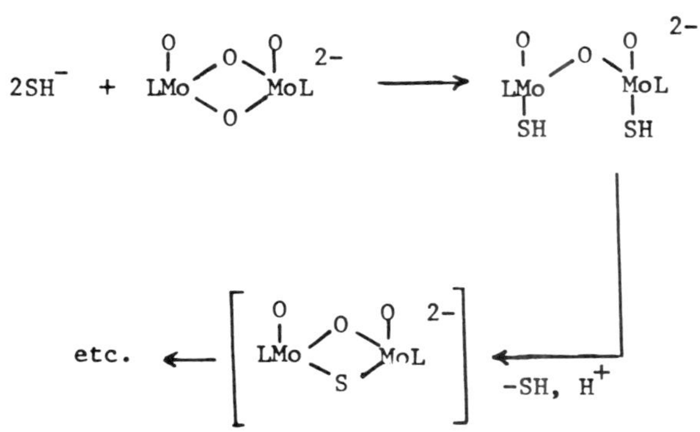

(56)

The base-catalyzed bridge-breaking reaction of $Mo_2O_4(cyst)_2^{2-}$ has been studied in some detail in this laboratory [48]. As mentioned in Section 3.1.3, when this complex is allowed to stand in basic solution (pH ≥7.5), one oxobridge opens [reaction (57)]. In

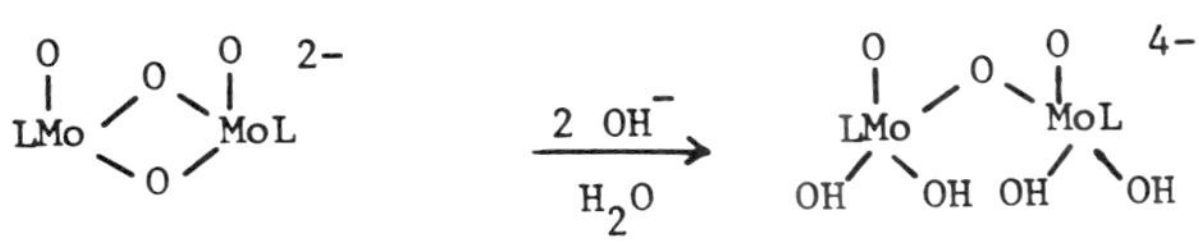

(57)

6 7

addition to bridge-opening, the weakly coordinating carboxyl group of cysteine is also displaced by OH^-. In the course of the reaction, a small ESR signal due to Mo(V) is developed. This signal while never more than ∿1% of total Mo, is a maximum at pH 11.0 [48] and is most likely due to the equilibrium shown in reaction

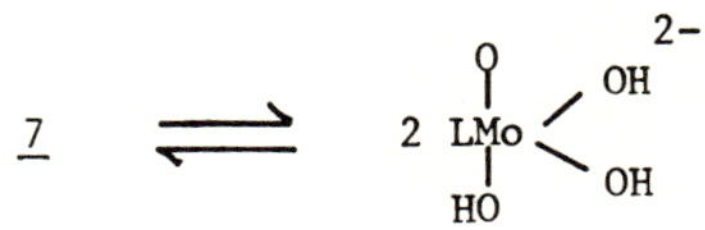 (58)

(58). In addition to OH^-, other bases such as NH_3 and RS^- also catalyze the reaction. In the presence of excess cysteine (RS^-), the reaction is accelerated, and preliminary kinetic studies indicate an intermediate involving the addition of an additional cysteine is formed [reaction (59)].

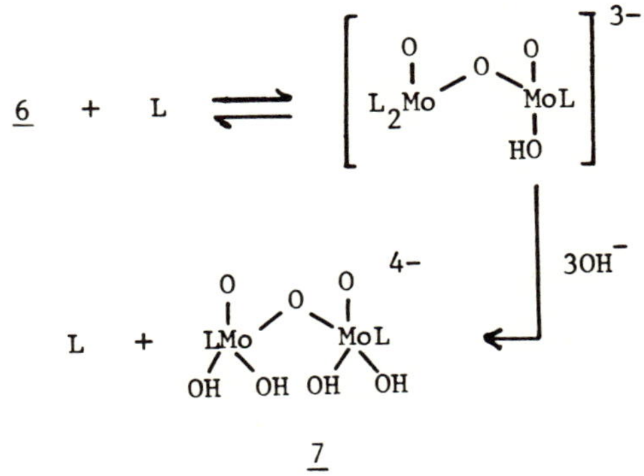 (59)

Further slow hydrolysis of complex 7 occurs, particularly at pH >10.0 to give reaction (60). As the pH is raised above 11.0, $MoO(OH)_3$ is formed.

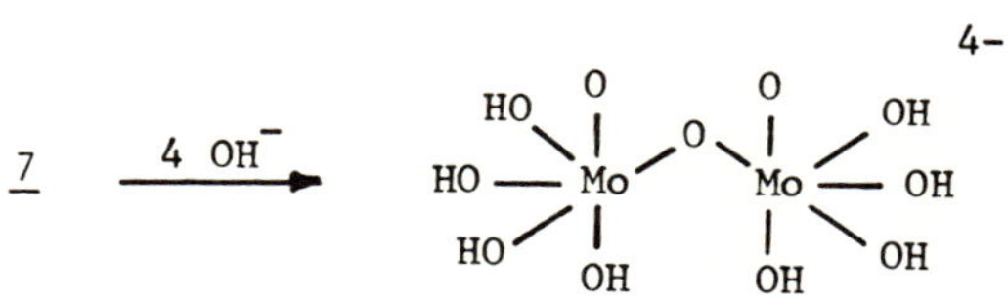 (60)

No such bridge-opening occurs with complexes without sulfur-containing ligands [57]. Furthermore, compounds containing disulfur bridges are stable even at pH 10 to 11, undergoing no bridge-opening [57]. Thus the presence of sulfur as a ligand labilizes oxobridges, but sulfur bridges are unreactive. As discussed previously (Section 3.1.3), the monooxobridged complex 7 is very reactive toward FMN, and observations in this laboratory indicate it is particularly sensitive to oxidation by O_2. Although most investigators favor an electron-transfer mechanism for xanthine oxidase involving Mo(VI) as the oxidation state present in the oxidized enzyme, the bridge-opening reactions suggest a possible alternative [reaction (61)]. The Mo(V) enzyme-substrate complex (either the dimer 13 or

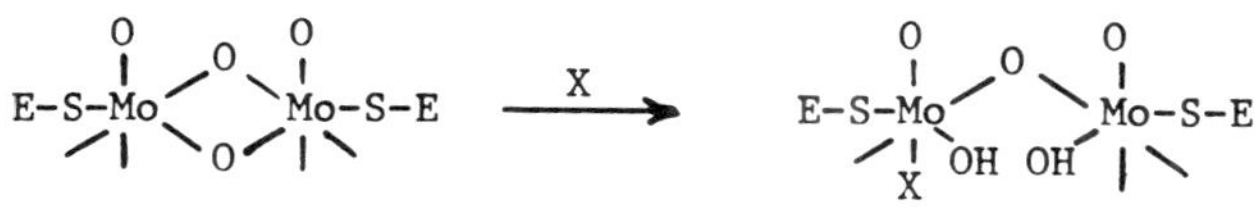

12 13

(61)

14

E = enzyme, X = Substrate

monomer 14 might then undergo electron transfer, either internally, or with FAD [reaction (62)]. The ESR signal due to Mo(V) would then be due to the monomer, or the mixed valence [Mo(IV)-Mo(V)] complex (15).

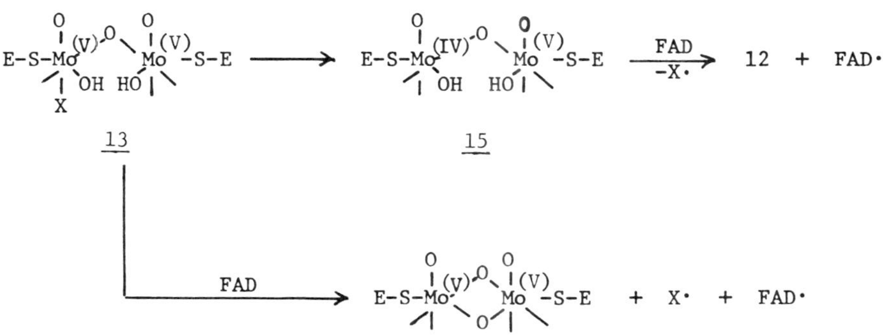

13 15 12

(62)

In any case, a monobridged Mo(V) dimer, because of its great reactivity, is an attractive model for the structure at the Mo(V) binding sites of the enzymes.

4. CONCLUSIONS

4.1. Enzymic Function of Molybdenum: Results from Model Studies

The results from the investigation of model reactions of molybdenum complexes, in general, strengthen the hypothesis that the metal is involved directly in the electron-transfer processes of the enzymes for which it is an essential cofactor [1].

In the reduction of nitrate, the model studies at pH 2.5 to 5.0 clearly indicate monomeric Mo(V), in equilibrium with dimer, is the reactive species. In strong acid solution, the results appear to be compatible with either a Mo(V) monomer or an Mo(IV) species. Although no detailed ESR studies concerning the sequence of steps have been made with the nitrate reductases, the most recent work, using the enzyme from *Micrococcus denitrificans* [68] gives unequivocal evidence for the presence of ESR-active Mo(V). In this case, the ESR signal is present in the resting enzyme and disappears in the presence of nitrate. A second ESR signal, differing somewhat in g values (80°K) was obtained upon reduction with $Na_2S_2O_4$. Although more work is certainly necessary, particularly to establish the sequence of electron transfers, the evidence from model studies clearly is in agreement with the proposed scheme for this enzyme [reaction (2), Section 2.2.3] and monomeric Mo(V) appears to be the key to nitrate reduction.

The results from the model nitrogenase studies most closely approximating biological systems clearly indicate some oxidation state lower than +5 is involved, since the reductants used ($Na_2S_2O_4$, $NaBH_4$, Ti^{3+}, V^{2+}, Cr^{2+}) are among the most powerful available in aqueous solution. In the absence of detailed experimental observations of the properties of these systems, it is difficult to draw any firm conclusions as to the species involved.

Based on arguments already presented (Section 3.1.2), the reactive Mo catalyst in the model systems is probably an Mo(III) complex. Such a species might also be present in nitrogenase. This hypothesis is supported by two facts. First, the natural reductants for nitrogenase (ferredoxins) have very negative reduction potentials [69] in the range of those of the chemical reductants used in the model systems. Second, no ESR signals attributed to molybdenum have been observed during reduction of N_2 by nitrogenase [9] and Mo(III) complexes are ESR-inactive in solution due to severe line-broadening. Before this tentative hypothesis can be accepted, however, much more work must be done, both with the model systems and nitrogenase.

The model studies of Mo(V,VI)-flavin reactions indicate electron transfer can occur between these two enzyme cofactors by both one- and two-electron steps, involving Mo(V) and Mo(IV) intermediates, respectively. The discovery that the monooxobridged Mo(V)-cysteine complex, formed in basic solution from the dioxo-bridged species, is the most reactive complex toward flavins yet reported (Section 3.1.3) is most interesting, since it suggests a similar structure may be present in enzymes such as xanthine oxidase, which contain two molybdenum atoms at the active site(s). Furthermore, the dibridged → monobridged reaction and the monomer-dimer equilibria observed with this complex suggest alternative interpretations to the generally accepted electron pathway of molybdenum in xanthine oxidase.

The limited number of studies reported concerning ligand displacement reactions of Mo complexes indicate both Mo(V) and Mo(VI) species undergo such reactions facilely and no conclusions concerning preferential substrate binding sites [Mo(VI) or (V)] can as yet be drawn.

4.2. Problems

Many of the model studies have raised questions which have already been discussed. In addition, there are a number of intriguing problems concerning the enzymic function of molybdenum that need solution.

Recent developments in xanthine oxidase investigations have been interpreted in terms of two separate Mo-sites, rather than one site at which Mo-Mo interaction exists [1]. This implies the presence, at least in the first stage of reduction, of monomeric Mo(V) complexes. Model studies, however, have indicated such complexes are not obtainable in aqueous solution except in very low concentrations or as reactive intermediates. Similar considerations apply to the currently popular proposal that the +4 oxidation state of Mo is involved during catalyses by xanthine oxidase [59]. In this case, with the possible exception of a poorly characterized thioglycolic acid species, no Mo(IV) complexes of biochemical interest that are stable in H_2O are known. There is even a question concerning the presence of enzyme-bound Mo(VI), since xanthine oxidase has maximum activity in slightly basic solution (pH $\sim$8) and Mo(VI) forms complexes only weakly, if at all, with most biological ligands under these conditions.

Clearly, such problems suggest a more intensive investigation of the coordination chemistry of molybdenum is needed, particularly with regard to complexes with more complex biochemical ligands, such as peptides and small proteins. Such ligands might provide the necessary structure to stabilize Mo complexes in several oxidation states. Since the findings also raise a question concerning the extent to which the binding site can be considered aqueous, these studies should be extended to nonaqueous solvents, which have been shown to make the preparation of unstable lower oxidation state complexes [Mo(IV), Mo(III)] feasible.

With the current interest in these problems by both biochemists and inorganic chemists, answers which give a deeper understanding of the biochemical function of this fascinating transition metal should soon be forthcoming.

ADDENDUM

Since the completion of the manuscript, a number of important studies concerning model reactions of molybdenum compounds have been published. This addendum makes no attempt to be either comprehensive or critical, but simply mentions some of the more relevant work.

Earlier investigations indicated the molybdenum in certain enzymes may be rather closely associated with the FAD cofactor. Recent ESR work with xanthine oxidase [74] indicates molybdenum may be remote from flavin in the enzyme, although the interpretation of the results on which this is based is somewhat questionable [75].

Although Mo(VI) was first assumed to be reduced to Mo(V) in the oxidases and dehydrogenases, it now seems possible Mo(IV) is the first reduction product, with the ESR active Mo(V) being formed by an equilibrium between Mo(VI) and (IV) [76].

Recent ESR studies of the nitrate reductases from *M. dentrificans* [77] and *E. coli* [78] have shown, in addition to an Mo(V) ESR signal, a second ESR signal, tentatively attributed to Mo(III), and Mo(III) ESR signals have recently been reported for model compounds [79].

The reduction of nitrate by Mo(V) complexes in nonaqueous solvents has now been reported [80-82]. Under these conditions NO_3^- is facilely reduced in a one-electron step to NO_2 (which disproportionates in water to nitrate and nitrite). The reduction of nitrite is much slower than the reduction of nitrate in nonaqueous media, in contrast to aqueous systems.

Treatment of the complex trans-$(Mo^{15}N_2)(dppe)_2$ with aqueous acid in N-methylpyrrolidone produces significant amounts of $^{15}NH_3$ [83,84]. The model system of Schrauzer has been extended and some details of the mechanism determined [85]. Recently, the reported stimulation of this system by ATP [43] has been questioned, with the effect being attributed to general acid catalysis [86].

Complete studies of the preparation of the monooxobridged

Mo(V) cysteine complex Mo_2O_3 (L-cysteine)$_2$(OH)$_4^{4-}$, and its reduction of flavins have now been published [56,87]. An electrochemical study of Mo(V)-flavin complexes in DMSO and an investigation of possible Mo(IV)-flavin complexes have recently appeared [88,89].

A study of the reduction of cytochrome c by Mo(V)-cysteine complexes has recently been completed in this laboratory [90]. The results indicate the reduction of cytochrome c by the monooxo-bridged cysteine complex, Mo_2O_3(L-cysteine)$_2$(OH)$_4^{4-}$, is extremely rapid (k_2 = 2.6 x 10^7 M^{-1} sec^{-1} at pH 7.0) and may be an excellent kinetic model for the enzyme sulfite oxidase.

Some ligand exchange reactions of the dioxobridged Mo(V)-cysteine complex, Mo_2O_4(L-cysteine)$^{2-}$, have been reported [91].

ABBREVIATIONS

A: ESR hyperfine splitting constant
ADP: Adenosine diphosphate
ATP: Adenosine triphosphate
Cyst: Cysteine
Cyt: Cytochrome
depe: 1,2-Bisdiethylphosphinoethane ($Et_2PCH_2CH_2PEt_2$)
dppe: 1,2-Bisdiphenylphosphinoethane ($Ph_2PCH_2CH_2PPh_2$)
DTDGA: Dithiodiglycolic acid
EDTA: Ethylenediaminetetracetic acid
F: Lumiflavin-3-acetic acid
FH_2: Dihydrolumiflavin-3-acetic acid
FAD: Flavin adeninedinucleotide
FMN: Flavin mononucleotide
$FMNH_2$: Dihydroflavin mononucleotide
Flavin: 7,8-Dimethylisoalloxazine
Fe-S: Nonheme iron-sulfur cluster
g: Spectroscopic splitting factor (ESR)
GSH: Glutathione (reduced)
GSSG: Oxidized glutathione
NADPH: Reduced nicotineadenine dinucleotide phosphate

Ox: Oxalate

Pi: Inorganic phosphate

Pyr: Pyridine

Q: 8-Hydroxyquinoline (oxine)

TGA: Thioglycolic acid

REFERENCES

1. R. C. Bray and J. C. Swann, *Structure and Bonding, 11,* 107 (1972).
2. S. T. Smith, K. V. Rajagopalan, and P. Handler, *J. Biol. Chem., 242,* 4108 (1967).
3. V. Massey, P. E. Brumby, H. Komai, and G. Palmer, *J. Biol. Chem., 244,* 1682 (1969).
4. V. Massey, H. Komai, G. Palmer, and G. B. Elion, *J. Biol. Chem., 245,* 2837 (1970).
5. P. C. H. Mitchell, *Coord. Chem. Rev., 1,* 315 (1966).
6. V. Massey and H. Ganther, *Biochemistry, 4,* 1161 (1965).
7. P. C. H. Mitchell, *Quart. Rev. (London), 20,* 103 (1966).
8. H. J. Cohen, I. Fridovich, and K. V. Rajagopalan, *J. Biol. Chem., 246,* 374 (1971).
9. H. Dalton, J. A. Morris, M. A. Ward, and L. E. Mortenson, *Biochemistry, 10,* 2066 (1971).
10. L. S. Meriwether, W. F. Marzluff, and W. G. Hodgson, *Nature, 212,* 465 (1966).
11. T. Huang and G. P. Haight, Jr., *J. Amer. Chem. Soc., 92,* 2336 (1970).
12. T. Huang and G. P. Haight, Jr., *J. Amer. Chem. Soc., 93,* 611 (1971).
13. J. T. Spence and H. Chang, *Inorg. Chem., 2,* 319 (1963).
14. R. L. Pecsok and D. T. Sawyer, *J. Amer. Chem. Soc., 78,* 5496 (1956).
15. J. T. Spence and E. R. Peterson, *Inorg. Chem., 1,* 277 (1962).
16. T. F. Emery, *Biochemistry, 6,* 3858 (1967).
17. J. T. Spence, *Coord. Chem. Rev., 4,* 475 (1969).
18. R. H. Garret and A. Nason, *J. Biol. Chem., 244,* 2870 (1970).
19. A. Nason, in "The Enzymes," Vol. 7 (P. D. Boyer, H. Lardy, and K. Myrbach, eds.), Academic Press, New York, 1963, p. 587.

20. G. P. Haight, Jr., P. Mohilner, and A. Katz, *Acta Chem. Scand., 16,* 221 (1962).

21. G. P. Haight, Jr. and A. Katz, *Acta Chem. Scand., 16,* 659 (1962).

22. G. P. Haight, Jr., *Acta Chem. Scand., 15,* 2012 (1961).

23. J. M. Kolthoff and I. Hodara, *J. Electroanal. Chem., 5,* 2 (1963).

24. E. P. Guymon and J. T. Spence, *J. Phys. Chem., 70,* 1964 (1966).

25. J. T. Spence and M. Heydanek, *Inorg. Chem., 6,* 1489 (1967).

26. D. H. Brown and J. MacPherson, *J. Inorg. Nucl. Chem., 33,* 4203 (1971).

27. J. T. Spence, *Arch. Biochem. Biophys., 137,* 288 (1970).

28. J. A. Frank and J. T. Spence, *J. Phys. Chem., 68,* 2131 (1964).

29. M. Hidai, K. Tominari, Y. Uchida, and A. Misono, *Chem. Commun.,* 1392 (1969).

30. J. E. Fergusson and J. L. Love, *Rev. Pure Appl. Chem., 20,* 33 (1970).

31. D. E. Harrison and H. Taube, *J. Amer. Chem. Soc., 89,* 5706 (1967).

32. E. E. Van Tamelen, *Accounts Chem. Res., 3,* 361 (1970).

33. J. E. Bercaw, R. H. Marvich, L. G. Bell, and H. H. Brintzinger, *J. Amer. Chem. Soc., 94,* 1219 (1972).

34. J. Chatt, G. A. Heath, and R. L. Richards, *Chem. Commun.,* 1010 (1972).

35. M. E. Vol'pin, A. A. Belyi, and V. B. Shur, *Izv. Akad. Nauk. SSR. Ser. Khim,* 2225 (1965).

36. M. E. Vol'pin and V. B. Shur, *Nature, 209,* 1236 (1966).

37. N. T. Denisov, O. N. Efimov, N. I. Shuvalova, A. K. Shilova, and A. E. Shilov, *Zh. Fiz. Khim., 44,* 2964 (1970).

38. A. Shilov, N. Denisov, O. Efimov, N. Shuvalov, N. Shuvalova, and A. Shilova, *Nature, 231,*460 (1971).

39. R. E. E. Hill and R. L. Richards, *Nature, 233,* 114 (1971).

40. J. R. Beneman, C. E. McKenna, R. F. Lie, T. G. Traylor, and M. D. Kamen, *Biochim. Biophys. Acta, 264,* 25 (1972).

41. A. R. Bowen and H. Taube, *J. Amer. Chem. Soc., 93,* 3287 (1971).

42. G. N. Schrauzer and G. Schlesinger, *J. Amer. Chem. Soc., 92,* 1809 (1970).

43. G. N. Schrauzer and P. A. Doemeny, *J. Amer. Chem. Soc., 93,* 1608 (1971).

44. G. N. Schrauzer, G. Schlesinger, and P. A. Doemeny, *J. Amer. Chem. Soc., 93,* 1803 (1971).

45. G. N. Schrauzer, P. A. Doemeny, G. W. Kiefer, and R. H. Frazier, Jr., *J. Amer. Chem. Soc., 94,* 3604 (1972).

46. G. N. Schrauzer, P. A. Doemeny, R. H. Frazier, Jr., and G. W. Kiefer, *J. Amer. Chem. Soc., 94,* 7378 (1972).

47. S. Ostrowetsky, *Compt. Rend., 251,* 1068 (1960).

48. P. Kroneck and J. T. Spence, *Inorg. Nucl. Chem. Lett., 9,* 177 (1973).

49. P. C. H. Mitchell and R. J. P. Williams, *Biochim. Biophys. Acta, 86,* 39 (1964).

50. P. Hemmerich and J. T. Spence in "Flavins and Flavoproteins," (E. C. Slater, ed.), Elsevier, Amsterdam, 1966, p. 82.

51. J. T. Spence, M. Heydanek, and P. Hemmerich in "Magnetic Resonance in Biological Systems" (A. Ehrenberg, B. G. Malmström, and T. Vänngard, eds.), Pergamon Press, Oxford, 1967, p. 269.

52. P. Hemmerich, F. Müller, and A. Ehrenberg in "Oxidases and Related Redox Systems," Vol. I (T. E. King, H. S. Mason, and M. Morrison, eds.), Wiley, New York, 1965, p. 157.

53. G. Colovos and J. T. Spence, *Biochemistry, 11,* 2542 (1972).

54. J. R. Knox and C. K. Prout, *Acta Cryst., B25,* 1857 (1969).

55. A. F. Isbell, Jr. and D. T. Sawyer, *Inorg. Chem., 10,* 2449 (1971).

56. P. Kroneck and J. T. Spence, *J. Inorg. Nucl. Chem., 35,* 3391 (1973).

57. P. Kroneck and J. T. Spence, unpublished results.

58. V. Massey and D. Edmonson, *J. Biol. Chem., 245,* 6595 (1970).

59. D. Edmonson, V. Massey, G. Palmer, L. M. Beachem, III, and G. B. Elion, *J. Biol. Chem., 247,* 1597 (1972).

60. J. F. Martin and J. T. Spence, *J. Phys. Chem., 74,* 3589 (1970).

61. J. F. Martin and J. T. Spence, *J. Phys. Chem., 74,* 2863 (1970).

62. M. Calvin in "Glutathione--A Symposium," (S. Colowick, A. Lazarow, E. Racker, D. R. Schwartz, E. Stadtman, and H. Waelsch, eds.), Academic Press, New York, 1964, p. 3.

63. P. F. Knowles and H. Diebler, *Trans. Faraday Soc., 64,* 977 (1968).

64. H. Gutfreund and J. M. Sturtevant, *Biochem. J., 73,* 1 (1959).

65. P. C. H. Mitchell, *J. Chem. Soc. (A),* 146 (1969).

66. A. Kay and P. C. H. Mitchell, *J. Chem. Soc. (A),* 2421 (1970).

67. B. Spivack and Zvi Dori, *Chem. Commun.,* 1716 (1970).

68. P. Forget and D. V. Dervartanian, *Biochim. Biophys. Acta, 256,* 600 (1972).

69. R. W. F. Hardy, R. C. Burns, and G. W. Parshall, *Advan. Chem. Ser., 100,* 219 (1971).

70. R. J. Downey, *J. Bacteriol., 105,* 759 (1971).

71. L. D. Solomonson and B. Vennesland, *Biochim. Biophys. Acta, 267,* 544 (1972).

72. K. V. Rajagopalan, I. Fridovich, and P. Handler, *J. Biol. Chem., 237,* 922 (1962).

73. K. V. Rajagopalan, I. Fridovich, and P. Handler, *J. Biol. Chem., 242,* 4097 (1967).

74. D. J. Lowe, R. M. Lynden-Bell, and R. C. Bray, *Biochem. J., 130,* 239 (1972).

75. R. C. Bray, *J. Less Common Metals, 36,* 413 (1974).

76. J. S. Olson, D. P. Ballou, G. Palmer, and V. Massey, *J. Biol. Chem., 249,* 4363 (1974).

77. P. Forget and D. V. Dervartanian, *Biochim. Biophys. Acta, 256,* 600 (1972).

78. P. Forget and D. V. Dervartanian, *Biochim. Biophys. Acta, 379,* 74 (1975).

79. P. C. H. Mitchell and R. D. Scarle, *J. Chem. Soc. (Dalton),* 110 (1975).

80. C. D. Garner, M. R. Hyde, F. E. Mabbs, and V. I. Routledge, *Nature, 252,* 579 (1974).

81. C. D. Garner, M. R. Hyde, F. E. Mabbs, and V. I. Routledge, in press.

82. R. D. Taylor and J. T. Spence, in press.

83. C. R. Brulet and E. E. Van Tamelen, *J. Amer. Chem. Soc., 97,* 911 (1975).

84. J. Chatt, A. J. Pearman, and R. L. Richards, *Nature, 224,* 1201 (1969).

85. G. N. Schrauzer, G. W. Kiefer, K. Tano, and P. A. Doemeny, *J. Amer. Chem. Soc., 96,* 641 (1974).

86. A. P. Krushch, A. E. Shilov, and T. A. Vorontsova, *J. Amer. Chem. Soc., 96,* 4987 (1974).

87. P. Kroneck and J. T. Spence, *Biochemistry, 12,* 5020 (1973).

88. D. T. Sawyer, J. N. Gerber, L. W. Amos, and L. J. DeHayes, *J. Less Common Metals, 36,* 487 (1974).

89. J. Selbin, J. Sherrill, and C. H. Bigger, *Inorg. Chem., 13,* 2544 (1974).

90. G. D. Lawrence and J. T. Spence, in press.

91. A. Kay and P. C. H. Mitchell, *J. Chem. Soc. (Dalton),* 1388 (1973).

Chapter 7

INTERACTION OF Cu(I) COMPLEXES WITH DIOXYGEN

Andreas D. Zuberbühler
Institute of Inorganic Chemistry
University of Basel
Basel, Switzerland

1. INTRODUCTION: Cu(I) IN BIOLOGICAL SYSTEMS

Metalloproteins engaged in biologically significant reactions of dioxygen ("molecular oxygen," "oxygen") may be divided into four classes: (i) Oxygen carriers, (ii) oxidases (electron-transfer oxidases), (iii) monooxygenases (mixed-function oxidases, hydroxylases), and (iv) dioxygenases (oxygen transferases) (cf. [1]). Copper has been established as the essential metal ion for one or several enzymes in each of these classes. Well-known examples are hemocyanin as an oxygen carrier [2], laccase and ascorbate oxidase as electron transfer oxidases [3,4], tyrosinase and dopamine-β-hydroxylase as monooxygenases [5,6], and quercetinase as a dioxygenase [7].

The interaction of Cu(I) with dioxygen plays an important role in three out of four of these classes of copper proteins: In some enzymes, and especially in the dioxygen carrier hemocyanin, the copper is now generally accepted to be permanently in the reduced state and reversibly interacting with O_2 [2]. For electron-transfer oxidases and monooxygenases, autoxidation (i.e., spontaneous oxidation by dioxygen at room temperature) of cuprous to cupric enzyme has been established or at least strongly indicated as an integral part of the reaction mechanism [8-10]. General types of such reactions are given by equations (1)-(3), which illustrate several possibilities for reduction of cupric enzyme by reduced substrate (SH_2) or cosubstrate (AH_2) followed by reoxidation with dioxygen (ping-pong mechanism).

$$\begin{gathered} 2Cu^{2+} + SH_2 \rightarrow 2Cu^+ + S + 2H^+ \\ 2Cu^+ + 2H^+ + O_2 \rightarrow 2Cu^{2+} + H_2O_2 \end{gathered} \qquad (1)$$

$$\begin{gathered} 4Cu^{2+} + 2SH_2 \rightarrow 4Cu^+ + 2S + 4H^+ \\ 4Cu^+ + 4H^+ + O_2 \rightarrow 4Cu^{2+} + 2H_2O \end{gathered} \qquad (2)$$

$$\begin{gathered} 2Cu^{2+} + AH_2 \rightarrow 2Cu^+ + A + 2H^+ \\ 2Cu^+ + SH_2 + O_2 + 2H^+ \rightarrow 2Cu^{2+} + SH(OH) + H_2O \end{gathered} \qquad (3)$$

Considering this biological significance of Cu(I), it is surprising, at least at first sight, that the number of studies on Cu(I) complexes of low molecular weight is practically negligible compared to the enormous literature on stability, structure, spectra, and reactivity of cupric compounds regarded more or less directly as models for biological systems. More specifically, the autoxidation of not a single Cu(I) complex had been described with detailed kinetics and a suitable mechanistic scheme until a few years ago. Thus many speculations, but few facts, could be put forward when questions such as correlation between nature or number of coordinating ligand atoms and stabilization of the Cu(I) state, rate of autoxidation, or ability of reversible dioxygen binding were discussed.

The aim of this chapter is to describe the kinetics and mechanisms of the interaction of low-molecular Cu(I) complexes with dioxygen and to discuss the results and their possible biological significance on the basis of the coordination chemistry of copper and the redox behavior of O_2. It is not intended to describe all copper enzymes that play a role in the biochemistry of dioxygen, or to give a complete bibliography of those biochemical systems which are mentioned.

2. REDOX PROPERTIES OF DIOXYGEN

The electronic structure of dioxygen, its physical properties, redox equilibria, interaction with transition metal ions, and fitness as a terminal oxidant in biological systems have been extensively reviewed [11-17]. Although the dioxygenyl cation O_2^+ can be obtained with extremely strong oxidants such as PtF_6 [18], the interaction of O_2 with coordination compounds normally is governed by its electron-accepting properties. Binding of dioxygen to transition metal complexes usually is described with a metal to ligand π bond (back donation), on which a weak ligand to metal σ bond is superimposed.

O_2 may coordinate to one or two metal ions. Structures Ia,

Ib [19] or Ic [20], and IIa [21] or IIb [22] have been suggested

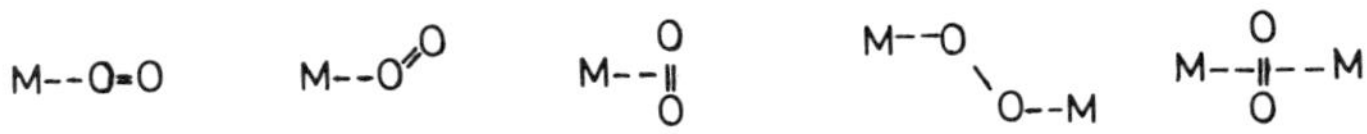

Ia Ib Ic IIa IIb

for these dioxygen adducts. Ic [23] and IIa [21] have been confirmed by x-ray analysis, but according to Goutermann and Zerner [24] the relative stability of the different structures may be strongly influenced by the nature of the other ligands coordinated to the central atoms, and indeed the x-ray analysis of oxygenated myoglobin seems to be better in line with the bent Pauling structure Ib than with Ic [25]. The reduction of dioxygen to its thermodynamically stable product, H_2O, can be described by equations (4)-(8) for one-electron steps and (9)-(11) for two- or four-electron reductions.

$$O_2 + e \rightleftarrows {}^{\bullet}O_2^- : E_0 = -0.4 \text{ V [12] } (-0.45 \text{ V [17]}) \tag{4}$$

$$O_2 + e + H^+ \rightleftarrows HO_2^{\bullet} : E_0 = -0.1 \text{ V [12] } (-0.32 \text{ V [17]}) \tag{5}$$

$$HO_2^{\bullet} + H^+ + e \rightleftarrows H_2O_2 : E_0 = +1.68 \text{ V [17]} \tag{6}$$

$$H_2O_2 + H^+ + e \rightleftarrows {}^{\bullet}OH + H_2O : E_0 = +0.8 \text{ V [17]} \tag{7}$$

$$^{\bullet}OH + H^+ + e \rightleftarrows H_2O : E_0 = +2.74 \text{ V [17]} \tag{8}$$

$$O_2 + 2H^+ + 2e \rightleftarrows H_2O_2 : E_0 = 0.688 \text{ V [26]} \tag{9}$$

$$H_2O_2 + 2H^+ + 2e \rightleftarrows 2H_2O : E_0 = 1.77 \text{ V [26]} \tag{10}$$

$$O_2 + 4H^+ + 4e \rightleftarrows 2H_2O : E_0 = 1.23 \text{ V [26]} \tag{11}$$

Two points are almost invariably mentioned when the redox properties of O_2 in biological systems are discussed: (i) the high redox potential of O_2, and (ii) its kinetic inertness. Both need some comment here.

(i) Classification of O_2 as a strong oxidant origins in the value of 1.23 V (11) which places it at about the same level as chlorine (E_0 = 1.36 V [26]). At neutral pH this value is reduced to 0.81 V; more important, however, is the fact that frequently H_2O_2 is the (metastable) endproduct of dioxygen reduction if no compounds with catalatic activity are present in the reaction mixture. In that case H_2O_2 can accumulate, and dioxygen should be looked at as a moderately strong oxidant in acid solution [E_0 = 0.688 V; eq. (9)], whereas it is rather weak at neutral pH (0.27 V).

(ii) Conceding the peculiar nature of dioxygen as a stable biradical, the restrictions imposed by its low potential as a one-electron reductant [eqs. (5) and (6)] and by the need for spin conservation [12], this low reactivity should not be overemphasized.

> It is often rather the inertness of bonds in organic matter than that of dioxygen which makes the simultaneous presence of both possible.
>
> Compared to really inert strong oxidants such as ClO_4^-, O_2 is very reactive, indeed. A striking example is the possibility of obtaining H_2O_2 as the practically sole reduction product of O_2. This means that in such systems H_2O_2 is less reactive against those reducing agents, and this, although its oxidation potential is almost 1 V above that of O_2 (in reactions leading to H_2O_2) at any pH.
>
> As we shall see later (cf. Sections 4 and 6), the autoxidation of some transition metal complexes should rather be classified as extremely rapid for redox reactions, because equilibria (12) and (13) are established with rates of $10^8\ M^{-1}\ sec^{-1}$, which are not very far from the diffusion controlled limit of about $10^{11}\ M^{-1}\ sec^{-1}$.

$$M^{n+} + O_2 \rightleftarrows M^{(n+1)+} + {}^{\bullet}O_2^- \quad (12)$$

$$M^{n+} + {}^{\bullet}O_2^- + 2H^+ \rightleftarrows M^{(n+1)+} + H_2O_2 \quad (13)$$

3. Cu(I): A REDOXACTIVE CLOSED-SHELL TRANSITION METAL ION

Cu^+ shares its closed-shell d^{10} electron configuration with eight other cations: Ag^+, Au^+, Zn^{2+}, Cd^{2+}, Hg^{2+}, Ga^{3+}, In^{3+}, and Tl^{3+}. Of these, only Cu^+, Ag^+, and Au^+ can be oxidized at all by chemical means, and Cu^+ is unique in being susceptible to autoxidation. Coordination compounds with easily oxidizable d^{10} central atoms have been obtained also for Ni(0), Pd(0), Pt(0), Co(-I), and others (cf. [27]), but strong electron-acceptor ligands and frequently aprotic media are needed, to stabilize these low valence states. On the other hand, Cu(I) readily forms complexes with NH_3, imidazole, Cl^-, and many other simple ligands, which are perfectly stable in aqueous solution if access of O_2 is excluded. As has been stated in Section 1, our knowledge of Cu(I) coordination chemistry is relatively limited. Several difficulties are connected with studies of Cu(I) complexes and may be used to explain that fact at least partially.

(i) It is hard to find sufficiently stable <u>and</u> soluble cuprous salts that could be used as a source of Cu(I). CuCl, the compound most commonly used in the past, is sparingly soluble in water ($S_{CuCl} = 10^{-6.73}$ or $10^{-5.92}$ [28] and highly complexed ($K^{Cu}_{CuCl} = 10^{3.63}$ [29], $\beta^{Cu}_{CuCl_2} = 10^{4.7} - 10^{6.0}$ [28]. The introduction of the water-soluble acetonitrile complex $Cu(CH_3CN)_4ClO_4$ [30,31] greatly helped to solve this problem.

(ii) The aquoion, Cu^+_{aq}, has a strong tendency to disproportionate into Cu^0 and Cu^{2+}: $[Cu^+]^2/[Cu^{2+}] = 10^{-6.0}$ [32]. Although Manahan [33] and Espenson [34] have prepared metastable cuprous perchlorate in dilute $HClO_4$, this instability of Cu^+_{aq} greatly hampers the study of weak Cu(I) complexes: The use of metallic copper-electrodes is excluded at low ligand concentrations, the polarographic half-wave potential can only be estimated indirectly, and often only the overall dissociation constants $\beta^{Cu}_{CuL_n}$ of the higher complexes are obtained, whereas the constants of the individual steps $K^{CuL_n}_{CuL_{n+1}}$ remain unknown or need the application of special techniques. Thus, in a recent study of Cu^+ in aqueous chloride, Ahrland and Rawsthorne [35] did not succeed in obtaining a value for the stability constant of the 1:1 complex. However, K^{Cu}_{CuCl} as well as the corresponding constant K^{Cu}_{CuAn} for acetonitrile (An) as ligand could be determined in a kinetic investigation of the autoxidation of Cu(I) in aqueous chloride or CH_3CN [29]. It is not surprising, therefore, that for ligands forming 1:1 and 2:1 or higher order cuprous complexes of comparable stability, no quantitative data are listed for the stability of the 1:1 complexes in the 1964 edition of "Stability Constants of Metal Ion Complexes" [36], with the single exception of $CuNH_3^+$, a system studied carefully by Bjerrum [37].

(iii) Even more severe limitations have their origin directly in the $3d^{10}$ electron configuration of Cu(I). With a diamagnetic ground state and a completely filled d subshell Cu(I) may be literally called invisible, as it is necessarily ESR-inactive and as visible absorption spectra from d-d transitions are excluded, too. Thus, in Cu(I) coordination chemistry the extremely important tools of ESR and visible absorption spectroscopy are of very little use. In marked contrast, the spectral properties of Cu^{2+} are probably better studied than those of any other transition metal ion. Here, spectrophotometry not only is very important for the determination of complex stability constants but also, as does ESR, helps to answer questions regarding the number and nature of coordinating ligand atoms (see, e.g., [10,38-41]).

Electrometric methods only give the number of ligands other than solvent molecules which are bound to the central atom. Due to the scarcity of spectral data, information about the structure of Cu(I) complexes in solution is difficult to obtain. Theoretical considerations, x-ray data from solids, and results from corresponding Ag(I) or Au(I) complexes are used as indirect evidence [42], although the question of solvent coordination should now become tractable by NMR.

As evidenced by x-ray analysis, and in common with Ag^+ and

Au^+, Cu^+ prefers low coordination numbers up to 4, whereas hexacoordination, the most frequent situation with transition metal ions, has not been observed in cuprous complexes at all.

The great majority of Cu(I) compounds shows tetrahedral coordination in solid state, but linear (e.g., in Cu_2O [43]) and trigonal arrangements are firmly established, too, the latter even in strictly monomeric species such as tris(trimethylphosphinesulfide)-copper(I) perchlorate [44] and tris(2-picoline)copper(I) perchlorate [45]. In some other cases the low coordination numbers are supplemented by one or two additional ligand atoms at considerably larger distances as in bis(thiourea)copper(I) chloride [46]. In solution the corresponding weak interactions are probably of minor significance, but one expects that addition of extra ligands is a relatively easy process and might play an important role in the reaction mechanisms of Cu(I) complexes.

Questions concerning the stereochemistry of cuprous species in solution have been discussed, e.g., by Orgel [47], Jörgensen [48], and Hemmerich [49]. Besides the tendency to low coordination numbers and the marked affinity for highly polarizable ligands, the reluctance of Cu(I) to form chelates with typical σ donors such as ethylenediamine is noteworthy. The generally accepted view that 2:1 complexes with monodentate ligands like NH_3 or imidazole (Im) are truly two-coordinate in solution is strongly supported by the properties of mixed systems containing two different ligands. For instance, if the complex formation is studied in solutions containing Cu(I) largely in the form of $Cu(An)_2^+$ ($\beta^{Cu}_{CuAn_2} = 10^{4.35}$ [31] or $10^{3.9}$ [50]), the addition of the first imidazole depends linearly on 1/[An], and the formation of $Cu(Im)_2^+$ on $1/[An]^2$ [49]. Quantitative treatment gives the same value for $\beta^{Cu}_{CuIm_2}$ in the presence or absence of CH_3CN, if its influence on $[Cu^+_{aq}]$ (14) is taken into consideration [51].

$$Cu(H_2O)_n^+ + 2An \rightleftarrows Cu(An)_2^+ + nH_2O \tag{14}$$

Thus, neither CH_3CN is bound to $Cu(Im)_2^+$ nor imidazole to

$Cu(An)_2^+$, and the same situation is met with other ligands. Contrary to previous suggestions [31] $Cu(An)_2^+$ probably is truly two-coordinate because one would expect easy replacement of any additional coordinated H_2O by either CH_3CN or imidazole in view of the very low affinity of Cu(I) for oxygen donors [49].

With CN^- [52], pyridine [53,54], Cl^-, CH_3CN [29,50,55], and other ligands, 3:1 complexes are easily formed. Using the above arguments, we conclude that three-coordination, very rare in transition metal chemistry, is quite common for Cu(I) complexes in solution. Tetracoordinated Cu(I) complexes, which are most numerous in the solid state, are preferentially formed with electron-accepting ligands like 2,2'-bipyridyl and 1,10-phenanthroline [53, 56].

4. AUTOXIDATION OF Cu(I) COMPLEXES

4.1. Theoretical Aspects

It will be assumed in the following that the autoxidation of Cu(I) complexes proceeds via direct interaction of dioxygen with the central atom. According to the accepted views of O_2 binding to metal ions [12-15], the extremely soft Cu^+ [49] should have reasonable affinity toward the "noninnocent" [57] π acceptor O_2. Further, H transfer, as opposed to electron transfer, has been disproved for the autoxidation of some other transition metal ions [58,59] and could be excluded for several Cu(I) complexes that do not contain suitably acid protons (see below).

Equations (15) and (16) describe two possibilities for the first step of dioxygen reduction by Cu(I) via one-electron transfer.

$$Cu(I) + O_2 \rightarrow Cu \ldots O_2 \rightarrow Cu(II) + {}^{\bullet}O_2^- \qquad (15)$$

$$Cu(I) + O_2 \rightleftarrows \begin{matrix} Cu(I)\text{-}O_2 \\ \updownarrow \\ Cu(II)\text{-}O_2^- \end{matrix} \rightleftarrows Cu(II) + {}^{\bullet}O_2^- \qquad (16)$$

Reaction (15) is bimolecular with one transition state and no intermediates. In reaction (16) a dioxygen adduct, corresponding to a minimum of the free energy of the system along the reaction axis, is formed. Distinction between reactions (15) and (16) is often difficult and becomes futile when the free energy of the intermediate is not much lower than that of the transition states [60]. Below we shall only speak of dioxygen adducts on the basis of direct chemical or kinetic evidence, although such species seem likely to be formed quite generally in the autoxidation of transition metal complexes.

In view of the scarcity of experimental data, it is not surprising that in the past different authors came to rather diverse conclusions concerning the susceptibility of Cu(I) complexes to autoxidation. For example, on the basis of semiquantitative experiments [61] Fallab [13,14] postulated that these reactions were relatively rapid with bimolecular rate constants of about 100 M^{-1} sec^{-1} and that the reactivity was little influenced by the nature of the ligands. The spheric electron distribution in the central atom was used as a theoretical explanation.

According to Hemmerich, π acceptors make Cu(I) inert toward dioxygen, whereas rapid autoxidation is expected with imidazole and sulfur-containing ligands [49], while according to Bayer and Schretzmann [15] the σ-donating power of the ligands is responsible for dioxygen binding, but the autoxidation proper is largely influenced by steric factors. Lastly, it was suggested by Orgel [62] that the π-bonding ability of the ligands was decisive for the reactivity of Cu(I) complexes.

Cu(I) and Cu(II) both have very high water exchange rates near 10^9 sec^{-1} [63]. Also, many Cu(I) complexes with their low coordination numbers should rather easily form unstable addition compounds with extra ligands (cf. Section 3). Binding of dioxygen and dissociation of the adduct into Cu(II) and $^{\bullet}O_2^-$ thus both could be rapid processes. On the other hand, as the coordination numbers and geometries of Cu(I) and Cu(II) complexes with the same ligand usually are completely different from each other, a relatively high free energy of reorganization might be needed to reach a transition state that is sufficiently "cupric-like" to allow easy oxidation of the central atom. From this point of view, the autoxidation of Cu(I) might be slower than that of other transition metal complexes, where both, the oxidized and reduced forms, often have the same (octahedral) geometry and a suitable transition state might be reached by simple molecular vibrations. It seems difficult to judge theoretically the relative importance of these and other conflicting factors in connection with the rate of Cu(I) autoxidation.

4.2. Hydrated Cuprous Ion

There exist several earlier reports on the autoxidation of CuCl in aqueous media containing excess chloride [64-68]. All of them were at least partially unsuccessful, either because of too high reaction rates or because of incomplete data. In no case was the nature of the reactive species established, but in the most detailed of those studies, Nord [68] was the first to demonstrate kinetically the formation of a dioxygen adduct with a low molecular cuprous complex. A mechanism avoiding such an adduct in Nord's system and with HOOCl as the redoxactive intermediate [reactions (17) and (18)] was postulated by Abel [67].

$$H^+ + O_2 + Cl^- \rightleftarrows HOOCl \tag{17}$$

$$HOOCl + Cu(I) \rightarrow Cu(II) + HO_2^{\bullet} + Cl^- \tag{18}$$

It is refuted, however, by the observation that the autoxidation is inhibited, rather than accelerated, by excess chloride [68].

More recently it was concluded from studies of $Cu(An)_4ClO_4$ in aqueous acetonitrile [29,69,70] that of the species Cu^+_{aq}, $Cu(An)^+$, $Cu(An)_2^+$, and $Cu(An)_3^+$, which are present in such solutions, the hydrated cuprous ion alone is reactive toward O_2. The strict parallelity between the polarographic reduction potential of Cu(I) and the rate of autoxidation was demonstrated in [70]. In the same paper, the pH dependence between pH 0 and 5 is given. The rate increases linearly with $[H^+]$ in the pH range 1.2 to 2.5 and is independent of the acidity below pH 0.5 and above pH 4.0.

The complete rate law is given by equation (19) [70], and the

$$\frac{-d[O_2]}{dt} = k_{Cu_{aq}} [Cu^+_{aq}][O_2] \frac{1 + k_H[H^+]}{1 + k'_H[H^+]} \tag{19}$$

$(k_{Cu_{aq}} = 3.5 \times 10^4\ M^{-1}\ sec^{-1} \quad k_H = 410\ M^{-1} \quad k'_H = 15\ M^{-1})$

equations (20)-(22) are needed to obtain an analogous expression by steady-state elimination of the dioxygen adduct CuO_2^+. (P = products.)

$$Cu^+ + O_2 \underset{k_{-20}}{\overset{k_{+20}}{\rightleftharpoons}} CuO_2^+ \tag{20}$$

$$CuO_2^+ + H^+ \xrightarrow{k_{+21}} P \tag{21}$$

$$CuO_2^+ \xrightarrow{k_{+22}} P \tag{22}$$

Obviously the velocity of autoxidation would be independent of pH if the first step between Cu^+ and O_2 was irreversible ($k_{-20} = 0$), as neither Cu^+, CH_3CN, nor O_2 have any significant proton affinity in water. Using the theoretical rate law obtained from equations (20)-(22) [70] and the experimental values $k_{Cu_{aq}}$, k_H, and k'_H, the rate constant for the formation of the dioxygen adduct can be calculated: $k_{+20} = 9.5 \times 10^5\ M^{-1}\ sec^{-1}$. Further, $k_{-20}/k_{+22} = 26.5$ [70], which means that at high pH dissociation of CuO_2^+ into Cu^+_{aq} and O_2 is about 25 times more likely than electron transfer to form Cu^{2+} and $^{\bullet}O_2^-$. For details of the reaction mechanism see Section 6.

Reexamination of the system $CuCl/Cl^-$ revealed its complete analogy to $Cu(An)_4ClO_4$ in aqueous An [29]. The inhibition of the autoxidation with increasing $[Cl^-]$, casually mentioned by Nord [68], can be quantitatively explained with Cu^+_{aq} alone contributing significantly to O_2 consumption in solutions with $[Cl^-] < 0.2$. Whether CuCl and $CuCl_2^-$ are completely unreactive toward dioxygen would have to be studied at higher $[Cl^-]$. The relevant kinetic and thermodynamic data of the two systems are summarized in Table 1.

TABLE 1

Thermodynamic and Kinetic data for the Systems Cu^+_{aq}/Cl^- and Cu^+_{aq}/An

	L = An	L = Cl^-
K^{Cu}_{CuL}	$10^{3.28}$ [29]	$10^{3.63}$ [29]
$\beta^{Cu}_{CuL_2}$	$10^{4.35}$ [31] $10^{3.9}$ [50]	$10^{5.19}$ [29] $10^{5.5}$ [55][a]
$\beta^{Cu}_{CuL_3}$	$10^{4.1}$ [50] $10^{4.39}$ [70]	$10^{5.19}$ [29][a]

$k_{Cu_{aq}}$ (M^{-1} sec^{-1})	3.5×10^4 4.0×10^4 3.0×10^4	[70] [29] [54]
k_H (M^{-1} sec^{-1})	410	[70]
$k_{Cu_{aq}} k_H$ (M^{-2} sec^{-1})	3.5×10^7	[69][b]
k'_H (M^{-1})	15	[70]

[a]For additional values see Ref. [36].

[b]At 30°C. All other kinetic values in Table 1 were obtained at 20°C.

4.3. Other Complexes with Monodentate Ligands

The autoxidation of Cu(I) complexes with the following monodentate ligands has been studied: Cl^-, CH_3CN, NH_3, Im, N-methylimidazole (Meim), and pyridine (Py). All these ligands form 1:1 and 1:2 complexes, whereas 1:3 complexes are obtained only with Cl^-, CH_3CN, and Py. In addition, ternary complexes $Cu(An)L^+$ are observed when $Cu(An)_4ClO_4$ in acetonitrile is used as the source of Cu(I) [51]. As has been already discussed in Section 4.2, complexes with CH_3CN or Cl^- are not directly oxidized by O_2.

In Table 2, the kinetic data for the other monodentate ligands are summarized. In each case, the velocity of autoxidation was

TABLE 2

Rate Constants for the Autoxidation of Cu(I) Complexes with Monodentate Nitrogen Donors

k \ L	NH_3	Im	Meim	Py
k_{CuL} (M^{-1} sec^{-1})	--[a]	380 [54]	--[a]	250 [54]
k_{CuL_2} (M^{-1} sec^{-1})	18 [71]	13.8 [54] 14.0 [71]	105 [71]	39 [54]
k_{CuL_3} (M^{-1} sec^{-1})	--	--	--	32 [54]
k'_{CuL_3}[b] (M^{-2} sec^{-1})	16000 [72]	6700 [54] 5500 [72]	18000 [71]	--
k_{CuLAn} (M^{-1} sec^{-1})	<2[c]	<2[c]	<20[c]	--[a]

[a]Not studied.

[b]$-d[O_2]/dt = k'_{CuL_3}[CuL_2^+][O_2][L]$.

[c]Under the assumption that 10% contribution to the O_2 consumption would be detectable.

proportional to $[Cu(I)]_{tot}$ and to $[O_2]$, and thus the bimolecular rate constants k_{CuL_n} (M^{-1} sec^{-1}) are the natural quantities for comparing the reactivity of the different systems.

As is shown in Table 2, k_{CuL_2} is known for all four ligands (L = NH_3, Im, Meim, or Py). The rate constants vary relatively little, ranging from 14 M^{-1} sec^{-1} for $Cu(Im)_2^+$ to 105 M^{-1} sec^{-1} for $Cu(Meim)_2^+$, and are smaller than $k_{Cu_{aq}}$ by several orders of magnitude. No influence of the pH, other than shifting the equilibria between the different complexes, was observed for any of these ligands [54,71,72].

$Cu(Py)_3^+$ is oxidized with a velocity that is almost identical with that of $Cu(Py)_2^+$ [54]. The reactivity of the 1:1 complexes and the mixed species $Cu(L)An^+$ was obtained from plots of k_{obs}:

$$-d[O_2]/dt = k_{obs}[Cu(I)]_{tot}[O_2] \tag{23}$$

vs pH [54,71]. Whereas $CuIm^+$ and $CuPy^+$ autoxidize rapidly (k_{CuIm} = 380 M^{-1} sec^{-1}, k_{CuPy} = 250 M^{-1} sec^{-1}) the mixed compounds do not contribute significantly to the O_2 consumption under the conditions studied, and the corresponding rate constants must be below 2 M^{-1} sec^{-1}. Figure 1 demonstrates the pH dependence of k_{obs} for the pyridine complexes formed in absence of acetonitrile.*

With NH_3, Im, or Meim the rate of autoxidation increases with [L] under conditions where the formation of the respective 2:1 complexes is practically complete [71,72]; no such enhancement by excess ligand was observed for $Cu(Py)_3^+$ [54]. The rate law cannot be described simply by introduction of an extra term proportional to $[O_2]$, $[Cu(I)]$, and [L] in addition to the bimolecular rate constants discussed above. The system with N-methylimidazole yielded the most

---

*Excess elemental copper was added to $CuSO_4$ in buffered solutions of Im or Py which had been freed from O_2. The flask was closed with a serum cap and shaken for equilibration. Positive pressure of N_2 was maintained permanently through a second inlet when aliquots of Cu(I) solution were withdrawn by means of a steel syringe put through the cap [54].

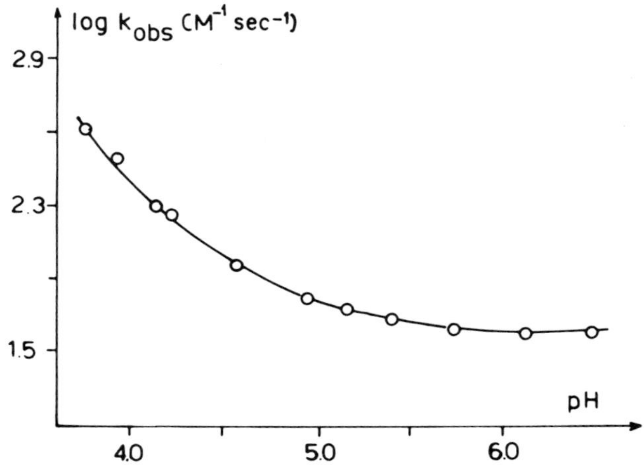

FIG. 1. Apparent bimolecular rate constant for the autoxidation of Cu(I) in aqueous pyridine as a function of pH. $[Py]_{tot} = 0.05$; acetate buffer, I = 0.2; T = 20°C. o Experimental points, (———) curve calculated with

$k_{Cu_{aq}} = 3 \cdot 10^4\ M^{-1}\ sec^{-1}$, $k_{CuPy} = 250\ M^{-1}\ sec^{-1}$,

$k_{CuPy_2} = 39\ M^{-1}\ sec^{-1}$, $k_{CuPy_3} = 32\ M^{-1}\ sec^{-1}$,

$K^{Cu}_{CuPy} = 10^{4.60}$, $\beta^{Cu}_{CuPy_2} = 10^{7.34}$, $\beta^{Cu}_{CuPy_3} = 10^{8.63}$.

Standard deviation of k_{obs}: 0.028 logarithmic units.

detailed insight into the autoxidation mechanism of Cu(I) complexes with monodentate nitrogen donors [71]. The experimental rate expression is given by equation (24), leading to linear dependence on [L] at high ligand concentration, and the individual steps given in equations (25)-(29) were shown to be necessary to explain it by steady-state approximation for $Cu(Meim)_2O_2^+$ and $Cu(Meim)_3^+$.

$$\frac{-d[O_2]}{dt} = [Cu(Meim)_2^+][O_2]\ \frac{A + B[Meim] + C[Meim]^2}{1 + D[Meim]} \tag{24}$$

($A = 105\ M^{-1}\ sec^{-1}$, $B = 1500\ M^{-2}\ sec^{-1}$, $C = 1.65 \times 10^5\ M^{-3}\ sec^{-1}$, $D = 9.3\ M^{-1}$ [71])

$$Cu(Meim)_2^+ + O_2 \underset{k_{-25}}{\overset{k_{+25}}{\rightleftharpoons}} Cu(Meim)_2O_2^+ \tag{25}$$

$$Cu(Meim)_2^+ + Meim \underset{k_{-26}}{\overset{k_{+26}}{\rightleftharpoons}} Cu(Meim)_3^+ \tag{26}$$

$$Cu(Meim)_2O_2^+ \xrightarrow{k_{27}} P \tag{27}$$

$$Cu(Meim)_2O_2^+ + Meim \xrightarrow{k_{28}} P \tag{28}$$

$$Cu(Meim)_3^+ + O_2 \xrightarrow{k_{29}} Cu(Meim)_2O_2^+ + Meim \tag{29}$$

It is suggested here that the same general mechanism which is shown in Figure 2 is valid for the autoxidation of the Cu(I) complexes with the four ligands NH_3, Im, Meim, and Py. The underlined species in Figure 2 all have been established either by equilibrium measurements or kinetically. It seems very likely that $Cu(L)O_2^+$ and $Cu(L)_3O_2^+$ are real intermediates, too, although the present results could also be explained under the assumption that they correspond to mere transition states in the course of the bimolecular reactions (28)-(30).

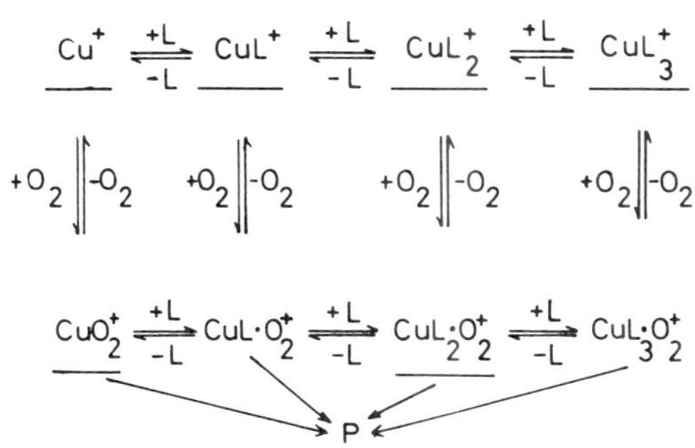

FIG. 2. General mechanism for the autoxidation of Cu(I) complexes with monodentate nitrogen donors.

$$CuL^+ + O_2 \xrightarrow{k_{30}} P \qquad (30)$$

The irreversible steps from the dioxygen adducts to the observable products Cu^{2+} and H_2O_2 or H_2O probably are rather complicated again and might contain additional rapidly playing equilibria. No further intermediates have been detected in any case, and no kinetic evidence for a two-electron transfer has been obtained, either.

4.4. Complexes with Chelating Ligands

As discussed in Section 3, Cu(I) normally does not form chelates unless the ligand has strong electron-accepting properties. Nevertheless, the influence even of chelators which lead to immediate disproportionation of Cu(I) in aqueous solution can be easily studied when the metal ion is sufficiently stabilized by acetonitrile. The kinetics of the autoxidation have been described for the complexes with the following ligands: 2,2'-bipyridyl (Bpy) [73], histamine (Ha), L-histidine (His), L-carnosine (Car), L-histidyl-L-histidine (Hishis) [74], oxalate (Ox), and malonate (Mal) [70]. As for Cu^+_{aq} and the complexes with monodentate ligands the velocity of autoxidation was always strictly proportional to the concentration of $[Cu(I)]_{tot}$ and $[O_2]$. With the single exception of oxalate, the influence of which was studied at low pH [70], no catalysis by H^+ or OH^- was observed. The kinetic data are summarized in Table 3.

The reactivity of the bis-chelated 2,2'-bipyridyl complex $Cu(Bpy)_2^+$ has been studied in detail by Pecht and Anbar [73]. O_2 and H_2O_2 both were used as oxidants. H_2O_2 was found to be almost 10 times less reactive than dioxygen ($k_{Cu(Bpy)_2} = 4.2 \times 10^2\ M^{-1}\ sec^{-1}$ and $3.2 \times 10^3\ M^{-1}\ sec^{-1}$,* respectively; cf. Table 3) and thus

---

*Pecht and Anbar [73] give the twice as high values 8.5×10^2 and $6.5 \times 10^3\ M^{-1}\ sec^{-1}$ for $d[Cu^{2+}]/dt = k[H_2O_2][Cu(Bpy)_2^+]$ and $k[O_2][Cu(Bpy)_2^+]$, respectively.

TABLE 3

Second-Order Rate Constants for the Autoxidation of Cu(I) Complexes with Chelating Ligands

Complex	k_{CuL_n} $(M^{-1}\ sec^{-1})$	Ref.
$Cu(An)Ox^-$, pH = 1.5	$>5 \times 10^5$ [a]	77
$Cu(Ha)Tris^+$	$>8 \times 10^4$	77
$Cu(An)Ox^-$, pH 4	$>5 \times 10^4$ [a]	77
CuHis	4.5×10^4	74
$CuHa^+$	2.6×10^4	74
$Cu(Bpy)_2^+$	6.5×10^3	73
	8.5×10^2 [b]	73
$Cu(An)Mal^-$	$>3 \times 10^3$ [a]	77
$Cu(HHa)_2^{3+}$	1.2×10^3	74
$Cu(HHis)_2^+$	1.0×10^3	74
Cu^+/Car, pH 6.7-8.7	130 - 600	74
Cu^+/Hishis, pH 6.3-8.5	100 - 310	74

[a] Using $K_{CuAn}^{Cu} = 10^{3.28}$ [29], k_{31}, and assuming that formation of 10% of the compound would have been detected by equilibrium measurements.

[b] With H_2O_2 as oxidant.

accumulates during the autoxidation of $Cu(Bpy)_2^+$ in presence of excess O_2. Although the rate was strictly proportional to $[O_2]$ and $[Cu(Bpy)_2^+]$, the authors suggested that the actual redox process took place within a dimeric species $[Cu(Bpy)_2]_2O_2^{2+}$ and that the formation of a monomeric dioxygen adduct $Cu(Bpy)_2O_2^+$ was rate-determining. A path involving one-electron steps was considered unlikely on the basis of the redox potentials for the couples $O_2/HO_2^{\bullet}$ and $Cu(Bpy)_2^{2+}/Cu(Bpy)_2^{+}$, for which they used the values -0.32 V [17] and +0.12 V [53], respectively.

Histamine and histidine both form three types of compounds III-V in aqueous CH_3CN [74]. Chelation as depicted in V cannot be observed directly, of course, but it is made very likely by the observation that the molecule of An, which is coordinated to the

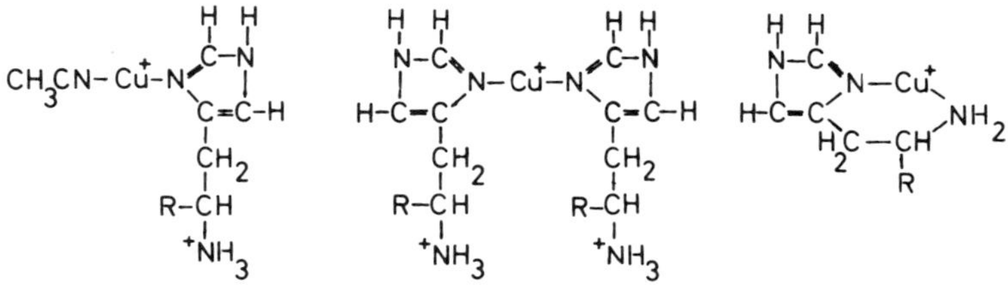

III IV V

Ha: R=H; His: R=COO^-

complexes $Cu(HHa)An^{2+}$ and $Cu(HHis)An^{+}$ with the monoprotonated ligands HHa^{+} and HHis, dissociates upon deprotonation of the ammonium group.

As can be seen from Table 3, the results for Ha and His are almost identical. Thus, the carboxyl group in His is not involved in complex formation with Cu(I), nor does it participate significantly in the transition state of the autoxidation. $Cu(HHa)An^{2+}$ and $Cu(HHis)An^{+}$ do not exhibit measurable reactivity toward O_2, in line with the results for the identically coordinated $Cu(Im)An^{+}$. Studies in absence of An have not been done with His and Ha, and nothing is known about the reactivity of $Cu(HHa)^{2+}$ and $Cu(HHis)^{+}$. The bimolecular rate constants for $Cu(Ha)^{+}$, $Cu(HHa)_2^{3+}$, and the corresponding histidine complexes are considerably higher than those of $Cu(Im)^{+}$ and $Cu(Im)_2^{+}$. The reason for this is not obvious, but one can assume that tetracoordinate "cupric-like" transition states (VI, VII) are reached easily. Involvement of an extra ligand in the autoxidation of $Cu(Ha)^{+}$ has actually been observed when

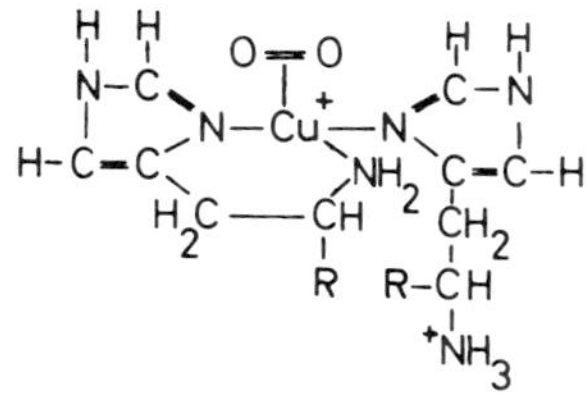

VI

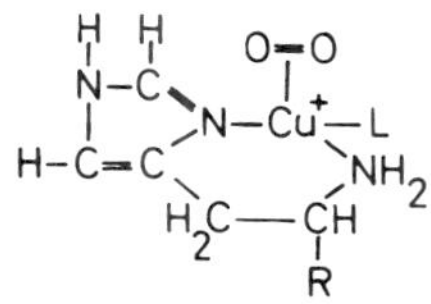

VII

Ha: R = H; His: R = COO$^-$

tris(hydroxymethyl)methylamine (Tris) was used as a buffer.

Carnosine and histidylhistidine form a variety of Cu(I) complexes in aqueous An [51]. No attempt has been made to assign individual rate constants to the different species, but it has been shown that none of them is extremely reactive toward O_2. Their autoxidation is considerably slower than that of the Ha or His complexes [74] under various conditions despite the fact that very stable Cu^{2+} chelates are formed with both ligands [75,76].

As Cu(I) has very little affinity toward oxygen donors [49], oxalate and malonate are not appreciably bound to Cu^+_{aq}. Thus, in solutions of Ox or Mal negative redox potentials are obtained for the Cu(II)/Cu(I) couple, and disproportionation of Cu(I) is practically complete. Upon addition of CH_3CN the cuprous state is stabilized, but measurable formation of complexes with Ox or Mal is still not observed in aqueous An. Nevertheless, the autoxidation rate of Cu(I) is greatly enhanced by these chelating dicarboxylic acids [70]. In both cases a reaction of formal third order was observed. By variation of [An] and pH the monoacetonitrile complex $Cu(An)^+$ and the dianions Ox and Mal were established as the reactive species, and in systems with oxalate at low pH a term proportional to $[H^+]$ was found as in the autoxidation of Cu^+_{aq}. The experimental rate law is given by equation (31). As Mal is a considerably stronger base than Ox, its influence could not be studied

$$-d[O_2]/dt = k_{31}[Cu(An)^+][O_2][L](1 + k'_{31}[H^+]) \qquad (31)$$

at sufficiently low pH in order to obtain the value for k'_{31} with Mal. In [70] $k_{31} = 3 \times 10^6\ M^{-2}\ sec^{-1}$ (L = Ox) and $1.7 \times 10^5\ M^{-2}\ sec^{-1}$ (L = Mal) were calculated from an estimated K^{Cu}_{CuAn}, but using the more recent experimental value of $10^{3.28}$ [29] these figures reduce to $5 \times 10^5\ M^{-2}\ sec^{-1}$ and $2.2 \times 10^4\ M^{-2}\ sec^{-1}$, respectively.

Three additional points of more general relevance might be mentioned here:

(i) Linear dependence on $[H^+]$, which has been observed for Cu^+_{aq} and $Cu(An)^+$ in the presence of oxalate, might be a general phenomenon of Cu(I) autoxidation, whenever it is possible to study a complex at sufficiently low pH. The corresponding values $k_H = 410\ M^{-1}$ [cf. (19)] and $k'_{31} = 320\ M^{-1}$ [70] of the two systems are surprisingly similar. On the other hand, catalysis by OH^-, a feature which is common to the autoxidation of most transition metal complexes [cf. (14)], has not been observed for any Cu(I) compound.

(ii) It is $Cu(An)^+$, and not Cu^+_{aq} which is redoxactive in the presence of Ox or Mal. This seems rather surprising considering the electron-accepting properties of CH_3CN and the fact that none of the other monoacetonitrile complexes $Cu(An)^+$, $Cu(An)Im^+$, $Cu(An)NH_3^+$, and so forth have shown any measurable reactivity toward O_2 at all. The transition state can be tentatively described by VIII involving tetracoordination of the central atom.

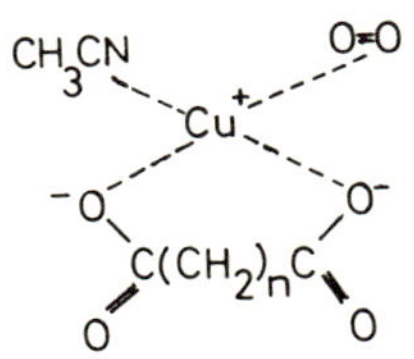

VIII n = 0 (Ox) or 1 (Mal)

Chelation seems to be indispensable, for otherwise it could not be explained why monocarboxylic or long chain dicarboxylic acids are without effect [70].

Of course the results do not exclude the possibility that $Cu(Ox)^-$ and $Cu(Mal)^-$ themselves in absence of An also autoxidize rather rapidly, as the concentration of Cu^+_{aq} was always small compared to that of $Cu(An)^+$. Anyway, inhibition by coordination of An is not especially effective in systems with Ox or Mal, and the bimolecular rate constants for the hypothetical ternary complexes $Cu(An)Ox^-$ and $Cu(An)Mal^-$ must be larger than those of the other cuprous CH_3CN complexes by many orders of magnitude.

(iii) Depending on the system and the initial ratio of [Cu(I)]: $[O_2]$ various amounts of H_2O_2 are produced. In the presence of Ox or Mal reduction of O_2 to H_2O_2 is exactly stoichiometric as long as an excess of O_2 over Cu(I) is present [70,77]. As a matter of fact, H_2O_2 was the major product in all rapid reactions (bimolecular rate constants above 10^2 M^{-1} sec^{-1}), and it has been suggested that the rates of oxidation by H_2O_2 depend much less on the nature of the Cu(I) complexes than those of autoxidation [77]. Furthermore, evidence has been obtained that at least part of the H_2O_2 consumption is due to the catalatic activity of copper complexes rather than to its reduction by Cu(I) [72]. On the other hand, $HO_2^{\bullet}$ or $^{\bullet}O_2^-$ which have been postulated as one-electron reduction products of O_2 [68-72], have not been identified by chemical or physical methods during autoxidation of Cu(I) complexes so far.

4.5. Studies in Nonaqueous Solvents

Little is known today about the autoxidation of Cu(I) complexes in nonaqueous solvents. In CH_3CN simple cuprous salts are stable indefinitely, but this is completely explained by the Cu(II)/Cu(I) potential of 0.8 V against Ag/0.01 M $AgNO_3$ [78], which makes their autoxidation thermodynamically impossible and even allows the preparation of $Cu(CH_3CN)_4ClO_4$ by simple heating of $Cu(ClO_4)_2$ in CH_3CN [79].

The autoxidation of a slurry of CuCl in glacial acetic acid containing various amounts of chloride has been studied by Henry [80]. The rate was proportional to $[O_2]$ and to $[Cu(I)]_{tot}^2$. The reactive species has not been directly identified but probably is the complex anion $CuCl_2^-$ and definitely not the solvated Cu^+, the concentration of which decreases rather than increases with increasing $[Cl^-]$. The rate law is given by equation (32) with $k_{32} = 0.11\ M^{-2}\ sec^{-1}$ when a value of 0.054 M/atm [81] is taken for the solubility of O_2 in acetic acid.

$$-d[O_2]/dt = k_{32}[Cu(I)]^2[O_2] \tag{32}$$

The same general behavior is observed for $Cu(CH_3CN)_4ClO_4$ in dimethylsulfoxide (Dmso) [79]. In Dmso acetonitrile does not appreciably bind to Cu^+ when $[An] \leqslant 0.1$, and the solvated ion is relatively stable against disproportionation ($[Cu^+]^2/[Cu^{2+}] = 0.5$ [82]). Proportionality to $[O_2]$ and $[Cu(Dmso)_n^+]^2$ (n is not known) is observed, and the termolecular rate constant was determined to be $7.1 \times 10^3\ M^{-2}\ sec^{-1}$ at 25°C in 0.1 M $LiClO_4$ [79]. Exactly 2 moles of Cu(I) are oxidized per mole of O_2, and thus dioxygen is reduced to the state of H_2O_2 under these conditions. The simple production of H_2O_2 and solvated Cu^{2+} is not likely, of course, in dry Dmso, and as a matter of fact strongly colored products of unknown nature are obtained.

The results from acetic acid and Dmso suggest the existence of a binuclear transition state $Cu(I)_2O_2$ for the redox step, contrary to the situation in water, where such a species has been postulated [73], but where no kinetic evidence is available for any system.

5. DIOXYGEN ADDUCTS OF LOW MOLECULAR Cu(I) COMPLEXES (?)

In the preceding section kinetic evidence has been presented for the formation of dioxygen adducts during the autoxidation of Cu^+_{aq}, $Cu(An)Ox^-$, $Cu(Meim)_2^+$, $CuCl_2^-$ in CH_3COOH, and Cu^+ in Dmso. These intermediates must be very unstable, however, and in no case could they be further characterized by their chemical or spectroscopic properties. The existence of stable or metastable low molecular dioxygen adducts has been postulated on the basis of spectroscopic evidence for the copper complexes of cysteine [61], cystamine [83], and oxalyldihydrazide in the presence of acetaldehyde [84]. None of these reports have been substantiated by the demonstration of actual binding or release of dioxygen, and for the last case, the absence of dioxygen has been demonstrated recently by x-ray analysis [85]. Likely the observed spectral changes are due to irreversible reactions involving the redoxactive ligands, namely, sulfhydryl groups or acetaldehyde.

To our knowledge not a single low molecular dioxygen adduct of copper in any valence state has been isolated or identified by its physical properties, and the hemocyanins and probably tyrosinase [86] remain the only such cases to the present day. In view of this lack of experimental data it might be appropriate to speculate a bit about the reasons why O_2 adducts of Cu(I) are so difficult to obtain, and what conditions, if any, would give the best chances of isolating such a complex.

Reversible binding of O_2 to transition metal complexes has been reviewed by several authors (cf. [12-16]). It may suffice here to mention the principal properties of a transition metal complex which have been stated to be essential for dioxygen carrying activity [15]:

> The complex should have filled π-bonding orbitals of intermediate energy to allow partial but not complete electron transfer from the metal atom to O_2.
>
> The coordination number and/or stereochemistry of the dioxygen adduct should be typical for the reduced state of the central atom but atypical or impossible for the oxidized state.
>
> The reduced complex must be able to increase its coordination number or have at least one very weakly bound ligand.
>
> The formation of $HO_2^{\bullet}$ or H_2O_2 should be suppressed.

Fulfillment of the last point does not seem feasible in aqueous solution, but using the other criteria one would presume that Cu(I) should be one of the fittest transition metal ions to form reversible dioxygen adducts. It has a whole set of low-lying filled π-bonding orbitals and correspondingly binds CO just as the biological dioxygen carriers, forms linear complexes that can be relatively easily converted into trigonal or tetrahedral compounds, and in practically no case Cu(I) complexes with a given set of ligands have any close resemblance to the corresponding cupric species. Nevertheless, dioxygen adducts of Cu(I), if formed at all, seem to be of very fleeting existence, indeed.

The situation becomes even more bewildering after confrontation with the fact that the overwhelming majority of the hitherto known dioxygen adducts have Co^{2+} as the central atom [87], an ion which does not fulfill the above criteria too well. Co^{2+} also has filled π-bonding orbitals. It is, however, not especially soft and does not bind CO. Most of its complexes are octahedral (often tetragonally distorted) in the 2+ as well as the 3+ state. Thus CO_2 complexes have no tendency to increase their coordination number and the coordination number of the coordination number of the central atom in the adduct is indeed possible, and even typical, of Co^{3+}.

Apparently, additional aspects have to be considered in order to explain this situation. It is suggested here that the kinetic lability or inertness of the metal ions in their oxidized forms is an important factor. This is sketched in Figure 3, assuming a

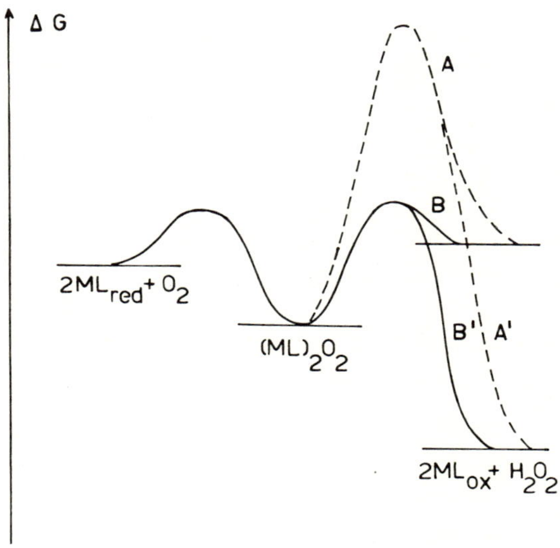

FIG. 3. Formation of stable (A, B), metastable (A'), and unstable (B') binuclear dioxygen adducts.

binuclear dioxygen adduct as it is found in oxyhemocyanin [88] and with the majority of Co(II) complexes. Reaction curves with low (A, A') and high (B, B'') ligand exchange rates of the oxidized central atoms are shown. For the sake of simplicity H_2O_2 is taken as the endproduct, and its thermodynamically favorable dismutation to H_2O and O_2 is neglected. Whether the intermediate should be called a dioxygen adduct or a peroxo complex is of no consequence in this context, and if both species were formed consecutively [14], the general situation still would be the same.

Figure 3 is practically self-explaining. In A and B truly reversible (with respect to O_2 and H_2O_2) dioxygen adducts are formed. For B, starting from ML_{red} or ML_{ox} should be equally feasible, whereas for A the latter path would have a prohibitively high free energy of activation. In A' and B' $(ML)_2O_2$ is unstable with respect to ML_{ox} and H_2O_2. The adduct can only be formed from ML_{red} and O_2. After B' $(ML)_2O_2$ will be a labile intermediate, which is perhaps not directly observable. On the other hand, the species related to A' will be metastable and pseudoreversible with all the phenomenological properties of a dioxygen adduct. Many cycles of oxygenation and deoxygenation might be possible before irreversible autoxidation takes place.

Substitution of A (A') and B (B') by Co(II, III) and Cu(I, II), respectively, suggests itself, implying that kinetic rather than thermodynamic reasons might be responsible for the easy availability of O_2 adducts with Co^{2+}. In view of the extreme lability of most low molecular Cu(II) complexes we might state that it will be extremely difficult, if not impossible, to find a nonbiological dioxygen adduct with Cu(I), <u>unless it is thermodynamically stable with respect to Cu(II) and H_2O_2</u>. However, as should be kept in mind, the catalatic activity of many copper complexes might make the observation of dioxygen adducts illusory even when this condition is fulfilled.

What should be done, then, in order to have the best chances of finding a Cu(I) dioxygen adduct?

(i) Rapid dissociation and rearrangement of the complex as well as addition of extra ligands to the dioxygen adduct producing a "cupric-like" stereochemistry must be excluded. This goal would probably have to be achieved with a rigid bi- or tridentate cyclic ligand shielding the back of the complex against attack by electron donors as shown in Figure 4.

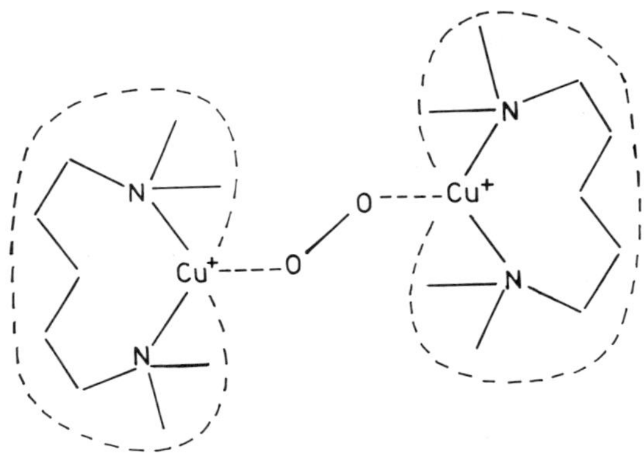

FIG. 4. Suggested coordination sphere of a hypothetical (meta) stable dioxygen adduct of Cu(I). Access of a fourth ligand is sterically suppressed. The two hydrophobic parts might have coalesced in a protein dioxygen carrier.

(ii) The complex should have a redoxpotential which corresponds closely to that of the couple O_2/H_2O_2 under the experimental conditions.

(iii) The system should have low catalatic activity, and the Cu(I) complex should not, or only slowly, be oxidized by H_2O_2.

(iv) The ligand itself must not be redoxactive.

(v) Suppression of H_2O_2 formation by aprotic conditions would perhaps allow the use of sufficiently electron donating ligands for stabilization of the adduct.

Point (i) seems to be rather difficult to fulfill and would probably involve some lengthy organic synthesis. Whether reversible binding of O_2 to a low molecular Cu(I) complex can be obtained without considering (i), remains to be seen.

6. DISCUSSION

6.1. Mechanism of Cu(I) Autoxidation

The identification of H_2O_2 as the product of O_2 reduction (cf. Section 4) excludes any step involving transfer of more than two electrons in Cu(I) autoxidation, and O_2 must behave as an oxidant of only moderate strength [see eq. (9)] when H_2O_2 is accumulating. Synchronous transfer of two electrons in a binuclear dioxygen adduct $[Cu(I)L]_2O_2$, leading directly to two cupric ions and peroxide, has been made likely by two studies in nonaqueous solvents [79,80]. Pecht and Anbar [73] assume the same general mechanism [reactions (33) and (34)] to be operative for the autoxidation

$$Cu(I)L + O_2 \underset{k_{-33}}{\overset{k_{+33}}{\rightleftharpoons}} [Cu(I)L]O_2 \tag{33}$$

$$[Cu(I)L]O_2 + Cu(I)L \underset{k_{-34}}{\overset{k_{+34}}{\rightleftharpoons}} [Cu(I)L]_2O_2 \xrightarrow{2H^+} 2Cu(II)L + H_2O_2 \tag{34}$$

of $Cu(Bpy)_2^+$ in H_2O. This mechanism is identical with that found for the autoxidation of most Co(II) complexes [13-16,87]. Without going into further details, the authors disqualify one-electron steps involving the formation of $^{\bullet}O_2^-$ or $HO_2^{\bullet}$ on the basis of the high free energy of those unstable intermediates. In the other reports on the autoxidation of Cu(I) complexes in aqueous solution the contrary view postulating consecutive one-electron steps has been taken [29,54,68-72,74].

Neither the formation of $HO_2^{\bullet}$ or $^{\bullet}O_2$ radicals nor that of a dimeric dioxygen adduct has been observed during the autoxidation of any Cu(I) complex in aqueous solution. The strict first-order dependence of the reaction rate on [Cu(I)] in all systems, in connection with kinetic evidence for unstable mononuclear dioxygen adducts [68-71], speaks for a stepwise formation of H_2O_2. In all cases, where binding of O_2 is not rate-limiting and the back-reaction to Cu(I)L (33) is significant (cf. [70]), the simple and suggestive mechanism given by Pecht and Anbar [73] cannot be true, as it would necessarily lead to a second-order dependence on $[Cu(I)]_{tot}$. Rather complicated reaction sequences, including at least two rate-determining rearrangements (k_{35}, k'_{35}) within the primary dioxygen adduct, would have to be invoked [reaction (35)] to reconcile two-electron steps with the rate law [eq. (19)] obtained for the autoxidation of Cu^+_{aq}, for instance.

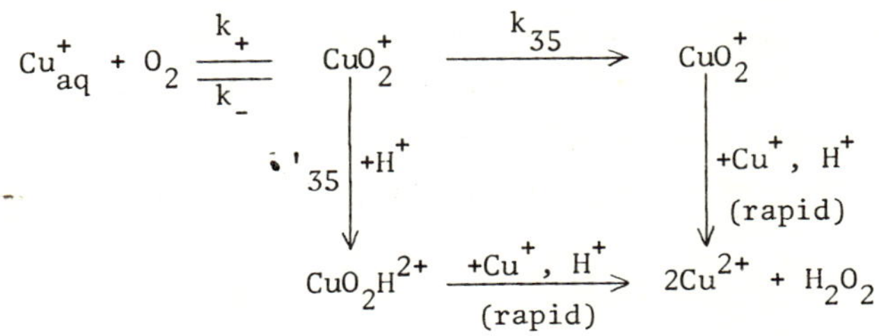

(35)

(CuO_2^{+*}: activated dioxygen adduct)

It cannot be denied, however, that one-electron steps involving $^{\bullet}O_2^-$ seem to be equally problematic in view of the relatively rapid autoxidation of many Cu(I) complexes. Using the standard potentials of -0.45 V and +0.159 V for the redox couples $O_2/^{\bullet}O_2^-$ [17] and Cu^{2+}_{aq}/Cu^+_{aq} [89] the equilibrium (36] can be calculated to lie

$$Cu^+_{aq} + O_2 \underset{k_{-36}}{\overset{k_{+36}}{\rightleftharpoons}} Cu^{2+} + {}^{\bullet}O_2^- \qquad K_{36} = \frac{[Cu^{2+}_{aq}][{}^{\bullet}O_2^-]}{[Cu^+_{aq}][O_2]} \qquad (36)$$

strongly on the side of Cu^+_{aq} and O_2: $K_{36} = k_{+36}/k_{-36} = 10^{7.4}$. A value for k_{-36} has been determined by Berdnikov et al. [90]. With $k_{-36} = 2.5 \times 10^8\ M^{-1}\ sec^{-1}$, k_{+36}, the rate constant for the autoxidation of Cu^+_{aq} at high pH, can be calculated: $k_{+36} = K_{36}k_{-36} = 11$ $M^{-1}\ sec^{-1}$. This is more than 3 orders of magnitude less than the experimental value of $3.5 \times 10^4\ M^{-1}\ sec^{-1}$ [70]. Even if the results of Berdnikov et al. [90] were disregarded and a diffusion-controlled rate were assumed for the back-reaction of (36), the observed value for k_{+36} remains too high. In several other systems the situation is similar.

By the above considerations, one-electron transfer from Cu^+ to O_2 and formation of free $^{\bullet}O_2^-$ is seemingly excluded. However, the generally used standard potential of -0.57 (-0.45) to -0.4 V [12,17] for the couple $O_2/{}^{\bullet}O_2^-$ can only be calculated indirectly from estimated thermodynamic data, and actually a much higher value of -0.2 ± 0.2 V has been suggested recently [91] for that redox system. This latter figure compares quite well with the value of -0.25 V, which can be calculated from the experimental rate constants k_{+36} and k_{-36}. All in all, the available evidence for or against stepwise reduction of O_2 to H_2O_2 should be called inconclusive.

A characteristic pecularity of Cu(I) autoxidation is its catalysis by H^+, which was found whenever a system could be studied below pH 3 [29,68-70,77]. In contrast, the autoxidation rate of other transition metal ions normally increases with $[OH^-]$, and pH-independent terms are found in strong acid [14,58,92,93]. The two different situations can be described schematically by equations (37) and (38). An explanation for the catalysis by either H^+ or

$$-d[O_2]/dt = k_{37}[M^{n+}][O_2](1 + k'_{37}[OH^-]) \tag{37}$$

$$-[O_2]/dt = k_{38}[M^{n+}][O_2](1 + k'_{38}[H^+]) \tag{38}$$

OH^- has been given in [70]: In the transition state IX the complete separation of the charged particles $M^{(n+1)+}$ and $^{\bullet}O_2^-$ needs a considerable amount of coulombic energy, which can be counterbalanced either by decreasing the positive charge on the central atom (A) or by neutralizing the negative charge on the dioxygen moiety (B). The low charge (+1) and high electron density, leading to a large value of δ, obviously predestinate Cu(I) to react according to B. No case is known where A and B both were operative for the same complex in solutions of different acidity.

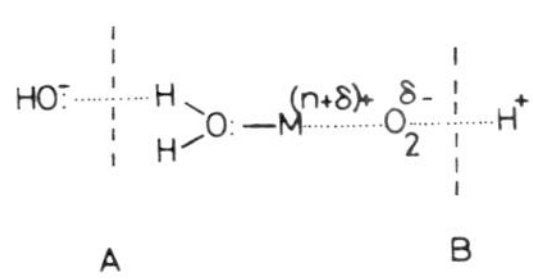

IX

6.2. Autoxidation Rate and Ligand Properties

Scrutinizing the values of the bimolecular rate constants in Tables 1-3 we find a rough correlation between autoxidation rate and thermodynamic equilibrium. The highest reactivity is observed with O ligands, followed by N chelators, monodentate N ligands, and π acceptors, which is also the general order of increasing standard potential for the redox couples Cu(II)L/Cu(I)L [48,49,53,56].

The parallelity of autoxidation rate and standard potential is by no means strict, however. For instance, $Cu(Bpy)_2^+$ autoxidizes by about 2 orders of magnitude more rapidly than the complexes with Car or Hishis, although $Cu(Bpy)_2^{2+}/Cu(Bpy)_2^+$ has a positive standard potential of 0.12 V [53], whereas the corresponding redox couples with Car and Hishis must be considerably more negative because these ligands form very stable chelate complexes with Cu^{2+} [75,76]. Also, the 2:1 complex with histidine autoxidizes less rapidly ($k_{Cu(HHis)_2} = 10^3\ M^{-1}\ sec^{-1}$ [74]) than Cu_{aq}^+ ($k_{Cu_{aq}} = (3.5\text{-}4.1)10^4\ M^{-1}\ sec^{-1}$ [29,70]), but the standard potential of the former (-0.17 V [53]) is about 0.3 V more negative than that of the aquoions. The situation with histamine is practically identical.

Finally, with several monodentate nitrogen donors (L = NH_3, Im, Meim) the potential of the half-reaction (39) falls off with $[L]^2$, but the rate of autoxidation depends linearly on [L]. Although the redox potentials of the different cupric/cuprous couples

$$CuL_2^+ + 2L \rightleftharpoons CuL_4^{2+} + e \tag{39}$$

seem to play an important role, they obviously are not the only decisive factors in Cu(I) autoxidation.

6.3. Autoxidation Rate and Steric Factors

For reasons already discussed (cf. Sections 3 and 4.1) the rate of ligand exchange reactions is not supposed to be of great significance in Cu(I) autoxidation. As can be easily concluded from the experimental data, steric factors do play an important role, however. In brief, it appears essential for a rapid overall reaction that the dioxygen adduct, which is formed prior to the actual electron transfer step, has or easily acquires a coordination sphere that is equally typical for Cu(II)-$^{\bullet}O_2^-$ and for Cu(I)-O_2. This is very frequently achieved via an increase in coordination number by one or even two entities as compared to the starting cuprous complex. It thus appears that a suitable geometry for electron transfer is more easily accomplished by molecular vibrations of a (unstable) tetra- or pentacoordinated complex than the oxidation of the central atom in a trigonal or digonal environment, which is typical for Cu(I).

At first glance, the rapid autoxidation of $Cu(Bpy)_2^+$ is surprising. Pentacoordination, which is most untypical for Cu(I) complexes, is likely involved in $Cu(Bpy)_2O_2^+$ as O_2 will not be able to displace one of the nitrogens of the very strong chelator 2,2'-bipyridyl. Many tetracoordinate Cu(I) complexes with chelating acceptor ligands are in fact rather inert toward O_2. On the other hand, $Cu(Bpy)_2^{2+}$ easily forms adducts with "soft" monodentate ligands [48,94]. Thus, the transition of $Cu(Bpy)_2O_2^+$ into a cupric complex of closely similar geometry might be an easy process. Also, we are left with a "complete" cupric compound even after splitting of the O_2-moiety, and for these two reasons the redox step might be very rapid once the adduct is formed, despite the strong electron-accepting properties of 2,2'-bipyridyl, which are directly demonstrated by a metal to ligand charge transfer band at 435 nm in $Cu(Bpy)_2^+$ [73].

In other systems the coordination number even is increased by at least two in the activated complex. NH_3, Im, and Meim do not form higher than 2:1 complexes CuL_2^+, but electron transfer preferentially takes place within a species $CuL_3O_2^+$ [71,72], which can be looked at as a typical cupric complex with $^{\bullet}O_2^-$ as the fourth ligand. In the absence of extra free L the autoxidation is strongly but not completely inhibited. Rather than assuming $CuL_2O_2^+$ to be the redoxactive species in that case, one may suggest that H_2O participates in the reaction. It is true that Cu(I) has little affinity toward oxygen donor ligands, but in the dioxygen adduct the electronic structure of the central atom will be significantly altered, so that the subsequent formation of $CuL_2(H_2O)O_2^+$ might easily be achieved.

The most typical example for the need of a "cupriclike" dioxygen adduct is given by the system Cu(I)/An in the presence of oxalate or malonate. Despite the reluctance of Cu(I) to form chelate complexes, the properties of the central atom apparently are sufficiently altered after binding of dioxygen to allow chelation (cf. Section 4.4). More surprising, however, is the observation that not $CuOx^-$ and $CuMal^-$, but $Cu(Ox)An^-$ and $Cu(Mal)An^-$, are the reactive species, although, thermodynamically, CH_3CN strongly stabilizes Cu(I) against autoxidation. Apparently, the need for a tetracoordinated complex VIII is stronger in this case than the inhibitory effect which is normally displayed by An.

Finally, the situation with carnosine may be commented upon From the fact that the redox potential must be rather low (cf. Section 6.2) we would expect a higher rate than the actually observed $k_{CuCar} = 130\text{-}600\ M^{-1}\ sec^{-1}$ at pH 6.7 to 8.7. The reason for the relatively slow reaction must lie in the very different structures of the cuprous (X) and the cupric (XI) complexes which were dominant under the conditions studied. It is not to be

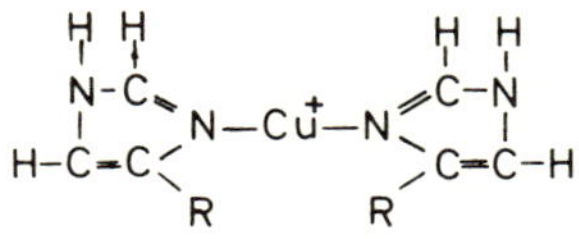

X

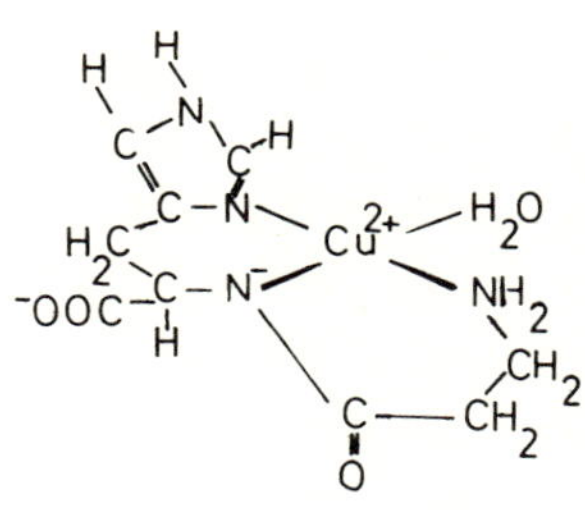

XI

$-R = -CH_2-CH(COO^-)NHCOCH_2CH_2\overset{+}{N}H_3$

expected that the complete rearrangement (X into XI) of the ligands will take place prior to the redox step, and as a matter of fact the consumption of O_2 is followed by a slower reaction that can only be observed spectrophotometrically and must correspond to the transition of a primary nonequilibrium cupric complex into XI [77].

For Hishis the same kind of arguments would apply, and this system thus is not discussed any further.

7. CONCLUSIONS

As we have seen, simple Cu(I) complexes autoxidize at very different rates, ranging from practically (or truly) zero for $Cu(CH_3CN)_3^+$ to $10^6\ M^{-1}\ sec^{-1}$ for Cu_{aq}^+ at low pH and for $Cu(An)Ox^-$. Stable dioxygen adducts have not been observed so far for any low molecular cuprous species, but transient binding of O_2 has been demonstrated kinetically in several systems, and inner sphere electron transfer from Cu(I) to O_2 is believed to be a step common to the autoxidation of most if not all Cu(I) complexes. In the organic solvents CH_3COOH and Dmso binuclear adducts $(CuL)_2O_2$ are probably formed, and O_2 is directly reduced to the oxidation state of H_2O_2. In water, no such binuclear complexes have been observed, and many pieces of evidence speak for a one-electron reduction of O_2 in this solvent, although all attempts to demonstrate directly the formation of $^{\bullet}O_2^-$ or $HO_2^{\bullet}$ have failed. Two factors seem to be of about equal importance with respect to the rate of autoxidation: The redoxpotential of the cupric-cuprous couple and the ease of obtaining a "cupric-like" transition state for the dioxygen adduct.

What could be called the biological significance of these results? Obviously, studies of the autoxidation of cuprous complexes cannot shed much light on the so-called "enzymic activation of dioxygen." However, this term is often used somewhat superficially. To our knowledge, direct electronic activation of O_2 has not been demonstrated for any copper enzyme, and a priori evidence would be rather difficult to obtain in view of the fleeting existence of the intermediates involved.

In fact, activation of dioxygen may frequently not be very important for enzymic reactivity: The dioxygenase quercetinase is a cupric enzyme with one copper atom per flavonol binding site [95,96]. No interaction with O_2 is observed, but the organic substrate is bound and probably activated. In mixed function oxidases, the situation is more complex. Binding of O_2 to Cu^+ or to a cuprous pair Cu_2^{2+} might be significant in enzymes such as tyrosinase, but no direct evidence for activation of dioxygen is available, the oxidation state of the redoxactive O_2 moiety is not known, and perhaps entropic factors (bringing together of substrate and O_2) and substrate activation are more important than activation of O_2 in these cases, too.

Reversible dioxygen adducts with other metal ions, especially Co^{2+}, are not very helpful in this context, either. In most of these complexes O_2 is even less easily reduced than in its molecular form. Recently, Takahashi et al. [97] described an interesting autoxidation of o-phenylenediamine to cis-cis-muconitrile catalysed by CuCl in pyridine, however. While the detailed mechanism is not yet known, the reaction surprisingly is not observed with cupric copper as a catalyst and possibly proceeds via "activated dioxygen" of some kind.

Autoxidation of low molecular Cu(I) complexes might bear some relevance for electron transfer oxidases and dioxygen carriers, although most questions concerning enzymic activity finally will have to be directly solved for those systems themselves. First, the high autoxidation rates found for some complexes are clearly in line with the working of the so-called ping-pong mechanism that has been invoked for the copper catalysed autoxidation of organic substrates in both enzymic and nonenzymic systems (cf. [13,98,99]). As the highest bimolecular rate constants for the autoxidation of Cu(I) complexes ($k_{CuL} \sim 10^6\ M^{-1}\ sec^{-1}$) are comparable to those of electron transfer oxidases (ceruloplasmin: $k = 10^5\ M^{-1}\ sec^{-1}$ [98], laccase: $k = 2.8 \times 10^6\ M^{-1}\ sec^{-1}$ [98]) we cannot maintain the idea that structures containing copper dimers or even tetramers are a conditio sine qua non for rapid reoxidation of the cuprous state. Special steric strains (distortion), imposed by the protein part of the enzyme, do not seem to be necessary either. Polynuclear active sites seem to be essential for the formation of water, rather than H_2O_2 as the reduction product of O_2, however. Also, the simultaneous reduction of more than one Cu(II) by organic substrate would avoid the formation of high-energy radical intermediates of the latter.

Even if more than one copper ion were present at the same active site, they would not necessarily have to be very close to each other. Instead of being reduced in a truly synchronous reaction, substrates might well "jump" from one site to the other without being released into the bulk solution. In fact, some enzymes such as dopamine-β-hydroxylase that are believed to contain 2 Cu^{2+} per active site have completely normal ESR signals [100, 101], an observation that is much less disturbing if the paramagnetic ions were some 0.5 or more nm apart. However actual copper-copper pairs seem to exist in laccase for instance [102], and the possibility of truly synchronous two- or four-electron reductions of dioxygen is of course by no means denied by the above considerations.

The results obtained with low molecular compounds also indicate the possibility of combining relatively high oxidation potential with rapid autoxidation, or the opposite. The former situation will be especially welcome in electron-transfer oxidases, whereas the latter is a prerequisite for dioxygen-carrying activity. The factors leading to complexes of the first or the second kind have been discussed in previous sections.

We conclude with the speculation that the binding of copper in hemocyanin might involve rather orthodox chemistry with two cuprous ions in an environment resembling the one in Figure 4 and with the protein part shielding the dioxygen adduct from access of extra ligands, thus leaving the metal in a Cu(I)-specific site.

ABBREVIATIONS AND DEFINITIONS

AH_2: Reduced cosubstrate

An: Acetonitrile

$\beta^{Cu}_{CuL_n}$: Overall formation constant $\dfrac{[CuL_n]}{[Cu^+][L]^n}$

Bpy: 2,2'-Bipyridyl

Car: L-Carnosine

$[Cu(I)]_{tot}$: Total (analytical) concentration of Cu(I)

Dmso: Dimethylsulfoxide

ESR: Electron spin resonance

Ha: Histamine

His: Histidine

Hishis: L-Histidyl-L-histidine

Im: Imidazole

k, k_H, k'_H, k_{obs}: Rate constants, defined by individual kinetic expressions

k_{CuL_n} : Second-order rate constant ($M^{-1}\ sec^{-1}$): $-d[O_2]/dt = k_{CuL_n}[CuL_n][O_2]$

k_n: Rate constant defined by equation (n)

$K^{CuL_{(n-1)}}_{Cul_n}$: Individual formation constant $\dfrac{[CuL_n]}{[Cul_{(n-1)}][L]}$

K_n: Equilibrium constant defined by equation (n)

L: Generalized ligand

Mal: Malonate dianion

Meim: N-Methylimidazole

$(ML)_2O_2$: Binuclear dioxygen adduct or peroxo complex

NMR: Nuclear magnetic resonance

Ox: Oxalate dianion

P: General product(s) of Cu(I) autoxidation

Py: Pyridine

S_{CuL}: Solubility product $[Cu^+][L]$

SH_2: Reduced substrate

Tris: Tris(hydroxymethyl)methylamine

ADDENDUM

Since the completion of this manuscript a number of studies have been published which clarify points raised in this chapter.

As discussed by Wood [103], a reliable value of -0.33 ± 0.01 V for the redox potential $E_{O_2}/{}^{\bullet}O_2^-$ has been determined by several independent groups, making the production of superoxide more feasible than the older data (cf. p. 352).

The bent Pauling mode of dioxygen coordination to transition metal complexes (see p. 326) has been firmly established by x-ray analysis [104-106] and the general advances in the coordination chemistry of dioxygen have been summarized [107,108].

The kinetics of the autoxidation of solvated Cu^+ in DMSO [109] and of the bis(1,10-phenanthroline)copper(I) complex in CH_3NO_2 [110] have been published. Both are strongly indicating a two-electron reduction of O_2 within a dimeric transition state $(CuL_x)_2O_2^{2+}$, while the rate of the autoxidation of cuprous-2,2',2"-terpyridine in H_2O [111] again was only directly proportional to [Cu(I)] (cf. p. 351).

The reversible uptake of O_2 (see p. 346) and CO by copper coordinated to a highly polymeric oxime ligand has been reported [112].

Finally, significant progress has been made in the study of the dioxygenation of aromatic compounds catalyzed by CuCl in pyridine by the identification of the "activated dioxygen" as cuprous peroxide [113] (cf. p. 358).

REFERENCES

1. O. Hayaishi, "Oxygenases," Academic Press, New York, 1962, p. 1.
2. R. Lontie, in "Metal Ions in Biological Systems," Vol. 3 (H. Sigel, ed.), Marcel Dekker, New York, 1974.
3. W. G. Levine, in "The Biochemistry of Copper" (J. Peisach, P. Aisen, and W. E. Blumberg, eds.), Academic Press, New York, 1966, p. 371.
4. Z. G. Penton and C. R. Dawson, in "Oxidases and Related Redox Systems," Vol. 1 (T. E. King, H. S. Mason, and M. Morrison, eds.), Wiley, 1965, p. 222.

5. H. Dressler and C. R. Dawson, *Biochim. Biophys. Acta, 45,* 515 (1960).

6. E. Y. Levin, B. Levenberg, and S. Kaufman, *J. Biol. Chem., 235,* 2080 (1960); S. Kaufman, W. Bridgers, F. Eisenberg, and S. Friedman, *Biochem. Biophys. Res. Commun., 9,* 497 (1962).

7. H. G. Krishnamurty and F. J. Simpson, *J. Biol. Chem., 245,* 1467 (1970).

8. L. Broman, B. G. Malmström, R. Aasa, and T. Vänngard, *Biochim. Biophys. Acta, 75,* 365 (1963).

9. W. E. Blumberg, M. Goldstein, E. Lauber, and J. Peisach, *Biochim. Biophys. Acta, 99,* 187 (1965).

10. R. Malkin and B. G. Malmström, *Adv. Enzymol., 32,* 177 (1970).

11. D. Samuel, in "Biochemie des Sauerstoffs" (B. Hess and H. Staudinger, eds.), Springer Verlag, Berlin, 1968, p. 6.

12. H. Taube, *J. Gen. Physiol., 49,* 29 (1965).

13. S. Fallab, *Z. Nat. Med. Grundlagenforschung, 1,* 333 (1963); S. Fallab, *Z. Nat. Med. Grundlagenforschung, 2,* 220 (1965).

14. S. Fallab, *Angew. Chem. (Int. Ed.), 6,* 496 (1967).

15. E. Bayer and P. Schretzmann, in "Structure and Bonding," Vol. 2 (C. K. Jörgensen, J. B. Neilands, R. S. Nyholm, D. Reinen, and R. J. P. Williams, eds.), Springer Verlag, Berlin, 1967, p. 181.

16. V. Ullrich, *Angew. Chem., 84,* 689 (1972).

17. P. George, in "Oxidases and Related Redox Systems," Vol. 1 (T. E. King, H. S. Mason, and M. Morrison, eds.), Wiley, New York, 1965, p. 3.

18. N. Bartlett and D. H. Lohmann, *Proc. Chem. Soc.,* 115 (1962).

19. L. Pauling and J. Weiss, *Nature, 203,* 182 (1964).

20. J. Griffith, *Proc. Roy. Soc., A, 235,* 23 (1965).

21. N. Vannerberg, *Acta Cryst., 18,* 449 (1965).

22. A. Vlcek, *Trans. Faraday Soc., 56,* 1137 (1960).

23. J. A. Ibers and S. J. La Placa, *J. Amer. Chem. Soc., 87,* 2581 (1965).

24. M. Gouterman and M. Zerner, in "Hemes and Hemoproteins," (B. Chance, R. W. Estabrook, and T. Yonetani, eds.), New York, 1966, p. 589.

25. H. C. Watson and C. L. Nobbs, in "Biochemie des Sauerstoffs" (B. Hess and H. Staudinger, eds.), Springer Verlag, Berlin, 1968, p. 37.

26. W. M. Latimer, "Oxidation Potentials," Prentice-Hall, Englewood Cliffs, N.J., 1952.

27. F. A. Cotton and G. Wilkinson, "Advanced Inorganic Chemistry," Wiley, New York, 1966.

28. See Ref. [36], p. 285.

29. A. Günter and A. D. Zuberbühler, *Chimia, 24,* 340 (1970).

30. B. J. Hathaway, D. G. Holah, and A. E. Underhill, *J. Chem. Soc.,* 2444 (1962).

31. P. Hemmerich and C. Sigwart, *Experientia, 19,* 488 (1963).

32. See Ref. [36], p. 13.

33. J. A. Altermatt and S. E. Manahan, *Inorg. Nucl. Chem. Lett., 4,* 1 (1968).

34. K. Shaw and J. H. Espenson, *Inorg. Chem., 7,* 1619 (1968).

35. S. Ahrland and J. Rawsthorne, *Acta Chem. Scand., 24,* 157 (1970).

36. L. G. Sillén and A. E. Martell (eds.), "Stability Constants of Metal Ion Complexes," The Chemical Society, London, 1964.

37. J. Bjerrum, *Kgl. Danske Videnskab. Selskab. Mat.-Fys. Medd., 12,* no. 15 (1934).

38. W. E. Blumberg, in "The Biochemistry of Copper" (J. Peisach, P. Aisen, and W. E. Blumberg, eds.), Academic Press, New York, 1966, p. 49.

39. A. S. Brill, R. B. Martin, and R. J. P. Williams, in "Electronic Aspects of Biochemistry" (B. Pullman, ed.), Academic Press, New York, 1964, p. 519.

40. C. K. Jörgensen, "Absorption Spectra and Chemical Bonding in Complexes," Pergamon Press, Oxford, 1962.

41. T. A. Kaden and A. D. Zuberbühler, *Helv. Chim. Acta, 51,* 1797 (1968).

42. C. B. Acland and H. C. Freeman, *Chem. Commun.,* 1016 (1971).

43. A. F. Wells, "Structural Inorganic Chemistry," Clarendon Press, Oxford, 1945, p. 316.

44. P. G. Eller and P. W. R. Corfield, *Chem. Commun.,* 105 (1971).

45. A. H. Levin, R. J. Michl, P. Ganis, and U. Lepore, *Chem. Commun.,* 661 (1972).

46. W. Spofford, III and E. Amma, *Acta Cryst. B, 26,* 1474 (1970).

47. L. E. Orgel, *J. Chem. Soc.,* 4186 (1958).

48. C. K. Jörgensen, in "The Biochemistry of Copper" (J. Peisach, P. Aisen, and W. E. Blumberg, eds.), Academic Press, New York, 1966, p. 1.

49. P. Hemmerich, in "The Biochemistry of Copper" (J. Peisach, P. Aisen, and W. E. Blumberg, eds.), Academic Press, New York, 1966, p. 15.

50. S. E. Manahan and R. T. Iwamoto, *J. Electroanal. Chem.*, *14*, 213 (1967).

51. T. A. Kaden and A. D. Zuberbühler, *Helv. Chim. Acta*, *49*, 2189 (1966).

52. R. M. Izatt, H. D. Johnston, G. D. Watt, and J. J. Christensen, *Inorg. Chem.*, *6*, 132 (1967).

53. B. R. James and R. J. P. Williams, *J. Chem. Soc.*, 2007 (1961).

54. A. Günter and A. D. Zuberbühler, unpublished results.

55. T. Hurlen, *Acta Chem. Scand.*, *15*, 1231 (1961).

56. C. J. Hawkins and D. D. Perrin, *J. Chem. Soc.*, 1351 (1962).

57. C. K. Jörgensen, *Z. Nat. Med. Grundlagenforschung*, *2*, 248 (1965).

58. F. B. Baker and T. W. Newton, *J. Phys. Chem.*, *60*, 1417 (1956).

59. T. A. Kaden and S. Fallab, *Chimia*, *19*, 176 (1965); T. A. Kaden, *Helv. Chim. Acta*, *49*, 1915 (1966).

60. G. S. Hammond, *J. Amer. Chem. Soc.*, *77*, 334 (1965).

61. L. Graf and S. Fallab, *Experientia*, *20*, 46 (1964).

62. L. E. Orgel, in "Metals and Enzyme Activity" (E. M. Crook, ed.), University Press, Cambridge, 1958, p. 8.

63. M. Eigen, *Ber. Bunsenges. Phys. Chem.*, *67*, 753 (1963).

64. G. W. Filson and J. H. Walton, *J. Phys. Chem.*, *36*, 740 (1932).

65. O. A. Chaltykyan, G. S. Chtyan, and G. Darbinyan, *Nauch. Trudy Erevan Gosudarst. Univ. Ser. Khim. Nauk.*, *53*, 95 (1956).

66. O. A. Chaltykyan and G. S. Chryan, *Nauch. Trudy Erevan Gosudarst. Univ. Ser. Khim. Nauk.*, *60*, 125 (1957).

67. E. Abel, *Monatsh. Chem.*, *87*, 354 (1956).

68. H. Nord, *Acta Chem. Scand.*, *9*, 430 (1955).

69. R. D. Gray, *J. Amer. Chem. Soc.*, *91*, 56 (1969).

70. A. D. Zuberbühler, *Helv. Chim. Acta*, *53*, 473 (1970).

71. A. D. Zuberbühler, *Chimia*, *23*, 416 (1969).

72. A. D. Zuberbühler, *Helv. Chim. Acta*, *50*, 466 (1967).

73. I. Pecht and M. Anbar, *J. Chem. Soc. Ser. A*, 1902 (1968).

74. A. D. Zuberbühler, *Helv. Chim. Acta*, *53*, 669 (1970).

75. R. B. Martin and J. T. Edsall, *J. Amer. Chem. Soc.*, *82*, 1107 (1960).

76. M. Doran, Diss. Clark Univ. (1958), cited from Ref. [36], p. 671.

77. A. D. Zuberbühler, Habilitationsschrift, Basel (1969).

78. D. A. Zatko, *Diss. Abstr. B, 28,* 524 (1967).

79. G. Rainoni and A. D. Zuberbühler, unpublished results.

80. P. M. Henry, *Inorg. Chem., 5,* 688 (1966).

81. Gmelins Handbuch der Chemie: "Sauerstoff," Syst. No. 3, Verlag Chemie, Weinheim, 1958, p. 490.

82. A. Foll, M. le Démezet, and J. Courtot-Coupez, *J. Elektroanal. Chem., 35,* 41 (1972).

83. H. Sakurai, A. Yokoyama, and H. Tanaka, *Chem. Pharm. Bull. Japan, 18,* 2373 (1970).

84. E. Frieden, J. A. McDermott, and S. Osaki, in "Oxidases and Related Redox Systems," Vol. 1 (T. E. King, H. S. Mason, and M. Morrison, eds.), Wiley, 1965, p. 257 ff.

85. C. R. Clark, B. W. Skelton, and T. N. Waters, *Chem. Commun.,* 180 (1972).

86. R. L. Jolley, L. H. Evans, and H. S. Mason, *Biochem. Biophys. Res. Commun., 46,* 878 (1972).

87. R. C. Wilkins, in "Bioinorganic Chemistry," *Advan. Chem. Ser. 100* (R. F. Gould, ed.), Amer. Chem. Soc. Publ., Washington, 1971, p. 111.

88. R. Lontie and R. Witters, in "The Biochemistry of Copper" (J. Peisach, P. Aisen, and W. E. Blumberg, eds.), Academic Press, New York, 1966, p. 455.

89. C. Berecki-Biedermann, G. Biedermann, and L. G. Sillén, Report to Analytical Section, JUPAC (1953).

90. V. M. Berdnikov, Y. N. Kozlov, and A. P. Purmal, *Khim. Vys. Energ., 3,* 370 (1969).

91. V. M. Berdnikov and N. M. Bazhin, *Russ. J. Phys. Chem., 44,* 395 (1970).

92. T. A. Kaden, D. Walz, and S. Fallab, *Helv. Chim. Acta, 43,* 1639 (1960).

93. P. George, *J. Chem. Soc.,* 4349 (1954).

94. G. A. Barclay, B. F. Hoskins, and J. H. L. Kennard, *J. Chem. Soc.,* 5691 (1963).

95. F. J. Simpson and T. Oka, personal communication (1972).

96. T. Oka, F. J. Simpson, and H. G. Krishnamurty, *Can. J. Microbiol., 18,* 493 (1972).

97. H. Takahashi, T. Kajimoto, and J. Tsuji, *Synt. Commun., 2,* 181 (1972).

98. T. Nakamura and Y. Ogura, *J. Biochem. (Tokyo)*, *64*, 267 (1968).

99. H. Nord, *Acta Chem. Scand.*, *9*, 442 (1955).

100. W. E. Blumberg, M. Goldstein, E. Lauber, and J. Peisach, *Biochim. Biophys. Acta*, *99*, 187 (1965).

101. S. Friedman and S. Kaufman, *J. Biol. Chem.*, *241*, 2256 (1966).

102. J. A. Fee, R. Malkin, B. G. Malmström, and T. Vänngard, *J. Biol. Chem.*, *244*, 4200 (1969).

103. P. M. Wood, *FEBS Lett.*, *44*, 22 (1974).

104. G. A. Rodley and W. T. Robinson, *Nature*, *235*, 438 (1972).

105. J. P. Collman, R. R. Gagne, C. A. Reed, T. R. Halbert, G. Lang, and W. T. Robinson, *J. Amer. Chem. Soc.*, *97*, 1427 (1975).

106. L. D. Brown and K. N. Raymond, *Inorg. Chem.*, *14*, 2595 (1975).

107. J. S. Valentine, *Chem. Rev.*, *73*, 235 (1973).

108. E. Ochiai, *J. Inorg. Nucl. Chem.*, *37*, 1503 (1975).

109. G. Rainoni and A. D. Zuberbühler, *Chimia*, *28*, 67 (1974).

110. J. A. Arce, E. Spodine, and W. Zamudio, *J. Inorg. Nucl. Chem.*, *37*, 1304 (1975).

111. A. L. Crumbliss and A. T. Poulos, *Inorg. Chem.*, *14*, 1529 (1975).

112. Z. Jedlinski, M. Rybczyk, J. Mainusz, and S. Janik, *Makromol. Chem.*, *176*, 891 (1975).

113. C. E. Kramer, G. Davies, R. B. Davis, and R. W. Slaven, *Chem. Commun.*, 606 (1975).

AUTHOR INDEX

Numbers in parentheses are reference numbers and indicate that an author's work is referred to although his name may not be cited in the text. Underlined numbers give the page on which the complete reference is listed.

H

I

J

K

L

M

Q

R

SUBJECT INDEX

B

N

T

U

V

X